Human Factors and Behavioural Safety

Human Factors and Behavioural Safety

Jeremy Stranks

LONDON AND NEW YORK

First published by Butterworth-Heinemann

First edition 2007

This edition published 2011 by Routledge
2 Park Square, Milton Park, Abingdon, Oxon OX14 4RN
711 Third Avenue, New York, NY 10017, USA

Routledge is an imprint of the Taylor & Francis Group, an informa business

British Library Cataloguing in Publication Data
A catalogue record for this book is available from the British Library

Library of Congress Cataloging-in-Publication Data
A catalog record for this book is available from the Library of Congress

ISBN-13: 978-0-7506-8155-1

Transferred to Digital Printing in 2013

Contents

Preface

People at work have experience, expectations, ambitions and skills. The problem with people is that they are inconsistent, they make mistakes, forget things, don't pay attention, don't understand things quite correctly and get their priorities wrong. In some cases, they wilfully disregard the safety rules, putting themselves and others at risk.

The last decade has seen considerable attention paid to the human factors aspects of health and safety at work. Much of this increased emphasis has been brought about as a result of the contribution of human failure to disasters, such as those at Bhopal in India, Moorgate, Kegworth and Longford, Victoria, South Australia, together with the Piper Alpha incident.

'Human factors' is an area of study concerned with people, the organizations they work for and the work they undertake. It is also concerned with communication systems within organizations and the training systems and procedures in operation, all of which are directed at preventing human error.

This book is about human factors and the behavioural aspects of safety. It examines psychological factors such as attitude, motivation and perception, theories of accident causation and the relationship of human reliability to accidents, together with the increasingly significant areas of ergonomics and stress at work.

One of the principal objectives of any organization is that of developing and promoting the right safety culture, an aspect which requires a significant human factors input if it is to be successful. This book looks at this aspect, along with important features in the development of a safety culture, such as communication, training and interpersonal skills.

I should stress that this book is not written by a psychologist for psychologists! It is targeted at health and safety practitioners, HR managers, trainers and managers in general, who need to have a broad understanding of the subject, together with those studying for NEBOSH and other qualifications in occupational health and safety.

I hope that all those who use this book will find it helpful.

Jeremy Stranks
2006

1 Human behaviour and safety

Human behaviour has a direct influence on safety in many aspects of life – at work, in the home, on the road, in the air and at sea. Evidence drawn from past disasters, such as the incidents at Flixborough, Kegworth and Moorgate, and the Piper Alpha incident, indicate that failures in, or inappropriate, human behaviour were a significant contributory factor.

What is meant, then, by behaviour?

Definition of 'behaviour'

'Behaviour' is variously defined as:

- how a person conducts himself;
- the demeanour and manners of an individual;
- an observable action of a person.

Behavioural sciences, therefore, are those sciences most concerned with the study of human and animal behaviour. This study allows the formation of general theories based upon the observation of specific events. The theories are subsequently used to explain observed events and, in some cases, to predict future events or outcomes.

The causes of human behaviour are associated with attitude, personality, motivation and memory, together with those physical and mental characteristics which constitute a person and his environment.

The study of human behaviour

Human behaviour is a wide area of study. The more significant areas of the study of human behaviour are outlined below.

Psychology

This is the science that studies the behaviour of human beings and animals. As a science, it is firstly empirical. Behaviour can be observed, recorded and studied resulting in the production of data derived from quantitative measurements. Secondly, it is systematic, in that, as a science, it endeavours to make sense of observations and summarizes these observations using established principles based on laws and accepted systems of classification.

Occupational psychology

Occupational psychology is concerned with the behaviour of people at work. It deals with areas such as the intelligence and aptitude required for work-related tasks, essential tests for the selection and placement of people, training and supervision requirements, the improvement of communications and, in some cases, the resolution of conflict involving employers and employees.

Human engineering (or engineering psychology) is another area of psychology which came to prominence during the Second World War. More commonly known as 'ergonomics', this area of human behaviour examines the design of equipment and the tasks of people who operate that equipment.

Sociology and social anthropology

This is a specialized area of psychology that is concerned with the behaviour of people as members of a group, at home, at work or in other circumstances. It studies the effects of group membership upon the behaviour, beliefs and attitude of an individual, together with the cultures and social structures of groups and societies existing together.

Anthropology

This is the study of the human race. It is the science of man and mankind with respect to physical constitution and condition.

Cognitive psychology

One of the changes that have taken place in recent years is the development of cognitive psychology, which stresses the importance of understanding. This has entailed a move away from 'behaviourism', where the individual was perceived merely to respond to external stimuli. Cognitive psychology sees people as active agents with ideas, plans and innovations. They are special people with internal mental structures who perceive the outside world within the framework of their own master plan, rather than responding to events and circumstances.

Perception, in its broadest sense, is an important element in this approach. This implies the formation of new attitudes, looking at the 'broad picture' and identifying both benefits and hazards arising from this approach.

Complexity of human behaviour

Human behaviour is a complex thing. People behave in different ways in different situations and no two people behave in the same way in a particular situation. This implies that all people are different in terms of their psychological make up.

Various theories have been proposed over the years, for example:

- The *Genes Theory* subscribes to the fact that all people have a particular mixture of genes which determine their behaviour.
- The *Nature v Nurture Theory*, on the other hand, seeks to ascertain whether the behaviour of people is due to their 'nature' (or genetics) or to their 'nurture', the environment in which they have been brought up.

Human behaviour is, however, associated with a range of factors which are considered below.

Factors affecting human behaviour

People behave in different ways in different circumstances. Moreover, individuals have their own particular modes and patterns of behaviour according to circumstances. These aspects of behaviour are associated with various psychological factors that contribute to the way a person behaves and include elements such as attitude, motivation, memory, personality and perception. Moreover, individual factors such as upbringing, past experience, a person's environment, the level of knowledge and understanding, emotions and stress, greatly influence the way people behave.

Elements and functions of human behaviour

Occupational psychology is concerned with how behavioural factors such as attitude, motivation, perception, memory and training, together with the mental and physical capabilities of people, can interact with work activities with particular reference to health and safety issues. In particular, it considers the individual differences in people, human reliability and the potential for human error, all of which may be contributory factors in accidents.

The various aspects of human behaviour are considered below.

Attitude

'Attitude' can be defined in a number of ways:

- a predetermined set of responses built up as a result of experience of similar situations;
- a tendency to behave in a particular way in a particular situation;

- a tendency to respond positively (favourably) or negatively (unfavourably) to certain persons, objects or situations;
- a tendency to react emotionally in one direction or another.

Rokeach (1968) defined 'attitude' as 'a learned orientation or disposition towards an object or situation which provides a tendency to respond favourably or unfavourably to the object or situation'.

Attitudes are acquired or learned as people progress through life. Part of the learning process involves *conditioning*. This is a restricted form of learning in which a single response is acquired.

Another aspect of learning and the development of attitude is the process of *reinforcement*. This is associated with two things: reward and punishment. Reward can take the form of praise for a job well done, financial reward in the form of a bonus for beating a sales target or company recognition of successful performance. Punishment, on the other hand, can include demotion within the organization or some form of financial penalty, such as a fine by a court. Both reward and punishment help to reinforce attitudes in many ways.

Attitudes are acquired in the same way that other responses are acquired, namely through classical and operant conditioning as part of the learning process.

Classical conditioning

This is the kind of learning originally studied in the 'classical' experiments of Ivan P. Pavlov, the famous Russian psychologist, with the behaviour of dogs. Pavlov introduced the concept of conditioning and established many of its basic principles. He established that animals can produce a conditioned response as a result of training and the provision of rewards. Pavlov trained dogs by sounding a bell immediately after presenting food and then measuring the amount of saliva produced by the dog. After pairing the sound of the bell with the provision of food a few times, the effects of the training were tested by measuring the amount of saliva which flowed when the bell was rung without the presentation of food.

Pavlov resumed the paired presentation of bell and food a few more times and then tested the dogs with the bell alone. He noted that, as training proceeded, the amount of saliva excreted in response to the bell alone gradually increased. The amount of increase over test trials could be plotted as a learning curve.

People, in the same way, can go through the process of classical conditioning at all stages in life. Children, for instance, will associate certain adults with fun and laughter so that, on sight of that person, they will start to laugh. In the work situation, because of past experiences, they may associate management with stress and harassment.

Operant conditioning

Operant conditioning is different from classical conditioning in a number of ways. Classically conditioned responses are, fundamentally, elicited responses as opposed to those learnt in operant conditioning, which are emitted responses. The elicited responses are relatively fixed, reflex-like responses, such as salivation or the bending of a limb. The emitted responses of operant conditioning, on the other hand, are variable responses, such as walking, talking and pushing.

Operant conditioning enables freedom of response, to shape behaviour through the appropriate use of reinforcement.

Features of attitude

Self-image

The 'self' represents the individual's awareness or perception of his own personality. This commences in early infancy and continues through life involving a range of personality traits or tendencies. When these traits are constantly applied, an individual accepts them as descriptions of himself.

Self-image is concerned with how a person likes to present himself to society. This may be characterized by factors such as his style of speech, self-expression, the manner of dressing, habits and personality traits. A person may wish to be perceived and described by others as 'cool', 'mean', 'honest', 'hard to read' or as 'not suffering fools gladly'.

Group influences

The influence of groups and the group norms prevalent at a point in time have a direct impact on attitudes held. Organizations endeavour to influence the attitudes of managers and employees through performance enhancing techniques such as Total Quality Management and staff appraisal schemes.

Whatever form of group is involved, continuing membership of the group implies complying with the group norms.

Beliefs, opinions and superstitions

Beliefs have a direct effect on attitudes. They may include religious, social and ethical beliefs, many of which are developed at school, in the home and subsequently, at work. Some people have the firm belief that all accidents are Acts of God, a matter over which they have no control.

An opinion is defined as 'a statement of something which may be subject to change'. People hold opinions on a range of matters which have been developed from childhood. A commonly held opinion is that only untrained, unskilled and careless people have accidents.

Superstitions, opinions for which there is no logical explanation, such as two crossed knives as a symbol of impending death, are passed down from generation to generation. Irrespective of the illogical nature of superstitions, they can have a profound effect on attitude.

Functions of attitude

According to Katz, there are four functions of attitude, outlined below.

Social adjustive function

This is concerned with how people relate and adjust to the influence of parents, teachers, friends, colleagues and their superiors. It is argued that by the age of 9 years, most attitudes are established. Behaviour is based, to some extent, on a philosophy of 'maximum reward, minimum punishment'.

Value expressive function

People use their attitudes to present a picture of themselves that is pleasing and satisfying to them. This is an important factor in that people see themselves as better and different, in some special way, from others around them (self-image). To promote this self-image, people may adopt extreme views of situations, dress in a particular way and adopt a particular political persuasion.

Knowledge function

Attitudes are used to provide a system of standards that organize and stabilize a world of changing experiences. On this basis, people need to work within an acceptable framework, have a scale of values and generally know where they stand.

Self-defensive function

The self-defensive function is concerned with the need to defend one's self-image, both externally, in terms of how people react towards us, and internally, to deal with inner impulses and our personal knowledge of what we are like.

Attitude surveys

People have beliefs which affect their attitudes. 'Belief' can be defined as 'information about an object, person or situation, which may be true or false, linking an attribute to it'. A typical belief in the field of health and safety may be that 'well-experienced operators don't need all this health and safety!'.

An attitude survey seeks to measure these beliefs which can roughly be classified as follows.

Strength of belief

The strength with which a person holds a particular belief is important. This can vary between total agreement and lack of agreement. This applies to specific aspects of a job, such as the need to add a hazardous substance to a process and the reasons why those aspects must be undertaken in a particular way.

Value

The actual value of a job varies from person to person according to how they perceive the significance of that job. The value can be measured on a scale from 'excellent' to 'very bad'.

Social beliefs

Social beliefs, or inner attitudes, about factors such as the causes of accidents and sickness absence, the skills required for jobs and the role of management, commonly feature in attitude surveys. Social beliefs are, however, a small part of human behaviour. Factors such as social pressures on the individual are also important.

Overall attitude

The relationship between various features of a job and their effects on a person's social life is considered in this type of survey.

Intended behaviour

A person's intentions in terms of future actions, for instance, whether he intends to stay with the organization or move on to something better, or seek promotion within the organization, whilst purely hypothetical, can correlate with actual behaviour at the time.

Motivation

A motivator is something which provides the drive to produce certain behaviour or to mould behaviour. For example, physical punishment was seen, for many years, to be an important motivator in terms of moulding the attitudes of school children in their formative years and as a means of deterring criminals from future criminal behaviour. For many people, money is an important motivator.

Everyone is motivated by the need, for example, to be safe, be warm, have sufficient food, belong to a group, such as a family or work group, to achieve things and to have an ordered life. Needs continually arise and in some cases these needs are satisfied. In other cases, the needs are not satisfied, which can result in stress, loss of motivation and loss of self-esteem.

The various theories of motivation are outlined below.

Theories of motivation

Taylor (1911) – Theory of management and organisation of work

Taylor said, 'Man is a creature who does everything to maximize self-interest'. In other words, people are primarily motivated by economic gain. 'What's in it for me?' is an important motivator for some people.

Maslow (1943) – 'Self-actualising man'

Abraham Maslow established a 'hierarchy of needs' in individuals (Figure 1.1). These are:

- Basic (survival) needs: physiological needs, such as air, food, drink, warmth, sex, sleep, etc.
- Safety and security needs: physical security, order, stability, limits, protection from the elements, etc.
- Belongingness and love needs: family, work group, relationships, affection, etc.
- Status needs: self-esteem and respect, achievement, independence, status, prestige, managerial responsibility, etc. and
- 'Self-actualization': self-fulfilment, achieving one's full potential; seeking personal growth and peak experiences; 'What a man can be, he must be!'.

This hierarchy was adapted in the 1970s to include:

- Cognitive needs: knowledge, meaning, etc. and
- Aesthetic needs: appreciation and search for beauty, balance, form, etc.

and again in the 1990s to include:

- Transcendence needs: helping others to achieve self-actualization.

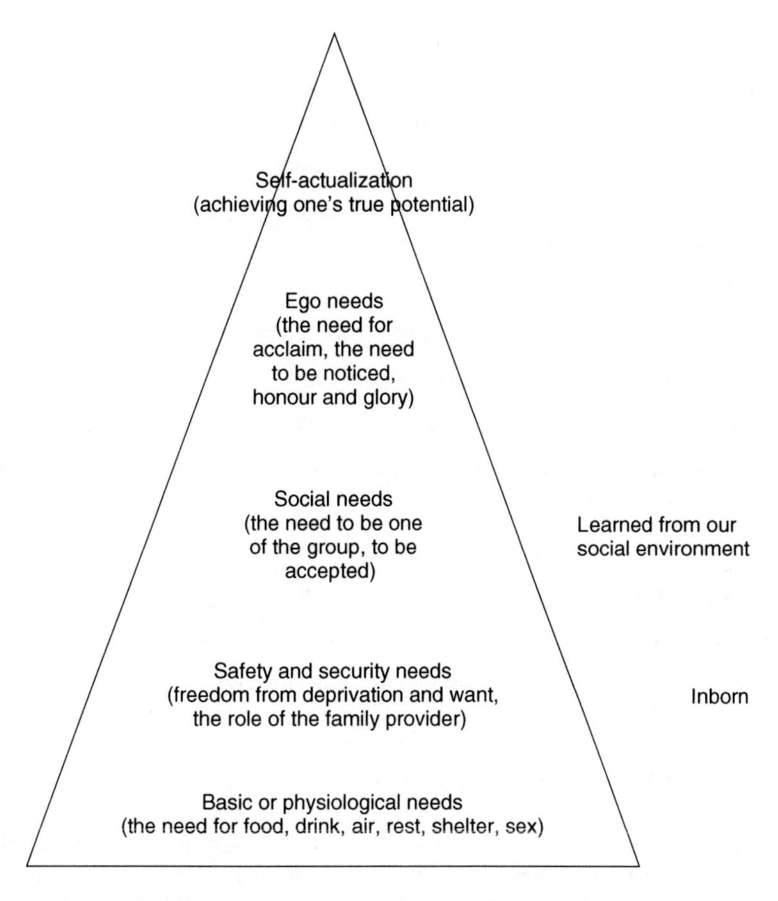

Figure 1.1 Maslow's hierarchy of needs

At any point in time, an individual is endeavouring to satisfy one of the various levels of the hierarchy. One of man's survival needs is to stay fit and healthy (physiological needs). People also have the need to be respected on the basis of their knowledge and skills and place in society.

Self-actualizing characteristics

One of the greatest motivators for many people is that of 'doing your own thing', as opposed to working for an organization.

Maslow stated a number of self-actualizing characteristics thus:

- keen sense of reality, aware of real situations, objective judgement, rather than subjective;
- see problems in terms of challenges and situations requiring solutions, rather than see problems as personal complaints or excuses;

- need for privacy and comfortable being alone;
- reliant on own experiences and judgement, independent, not reliant on culture and environment to form opinions and views;
- not susceptible to social pressures, non-conformist;
- democratic, fair and non-discriminating, embracing and enjoying all cultures, races and individual styles;
- socially compassionate, possessing humanity;
- accepting others as they are and not trying to change people;
- comfortable with oneself, despite any unconventional tendencies;
- a few close intimate friends rather than many surface relationships;
- sense of humour directed at oneself or the human condition, rather than at the expense of others;
- spontaneous and natural, true to oneself, rather than being how others want;
- excited and interested in everything, even ordinary things;
- creative, inventive and original;
- seek peek experiences that leave a lasting impression.

Herzberg (1957) – Two factor theory

Fred Herzberg undertook an extensive study throughout the United States, Canada and the United Kingdom seeking to identify among workers in many organizations the factors that produced job satisfaction and dissatisfaction.

Herzberg examined homeostatic needs (*hygiene factors* or *maintenance factors*), which are concerned with avoiding pain and dissatisfaction, and growth needs (*motivators*), which are concerned with actively seeking and achieving satisfaction and fulfilment (Figure 1.2).

Hygiene factors	Motivators
Money	Challenge
Working conditions	Responsibility
Safety arrangements	Advancement
Quality of supervision	Interest and stimulation created by the job
Administrative procedures	Achievement
Interpersonal relationships	Recognition
Status	Possibility of growth
Security	

Figure 1.2 Hygiene factors and motivators (Herzberg, 1957)

He asked many people in different jobs at different organizational levels two questions:

> What factors lead you to experience extreme dissatisfaction with your job?
>
> What factors lead you to experience extreme satisfaction with your job?

He established that there was no one factor that determined the presence or absence of job satisfaction. However, he points out that before satisfaction with work can be improved, the factors that cause dissatisfaction must be dealt with by management. As a result of his research, he identified two specific motivational factors:

- the *dissatisfiers (hygiene factors)* – those factors which produced dissatisfaction, namely the wages paid, working conditions, including safety standards, and the quality of supervision; and
- the *satisfiers ('motivators')* – those factors which produced satisfaction, in particular, challenge, responsibility, advancement and the interest and stimulation created by a job.

In the typical work situation, hygiene or maintenance factors involve the total environment affecting the employee, namely physical working conditions, pay, safety, security, social factors and interpersonal relationships. On the other hand, motivational needs or motivators that lead to positive happiness are the needs for growth, achievement, responsibility, accountability and recognition. These needs can only be met by undertaking the actual work itself.

Herzberg's view was that the job itself can provide a potentially more powerful motivator than any externally introduced incentives. Employees can be actively satisfied only when the work done is perceived by the worker as being meaningful and challenging, thereby fulfilling their motivational needs.

People expect the hygiene factors to be present and satisfactory. If non-existent or poorly managed, this will bring about dissatisfaction. On the other hand, whilst motivators give rise to job satisfaction, they will not necessarily result in job dissatisfaction if absent.

Herzberg concluded that before employers can increase satisfaction with the work, they must, firstly, reduce dissatisfaction. In other words, if management is to motivate people to take greater responsibility and stimulate job interest, they must get the hygiene factors right. This means the provision of better working environments and safety standards in particular.

Improving motivation

Herzberg suggested the concept of *job enrichment* as a solution to the problem of meeting motivational needs. Care must be taken, however, to ensure that job enrichment techniques are not introduced in such large amounts or at such speed as to create excessive alarm or fear among workers. When introduced on a planned and phased basis, and with consultation, job enrichment makes it possible for the growth and achievement needs of people to be met as a result of their efforts at work.

The significance of the job enrichment concept lies in the clarity with which it focusses attention on the motivational distinction between:

1. task and environment; and
2. intrinsic and extrinsic factors.

According to Herzberg, *task* impoverishment, namely removing the individual interest, challenge and responsibility from a job, results in deteriorating motivation. Conversely, no amount of environmental improvement can compensate for this impoverishment. Clearly, there is a need, therefore, to examine the tasks that people carry out with a view to identifying the factors that provide interest, challenge and responsibility for workers. These higher levels of motivation can be achieved through job enrichment, job enlargement and job rotation, accompanied by various worker participation schemes:

- **Job enrichment**. An increase in satisfaction and the responsibility attached to a job is achieved either by reducing the degree of supervision or by allocating each worker a unit of work in which they have freedom to select their work method and the sequence of operations.
- **Job enlargement**. In this case, the worker is required to progressively increase the actual number of operations that he undertakes.
- **Job rotation**. Whilst this concept has been unpopular, to some extent, with workers, the objective is to give more variety on simple, repetitive and usually automated tasks.
- **Worker participation**. Worker participation in varying degrees can be achieved in the following areas:
  - *Human resources-related decisions*. Transfers to other jobs, disciplinary matters, various forms of training and instruction;
  - *Social decisions*. Welfare arrangements, health and safety procedures and systems of work, regulation of working hours, rest periods;
  - *Economic decisions*. Methods of production, production planning and control, rationalization, changes in plant organization, expansion and contraction.

The fundamental objective is to reduce or eliminate the authoritarian approach by management and replace it with a more participative style of management. The participative leader is one who plans work and consults his subordinates as to the best course of action. Such leaders are skilled in reconciling conflicts so as to achieve group cohesion and effectiveness. This sort of person is also interested in individual employees and their particular problems, whereas the authoritarian uses rewards and punishment of the traditional sort, exercises close supervision and is more interested in the activities of those above than those below them.

Motivation and safety

Important factors for consideration in motivating people to better levels of safety performance include the following.

Joint consultation

Consultation with workers in planning the organization of work is one of the greatest motivators from a health and safety viewpoint. Consultation is best undertaken through a formally established health and safety committee with clearly defined objectives that is representative of all the people concerned. This committee should meet regularly, publish agendas and minutes, and implement decisions consistently and in an expeditious manner. Above all, it must have credibility with the workforce.

Trade union safety representatives and worker-elected representatives of employee safety have, further, an important role to play in the joint consultation process. It is important that the role and functions of such people are clearly identified and that they be adequately trained in the various aspects of occupational health and safety in order to make as constructive a contribution as possible.

The use of working parties

These can be used to define objectives, including those for health and safety. In certain cases, small working parties can examine a particular situation, reporting their findings to the health and safety committee.

Attitudes currently held

Those attitudes currently held by management and workers with regard to safe working are true indicators of the importance attached to health and safety at work. It is important that both groups are positively motivated towards success in improving levels of health and safety performance.

The communications system

Communications systems within the organization should provide information that is comprehensible and relevant to all concerned. Many people become demotivated by communications that they find difficult to understand or interpret.

The quality of leadership

Leadership, as with all areas of management, should come from board level if people are to be adequately motivated towards improved standards of health and safety performance.

Individual needs and safety incentives

The individual needs of people and the safety incentives necessary to satisfy these needs can be summarized as indicated in Table 1.1.

Perception

People perceive and gain information through what they see, hear, touch, taste and smell. Perception is the complex mental function giving meaning and significance to sensations. An individual is constantly responding in some way to incoming stimuli. These stimuli can be accepted, rejected, ignored or distorted. It all depends upon whether the stimulus supports or contradicts the individual's beliefs, values and attitudes. This process forms the basis of perception, i.e. the

Table 1.1 Individual needs and safety incentives

Motives	May be satisfied by
Financial gain through increased departmental or company profits	Monetary awards through suggestion schemes, profit-sharing plans, promotions, increased responsibility, propaganda
Fear of painful injury, death, loss of income, family hardship, group disapproval or ridicule, criticism by managers/supervisors	Visual material – posters, films, videos, public reports of accidents
Participation, that is, the need to be 'one of the gang'	Group and individual activities, e.g. safety committees, working parties, safety campaigns
Competition – a desire to beat others	Health and safety competitions and award schemes
Pride in safe workmanship, both individual and group	Recognition for individual and group achievement – trophies, awards, publicity
Recognition – desire for the approval of others in the group and family, for praise from supervisor	Publicity – photographs and articles in company and community newspapers, use of notice boards

way the individual interprets incoming information. To summarize, perception could be defined as 'how the environment presents itself to the individual'.

Perception is developed in order to satisfy an individual's needs so that he or she can cope with reality. Each individual is different due to factors such as heredity and environment. This is what makes each individual unique and each individual's 'reality' unique.

Further consideration of perception is provided in Chapter 2, Human sensory and perceptual processes.

Memory

Memory is the process of retaining, recognizing and recalling experience (remembering) and is particularly concerned with how people learn. It is, essentially, the information storage system of human beings. It is an important feature of human behaviour, particularly where people are frequently exposed to hazards, and is directly affected by learning, past experience, feedback from events of particular significance and an individual's capacity for storing information. Poor memory skills can be significant in accident causation.

People have both short-term and long-term memory components.

Short-term memory

Short-term or immediate memory refers to the temporary storage of information for a few seconds as, for example, with a telephone number or postal code. Short-term memory is also concerned with the amount of material that an individual can take in and retain. In many cases it is a limiting factor in individual ability and individual safety. The limited storage capacity of short-term memory is also shown by the fact that memory span for a single repetition is about seven items long. This means that, without regrouping or reorganizing the information as we receive it, most people cannot retain more than about seven items, for example, numbers, after one exposure to them. The short-term memory is susceptible to interference from other sources.

The existence of short-term memory has been demonstrated in experiments with people memorizing numbers, the ability to accurately retain such numbers reducing as the number of digits increases. Short-term memory is, thus, limited in capacity and highly susceptible to disruption or interference as compared with long-term memory.

However, people have learned to recode the information into chunks or 'bits' and the trick is to recode the information as it is received. Most people can only take in and retain around 3.1 'bits' of information at any point in time. Once this level is exceeded, the ability to recall the information reduces proportionately.

This fact is aptly demonstrated when seeking travel directions from people. In many cases, and in trying to be helpful, people provide so much extraneous information to a person asking the way that the short-term memory becomes overloaded. By the time the person providing the information has finished, the first part or parts of the directions are confused or lost!

Short-term memory and its limitations can be a significant factor in the causes of accidents at work and one that is frequently associated with human error or poor memory skills. Important on-the-spot instructions should, therefore, be repeated several times to ensure that the recipient fully understands and can recall those instructions accurately.

Thus, there is a direct connection between short-term memory and the human errors or omissions which result in accidents.

Long-term memory

Long-term memory is concerned with the individual's ability to store, and subsequently recall, information. It is a vast store of information that is organized in some form of classification. On this basis, any new information is perceived in terms of these categories and forced into the classification system even when it does not fit exactly. In this process there is a chance that it may become distorted.

Long-term memory is developed largely through the repetition of items and codifying them to produce a meaning. A mnemonic is an example of a codifying system. There is a characteristic drop in memory over a period of time, associated with the ageing process.

Interference with long-term memory can be caused by:

- events of close similarity which tend to confuse; and
- the effect of recall on the subsequent memory that can again cause confusion, resulting in the individual forgetting.

Limitations in memory recall, or remembering, can frequently be overcome by recalling the circumstances in which the original memory was stored or approaching it via memories that we know were associated with it. Unavailability of memory may be significant in certain emergency situations or where a quick response is required from an individual. This unavailability can be overcome by recalling and reusing the memories (knowledge and skills) at regular intervals. Various forms of refresher training, fire and emergency drills and practical sessions all assist in reducing unavailability of memory.

Generally, memory skills vary considerably from one person to another. Variables that influence the amount of material retained in long-term memory are:

- the meaningfulness of the material;

- the degree of learning of the material; and
- interference with the learning material.

In many cases, memory may undergo significant change over a long period of time. This is particularly common where memory may be unpleasant and it is likely that distortion will occur at each recall. Thus, these various memory recalls, distorted to various degrees, will eventually be remembered rather than the original version. A stage is reached where the individual is unable to distinguish between the correct facts and those that were introduced as part of subsequent recalls. Such a phenomenon is commonly encountered in accident investigation whereby a witness may make a particular statement following an accident but, 3 months later, after he has endeavoured to recall the situation on numerous occasions, may have a totally different version of the events leading to the accident.

Memory defects are sometimes associated with the phenomenon of accident proneness (see below).

Personality

Allport (1961) defined 'personality' as 'the dynamic organisation within the individual of the psychophysical systems that determine his characteristic behaviour and thought'. Personality is directly related to people's behaviour, e.g. rigid, honest, overbearing, bumptious, etc.

'Dynamic' implies that personality is composed of interacting parts and that this interaction produces flexibility of response, i.e. a person who is subject to change. The degree of subjection to change is important, particularly in the selection of people who may be exposed to continuing forms of danger.

Theories of personality

Aspects of individual personality have been the subject of considerable study and debate over the last century and there has been no general agreement by psychologists on one single theory. Some of the more prominent theories of personality are outlined below.

Psychoanalytic theory

Sigmund Freud, who was the founder of psychoanalysis, based his theory of personality on psychoanalytic theory. Freud believed that personality has a three-part structure. These parts, the *Id*, the *Ego* and the *Superego*, work together to produce all complex behaviours. All three components need to be in balance in an individual for him to have a good amount of psychological energy available and a well-adjusted mental health.

1. **The *Id*.** The *Id* is the primitive mind and functions in the irrational and emotional part of the mind. In infancy a baby's mind is totally *Id* and contains all the basic needs and feelings for food, love and attention. This continues, to some extent, through childhood and is concerned with seeking satisfaction in terms of food, warmth, love and attention, the 'pleasure principle'. The *Id* can be viewed as a store of biologically based motives and 'instinctional' reactions satisfying motives. The major motives or 'instinctional drives', in psychoanalytical theory, are the sexual and destructive urges. The libido is the energy of these motives. On its own, the *Id* energy or libido would satisfy basic motives as they arose without any consideration of the realities of life. The *Id* is, to some extent, tied in with the *Ego*.
2. **The *Ego*.** The *Ego* functions within the rational part of the mind and develops out of growing awareness that needs and feelings cannot always be satisfied. It comprises elaborate ways of behaving and thinking that have been learnt for dealing effectively with situations. It recognizes the need for compromise (the 'reality principle') and negotiates between the *Id* and the *Superego*. It delays the satisfaction of motives and channels motives into socially acceptable outlets. In particular, the *Ego* enables a person to get along with other people, to earn a living and adjust to the realities of life. Freud characterized the *Ego* as working 'in the service of the reality principle'.
3. **The *Superego*.** This is the last part of the mind to develop and corresponds closely with a person's 'conscience' or the moral part of the mind. It incorporates restraints which have been acquired in the course of personality development on the *Ego* and the *Id*. It stores and enforces rules and constantly strives for perfection, even though this perfection objective may be unachievable in reality. The *Superego* has two subsystems:
   (a) **The *Ego* ideal.** This provides rules for good behaviour and standards of excellence towards which the *Ego* must strive. The *Ego* ideal is what a child's parents approve of or value.
   (b) **Conscience.** These are the rules about what constitutes bad behaviour, those aspects of behaviour that a child feels his parents may disapprove of and for which punishment may be given.

Whilst any clear link between personality and safety performance is difficult to identify, certain *personality traits*, such as a lack of regard for rules, carelessness with respect to practical matters and a lack of self-discipline on the part of an individual, could well be a contributory factor in accidents.

In the case of people involved in high-risk activities, the use of this test may be appropriate. Moreover, in the assessment by an employer of human capability prior to the allocation of tasks, which is required under the Management of Health and Safety at Work Regulations, personality testing could become quite a significant feature of such assessment with respect to certain tasks.

Developmental psychology is a branch of psychology studying changes in behaviour that occur with changes in age. This is particularly significant in the case of personality. Freud viewed personality development as taking several overlapping stages during the period from a baby to an adult, as follows.

- **The oral stage**. This stage, which occupies most of a baby's first year, is where the baby gets pleasure from the suckling process and other activities involving his mouth. Where he is not allowed to suck, or may become anxious about sucking, the baby may acquire an oral fixation. As he gets older the oral syndrome may include excessive oral behaviour and certain adult personality traits (or character traits) such as dependence, greediness and passivity.
- **The anal stage**. During early childhood, this stage is common. According to Freud, the young child becomes highly conscious of anal activities, in most cases due to attempts by parents at toilet training. If the training is too strict and creates anxiety in the child with respect to these activities, it may result in compulsiveness, excessive conformity or self-control due to the adult anal syndrome.
- **The phallic stage**. When the stage of toilet training has been completed, the child develops an interest in his sexual organs. At this pregenital phallic stage, a child will develop loving, and even romantic, feelings towards the parent of the opposite sex, that is, a boy towards his mother and a girl towards her father. In the case of boys, Freud called this the Oedipus Complex, associated with the story of Oedipus who killed his father and subsequently married his mother when he became King of Thebes. According to Freud, this stage, accompanied by its Oedipal Complex, is the most significant stage in human development. At this stage, the child believes that he may be punished for the romantic attachment to the particular parent. For boys, this punishment could be castration by the jealous father which, in turn, creates anxiety and defences to this anxiety. Ultimately, the defence which emerges is identification with the threatening father and the boy endeavours to become like his father. In this process, the boy adopts the behavioural patterns and traits of his father, particularly with reference to what is right and wrong, which are a feature of his culture. In the case of girls, while the situation is slightly different, the principle is the same. Identification, and as a result of identification, the girl becomes indoctrinated into the culture as a result of the outcome of the resolution of the Oedipal conflict.
- **The latency period**. This period commences around the age of 6 years, following the pregenital stages. In this period, to the onset of puberty, no significant new psychological mechanisms emerge. The ego develops as the child is exposed to a range of learning situations at school, in particular. It entails a period of strengthening and elaborating of the defence mechanisms which developed in the earlier stages against conflict-producing anxiety.
- **The genital stage**. This stage commences at puberty. Instead of devoting attention to himself as was the case in earlier stages of development, normal heterosexual interests arise, interest is developed more towards other people and on considering entry to adulthood.

Personality traits

Gordon Allport, often called the father of personality theory, was very much a trait theorist. He believed in the individuality and uniqueness of the person and that people have consistent personalities. Allport believed that people have a range of traits or characteristic features:

- **Individual** – traits possessed by one person;
- **Common** – traits possessed by many people;
- **Cardinal** – a single trait that dominates one person;
- **Central** – a small number of important traits that may affect many people;
- **Secondary** – many consistent traits which are not often exhibited;
- **Motivational** – very strongly-felt traits; and
- **Stylistic** – less strongly felt traits.

Allport believed that through the writing of diaries, autobiographies and letters, an understanding of an individual's personality could be gained.

Supertraits

Hans Eysenck believed that all people could be described in terms of two supertraits which had a biological basis, namely:

- **Introversion–extroversion**: continuum of sociability, dominance, liveliness, etc; and
- **Emotionality–stability**: neuroticism, continuum of upset and distress.

A third supertrait, psychoticism, was added later. This is a predisposition towards becoming either psychotic or sociopathic (psychologically unattached to other people) and a tendency to be hostile, manipulative and impulsive.

The Big Five personality factors

In the last 20 years, a strong concensus on basic traits has emerged, namely five superordinate factors, referred to as 'The Big Five' or the 'Five Factor Model'. The Big Five are:

- **Neuroticism**–emotional stability;
- **Extraversion**–introversion;
- **Openness to experience**–closedness to experiences:
- **Agreeableness**–disagreeableness;
- **Conscientiousness**–lack of conscientiousness.

Each supertrait is measured by six facets, or subordinate traits, shown in Table 1.2.

Table 1.2 The Big Five personality traits

Neuroticism	Extraversion	Openness	Agreeableness	Conscientiousness
Anxiety	Warmth	Fantasy	Trust	Competence
Angry hostility	Gregariousness	Aesthetics	Straightforwardness	Order
Depression	Assertiveness	Feelings	Altruism	Dutifulness
Self-consciousness	Activity	Actions	Compliance	Achievement striving
Impulsiveness	Excitement-seeking	Ideas	Modesty	Self-discipline
Vulnerability	Positive emotion	Values	Tendermindedness	Deliberation

Measuring personality

It is possible to measure personality through the use of the STEN (Standard 10 Personality Factor Test).

Cattell (1965) considered responses to questionnaires from people on their various beliefs and preferences, subsequently developing a list of 16 'personality factors' (see Table 1.3). These factors are shown as 16 dimensions for which a person's level in each case can be recorded to produce a 'personality profile', which is unique to the person completing the questionnaire.

Table 1.3 Features of personality

	Features of personality 1 <——> 10	
1.	**Reserved**, detached, critical aloof	**Outgoing**, warmhearted, easy-going, participating
2.	**Less intelligent**, concrete thinking	**More intelligent**, abstract-thinking, bright
3.	**Affected by feelings**, emotionally less stable, easily upset	**Emotionally stable**, faces reality, calm, mature
4.	**Humble**, mild, accommodating, conforming	**Assertive**, aggressive, stubborn, competitive
5.	**Sober**, prudent, serious, taciturn	**Happy-go-lucky**, impulsively lively, gay, enthusiastic
6.	**Expedient**, disregards rules, feels few obligations	**Conscientious**, persevering, staid, moralistic
7.	**Shy**, restrained, timid, threat-sensitive	**Venturesome**, socially bold, uninhibited, spontaneous
8.	**Tough-minded**, self-reliant, no-nonsense	**Tender-minded**, clinging, overprotected, sensitive
9.	**Trusting**, adaptable, free of jealousy, easy to get along with	**Suspicious**, self-opiniated, hard to fool

(continued)

Table 1.3 (Continued)

	Features of personality 1 <——> 10	
10.	**Protective**, careful, conventional, regulated by external realities, proper	**Imaginative**, wrapped up in inner urgencies, careless of practical matters, Bohemian
11.	**Forthright**, natural, artless, unpretentious	**Shrewd**, calculating, worldly, penetrating
12.	**Self-assured**, confident, serene	**Apprehensive**, self-reproaching, worrying, troubled
13.	**Conservative**, respecting established ideas, tolerant of traditional difficulties	**Experimenting**, liberal, analytical, free-thinking
14.	**Group-dependent**, a 'joiner' and sound follower	**Self-sufficient**, prefers own decisions, resourceful
15.	**Undisciplined**, self-conflict, follows own urges, careless of protocol	**Controlled**, socially precise, following self-image
16.	**Relaxed**, tranquil, unfrustrated	**Tense**, frustrated, driven, overwrought

Under the Management of Health and Safety at Work Regulations, when entrusting tasks to his employees, an employer is required to take into account their capabilities as regards health and safety. He must ensure that the demands of the job do not exceed an employee's mental and physical capabilities to undertake the work without risks to himself or others. Personality testing could become quite a significant feature of such an assessment of a person's capabilities in this respect.

Ancestry and social background

A person's upbringing and social background are significant in the formation and retention of attitudes and in the development of personality. It is not uncommon for people to be classified as 'working class', 'upper class', 'blue collar', 'professional class', etc., whereby a certain standard of behaviour is expected in each case.

Ancestry is associated with a person's forefathers, where they came from, what they did and how they managed their lives. Attitudes and personality traits are commonly derived from parents and other family members as part of the growing up process.

Similarly, a person's social background in terms of the people with whom he associates in the home, at work and within a community has a direct effect on his behaviour. People are described as having a 'working class' background, implying their belonging to a particular class of society and, on this basis, are expected to have certain attitudes to, for instance, employers, the law and society in general.

Experience, intelligence, education and training

These aspects are significant in the development of individuals, their behaviour and potential for accidents.

Experience

Past experience of situations and events has a direct effect on human behaviour. The process of gathering experience is a continuous one from a child's early days at school through to the stage of adult maturity. People learn by their experiences in all sorts of ways. Positive experience may arise through actively learning to undertake a task correctly and safely. Negative experience, on the other hand, may be provided through learning from mistakes made, 'near miss' situations, such as those arising from careless driving, and the outcome of other people's accidents.

Intelligence

Intelligence can be defined as 'the ability to learn, manipulate concepts in the brain and solve problems'.

Intelligence can be measured by means of intelligence tests. Binet and Simon (1905) developed the concept of intelligence quotient (IQ) to describe his findings that, in children, greater intelligence seemed to result in a child being able to do certain tasks sooner than the average child.

$$\text{Intelligence quotient (IQ)} = \frac{\text{mental age}}{\text{chronological age}} \times 100.$$

By definition an average IQ is 100.

Theories of intelligence

A number of theories exist, but most are based on the subdivision of intelligence into a number of different *aptitudes*. A commonly used classification divides these aptitudes into six different categories, thus:

Verbal

Numerical

Spatial

Mechanical

Manual

Musical

Tests are available to deal with the above aptitudes. However, they need to be selected, administered and interpreted by specialists to be valid and meaningful.

Education

'Education' literally means a drawing out or leading out of the individual. Maritain (1952) defined education as 'a human awakening'. Education continues throughout life in many ways.

Training

Training, on the other hand, is concerned with the systematic development of attitude, knowledge and skill patterns required by an individual to perform adequately a given task or job (see Chapter 14, Systematic training).

Attitudes and behaviour

Many people endeavour to bring about changes of attitude on the part of others. For instance, a doctor may endeavour to change a patient's attitude towards a healthier way of life. Similarly, a health and safety practitioner, through training and consultation, may try to change the attitudes of operators towards operating the safety procedures that have been established. Changing attitudes is a difficult task, particularly as people commonly revert to their originally held attitudes once the pressure for the change is removed or lessened.

To be successful, attitude change must take place in a number of well-controlled stages:

- by attracting the attention of the individual to the fact that a change of attitude is needed; and
- by convincing that individual that his currently held attitude is wrong or inappropriate to the situation.

Cognitive dissonance

This is one of the barriers to attitude change, the conflict situation where a person takes an attitude which is not compatible with the information presented. People display cognitive dissonance in all sorts of situations. For instance, prior to the legislation requiring the wearing of seat belts by drivers and front seat passengers, many people expressed the view that such a requirement was an infringement of their civil liberties or that seat belts could be dangerous in an accident due to the driver becoming entangled. Whilst the research overwhelmingly showed that the wearing of a seat belt saved lives, these people endeavoured to rationalize their own particular resistance to the requirement.

People frequently display cognitive dissonance when required to wear personal protective equipment, such as hearing protection or head protection. Common responses are:

The ear muffs are uncomfortable.

The hard hat gets in the way when I am working.

Where cognitive dissonance is encountered amongst groups of workers, the ideal group attitude change process should take place as a series of specific considerations by each individual in the group, thus:

1. I don't use hearing protection. You can't hear what people say.
2. They say you can go deaf if you don't wear it all the time.
3. I could go deaf eventually.
4. If I wear the protection, it should stop me going deaf.
5. OK! I'll wear the ear muffs from now on.

Approaches to change

Organizations, in most cases, must bring about change in order to survive, to introduce new work processes or management systems, to increase profitability and to improve service. How an organization approaches the introduction of change is crucial to successfully bringing about this change. This has been seen with, for example, the introduction of Quality Management systems in organizations.

The majority of people are averse to change. Inevitably, this requires a change of attitude on the part of management and employees. Organizations must, therefore, manage effectively any changes they wish to install (see Chapter 17, Change and change management).

Changing attitudes

A number of factors must be considered when endeavouring to change attitudes towards safer working practices, in particular:

- **The individual** can be affected by longstanding opinions, past experience of the work, the level of intelligence and education, actual motivation towards change and the extent that people are prepared to blindly follow those whose views and opinions they value.

- **Attitudes currently held** may be affected by cognitive dissonance, self-image, group norms, the potential for financial gain, the opinions of other people, individual skills available.
- **The situation** i.e. group situations, where membership of the group relies heavily on complying with the group norms or standards, the pressures exerted by change agents, such as officers of the enforcement agencies, health and safety practitioners and insurance companies, the sanctions imposed or that could be imposed, the prestige attached to the change required and the climate for change within an organization; and
- **Management example** which is the strongest of all motivators to bring about attitude change.

The human factors issues

Current legislation requires organizations to take a human factors-related approach to occupational health and safety, which means that managers must consider human capability from a health and safety viewpoint when entrusting tasks to their employees and clients.

Human capability

What is meant by 'human capability'? Various terms are found in the average dictionary: able, competent, gifted and having the capacity. Perhaps the last term is the most significant from a health and safety viewpoint. Capacity implies both mental and physical capacity, for instance, the mental capacity to understand why a task should be undertaken in a particular way, and physical capacity, in terms of the actual physical strength and fitness to undertake the task in question.

Human factors

Human factors, in its application to occupational health and safety, has been defined as

> a range of issues including the perceptual, physical and mental capabilities of people and the interactions of individuals with their jobs and working environments, the influence of equipment and system design on human performance and those organizational characteristics which influence safety-related behaviour at work.

Although most health and safety legislation places the duty of compliance firmly on the employer or body corporate, this duty can only be discharged by the effective actions of its managers. For instance, it is management's job to report certain incidents to the enforcement authority and to investigate accidents and ill health arising from work, but frequently the causes are written down to 'operator carelessness', 'not looking where he was going' or, quite simply, 'human error', indicating that nothing can be done and no further action should be taken.

Perception of risk

People perceive risk in different ways and no two people necessarily perceive the same risk in the same way. How people perceive risk may be associated with, for example, the skills available to the individual, motivational factors, past experience, the context in which a stimulus is produced, ergonomic factors, such as the layout of controls and displays to machinery and vehicles, a person's level of arousal and the degree of competence in a particular task.

In driving situations, factors such as prevailing weather conditions, the state of the road, the width of the road, density of traffic and the driver's past experience have a direct effect on that driver's perception of risk (see Pre-accident and post-accident strategies below).

Risk homeostasis

'Homeostasis' is derived from the Greek words for 'same' and 'steady'. In its biological application, it refers to the ways the body acts to maintain a stable internal environment in spite of environmental variations, such as temperature and blood sugar levels, and disturbances which may be of a stressful nature.

Risk homeostasis was first propounded by Gerald J.S. Wilde (1994) who said:

> Give me a ladder that is twice as stable and I will climb it twice as high, but give me a cause for caution, and I'll be twice as shy.

Wilde's Risk Homeostasis Theory is based on a number of premises. The first is the notion that people have a target level of risk, that is, the level of risk they accept, tolerate, prefer, desire or select. This target level of risk depends upon perceived benefits and disadvantages of safe and unsafe behaviour alternatives, and it determines the degree to which they will expose themselves to health and safety hazards.

The second premise is that the actual frequency of lifestyle-dependent death, disease and injury is maintained over time through a closed-loop, self-regulating control process. Fluctuations in the degree of caution people apply in their behaviour determine the variations in the loss to their health and safety. Furthermore, the ups and downs in the amount of actual lifestyle-dependent loss determine the fluctuations in the amount of caution people exercise in their behaviour.

Finally, the third premise holds that the level of loss to life and health, in so far as this is due to human behaviour, can be decreased through interventions that are effective in reducing the level of risk people are willing to take by means of programmes that enhance people's desire to be alive and healthy.

Behavioural safety – what does it mean?

'Behavioural safety' is the systematic application of psychological research on human behaviour to the problems of safety in the workplace (Cooper D., 1999).

Fundamentally, it is a means of gaining further improvements in safety performance through promoting safe behaviours at all levels in the workplace. However, it is not a substitute for the important and traditional approaches of, for example, risk assessment and the implementation of 'safe place' and 'safe person' strategies, but rather adds a new dimension to emphasize employee involvement and personal responsibility. It is a natural progression of safety management from

- the highly regulated and disciplined early approach to health and safety of prescriptive legislation and punishment; through to
- the safety management systems approach which most organizations have been operating for some time; to
- a system which recognizes employees as mature human beings with a genuine interest in their own well-being, who contribute best when they can see that they, themselves, can have an influence on their own safety.

In order to achieve this transition, it is necessary to change the 'culture' and attitudes of all levels of the organization, a process which does not achieve results quickly.

Behavioural safety techniques are based primarily on observation, intervention and feedback as a means of changing behaviour. They incorporate a number of features:

- identification of behaviours which could contribute to, or have contributed to, accidents;
- a system of on-going observation and feedback (intervention); and
- the use of the information obtained to identify corrective actions.

Behavioural safety programmes

According to Sulzer-Azeroff and Lischeid (1999), a behavioural safety programme should comprise a number of essential elements:

- significant workforce participation;
- the targeting of specific unsafe behaviours;

- the collecting and recording of observational data via peer-to-peer monitoring;
- decision-making processes being data-driven only;
- a systematic, scheduled observational improvement intervention;
- the provision of regular focussed feedback about on-going safety performance;
- all levels of personnel involved;
- demonstrable and visible on-going support from management and supervisors.

The implementation of a behavioural safety programme should take place in a number of clearly defined stages.

1. Obtain the involvement and commitment of the workforce through consultation, discussion and the provision of information as to what behavioural safety entails, what it means to them and what they are expected to do to make the system work in conjunction with management.
2. Install a project team or steering committee to implement and monitor progress in the development of the programme.
3. Analysing recent accident and near miss records to identify those unsafe behaviours responsible for a substantial amount of the organization's accidents.
4. Develop workgroup-specific check lists that incorporate the particular behaviours identified above.
5. Train personnel from each work group in observation techniques and the means of providing feedback to the people concerned.
6. Establish a baseline, that is, monitor behaviour in the workplace for a period of, say, 4 weeks to ascertain the current average levels of safe behaviour in each department or work area.
7. Establish with each department or work group a safety improvement target, using their baseline average as the comparison starting point.
8. Monitor progress on a daily basis and provide detailed feedback to each department or work group on a weekly basis.
9. Review performance trends to identify any barriers to improvement.
10. Provide feedback, by way of briefing sessions, on the success or otherwise of the programme at regular intervals, seeking the views, recommendations and opinions of all personnel.

Perception of risk and its influence on risk-taking behaviour

People take risks in all sorts of situations, for example, whilst operating machinery, driving vehicles and in working at height. As stated earlier in this chapter, perception of risk is associated with a number of psychological mechanisms, such as motivation, the level of arousal, past experience and the skills available to an individual.

Various theories relating to personal risk-taking have been put forward and these are summarized below.

Personal risk-taking

What is a risk?

Various definitions of the term can be noted:

- chance of loss or injury;
- to expose to a hazard;
- the probability of harm, damage or injury;
- to expose to mischance;
- the probability of a hazard leading to personal injury and the severity of that injury.

Risk-taking theories

Mathematical models of people's behaviour

The American school of psychologists, e.g. Ward Edwards, said that one can consider risk-taking as an attempt by the individual to maximize some function of probability and the value of the outcome of his decision. The logical way is to calculate probability and the value of all possible outcomes and rank them according to degree. His experiments with gambling showed specific preferences for betting situations, i.e. probability (p) × value (v), but this simple equation does not predict people's behaviour and is very crude statistically. For instance, it does not explain why people will gamble on only very low probabilities, e.g. on the football pools.

Other factors affecting risk and the value of an outcome are the urge to win and the severity of injury or loss resulting from a risk-taking activity. In gambling situations, individual differences between the sorts of risk people prefer to take can be noted. Cohen studied compulsive gamblers, showing that the probability of winning was the overriding factor in their taking risks, the value of the outcome being insignificant, i.e. the greater the risk, the happier they were – a 'people battle' between the gambler and 'fate'.

Risk-taking in relation to arousal

Generally, if people can avoid taking risks, they will do so. Some people, however, enjoy taking risks as in the case of Cohen's study of compulsive gamblers. The question of arousal and the level of personal risk-taking is based on the fact that when people take risks, e.g. physical, financial risks, it increases their level of arousal up to an optimum level. Once this level of arousal was reached, however, people tend to take fewer risks.

Taylor examined the aspect of galvanic skin response (the basis for the lie detector test) in driving experiments and found a relationship between road conditions and the level of arousal. Independent evidence to support Taylor's findings was produced by Cohen who studied the effects of alcohol on individual risk-taking. It was noted in these studies that drivers were more willing to drive their buses through narrower gaps when affected by alcohol.

Risk thresholds

Steiner *et al.* (1970) tried to measure people's risk thresholds, i.e. the levels that people would prefer to operate at in terms of risk. They considered physical risk-taking (e.g. walking in front of a bus), financial risk-taking (e.g. insurance, playing the stock market, gambling on horses) and social risk-taking (e.g. activities which could result in loss of face or reputation and social status). Whilst their results are not clear-cut, there emerges a broad threshold across these areas for some people, combined with a personality factor or variant, the latter being the more important.

Achievement motivation

Atkinson (1964) related risk-taking to the concept of achievement motivation, which is derived from motivation theory. He showed that there are many sorts of motivation affecting people, namely fear of failure (negative) and motivation towards success (positive). Such factors can be measured by a well-validated questionnaire. Atkinson's ideas stemmed from a larger study of business risk-taking by McClelland *et al.* (1953), who examined the factors that motivated people to establish their own businesses, which involves some degree of financial risk, as opposed to the relative job security of the large organization. He showed that achievement motivation, the desire to 'do your own thing' was a strong factor in the first case, i.e. they actually enjoyed the financial risk-taking involved.

Atkinson went further to produce a hypothesis on people behaviour relating to throwing balls into a basket. With this type of task, people high on the 'fear of failure' rating tried to get out of the task, whereas those high on the 'motivation towards success' rating readily co-operated.

Skill versus chance

There is a great difference in people's behaviour in the skill versus chance situation. With skilled tasks, people find the risk challenging (Atkinson, 1964). They feel their behaviour is influencing the situation. However, when people perceive their behaviour does not affect the situation, i.e. a chance situation, then their behaviour is determined largely by the value of the outcome and the probability value. In other words, they act more directly towards the mathematical model. However, when skill enters the situation, the value of the risk is important – a challenge to the person's behaviour.

The skill versus chance situation is important in the evaluation of occupational risks as shown in Table 1.4.

Another way is to get people to judge the risks in various work situations from photographs and correlate the risks with accident rates. People who have many accidents tend to judge the risks higher and vice versa, but they judged the consequences of taking the risk as being much less serious, i.e. probability high, severity low.

Spaltro (1969) used people's judgements of situations. Similar results emerged from these studies, i.e. high accident people overestimated the risks but underrated the outcome. High accident people would also not change their behaviour to reduce the pain and suffering from accidents. Latent aggression is important in this case. It increases in order to sustain people in their work.

Summary of risk-taking concepts

Accidents can be caused for a number of reasons.

- The information received is incorrect – the objective danger is higher than the subjective perception of risk involved.
- The skill is not available – a difficult situation is where the build-up of danger is so small as to be imperceptible.
- The individual's motivation is inappropriate to the circumstances.
- The probability matrix is incorrect.
- Feedback is ineffective – where people do not learn by their experiences.

Table 1.4 The skill versus chance situation

Value of outcomes	Probability (value of risk)
Affected by:	
Punishment	Personality
Financial gain	The situation
The severity of harm or injury	
Other people's opinion	

In endeavouring to change the behaviour of people with regard to attitudes, an indication as to which of the above five factors is incorrect or wrong is important.

Accident causation

What are the causes of accidents?

Why do some people have more accidents than others?

Is there such a thing as accident proneness?

The above questions feature prominently in the study and investigation of accidents and the causes of occupational ill-health. Hazards may result in immediate physical injury, long-term physical injury, long-term damage to health, including psychiatric injury associated with stress at work. It is possible to construct accident causation models, such as that produced by Hale and Hale (1972) (see Chapter 5, Perception of risk and human error).

However, there is a basic chain of causation in any accident situation, commencing with a set of events or circumstances which leads to the event resulting in the accident or ill-health condition. The objective of any accident prevention strategy is, therefore, the breaking of this sequence by some form of pre-accident strategy.

Classification of hazards

Hazards at work can be classified as follows.

- **Physical hazards** Physical hazards are those associated with physical agents, such as heat, noise, vibration and humidity, friction and pressure, together with inadequate lighting and ventilation. People may suffer conditions such as heat stroke, noise-induced hearing loss, bursitis, visual fatigue and hand-arm vibration syndrome as a result of exposure.
- **Structural hazards** These include hazards associated with the structure of buildings, such as floors, staircases, windows, doors and gates. People are exposed to the risk of falling from a height, falling through open windows, falls on the same level and being struck by a door.
- **Chemical hazards** Many people are exposed to hazards from chemical substances used at work, such as acids, solvents, alkalis and dangerous substances, such as asbestos and lead. Exposure can result in many ill-health conditions, such as lead poisoning, acid burns, occupational cancers, asbestosis and occupational dermatitis, the most common of all occupational diseases. In considering chemical hazards, not only must the nature of the substance be considered, but also its form, e.g. a gas, fume, dust, etc.

- **Biological hazards** Harmful exposure to various forms of microorganism, such as bacteria and viruses, may arise in some occupations. These organisms which cause disease may be
  - animal-borne in the case of zoonoses, such as brucellosis;
  - vegetable-borne, as with aspergillosis (farmer's lung); or
  - associated with humans, as in the case of viral heptatitis.
- **Psychological hazards** Over the last decade, considerable attention has been paid to stress at work and the risk of employees contracting psychiatric injury (see Chapter 18, Stress and stress management). Stress-induced injury may arise as a result of too much or too little responsibility, conflicting job demands, deficiencies in interpersonal skills or incompetent superiors. Bullying and harassment at work may be a contributory factor in stress-induced ill-health.

Risk assessment

The risk assessment process endeavours to predict the probability or likelihood of the above accident and ill-health situations arising together with the severity of injury or ill-health that could arise, taking into account the requirements and prohibitions of current health and safety legislation at the same time.

Risk assessment is directly tied up with accident causation theory and it is important, therefore, to consider a number of theories of accident causation when endeavouring to answer the questions above.

Accident causation theories

A number of theories have been put forward over the last century with respect to the causes of accidents.

The Pure Chance Theory

This theory states that everyone in the population has an equal chance of sustaining an accident. It suggests that no discernible pattern emerges in the events that lead up to an accident. An accident is usually treated as an Act of God, leaving one to accept the fact that prevention is non-existent.

The Biased Liability Theory

This theory considers that once a person sustains an accident, the probability that the same person will sustain a further accident in the future has either decreased or increased when compared with the rest of the population at risk. If the probability has increased, the phenomenon is referred to as the *Contagion Hypothesis*. If the probability has decreased on the other hand, it is commonly called the *Burned Fingers Hypothesis*.

The Domino Theory

One of the more colourful theories of accident causation was formulated by Heinrich in 1959, and is known as the Domino Theory. This theory explains the accident process in terms of five factors:

1. ancestry and social environment;
2. fault by the person;
3. the unsafe act and/or mechanical or physical hazard;
4. the accident;
5. the injury.

These factors are in a fixed and logical order. Each one is dependent on the one immediately preceding it, so that if one is absent, no injury can occur. The theory can be visualized as five standing dominoes in which the behaviour of these dominoes is studied when subjected to a disturbing force. When the first, social environment, falls, the other four follow automatically unless one of the factors has been corrected, i.e. removed, thereby creating a gap in the required sequence for producing an accident. The five factors in Heinrich's Domino Theory are described in Table 1.5.

The Domino Theory was subsequently extended by Bird and Loftus (1984) to include the influence of management as part of the causes and effects of accidents. A modified sequence of events is shown in Table 1.6.

This modified accident sequence is applicable to most types of accident and can be shown in diagrammatic form (Figure 1.3).

Multiple Causation Theory

Heinrich's theory (1931) is very much a theory of single causation. However, very rarely is there one single cause of an accident. Multiple causation (or causality) refers to the fact that there may be more than one cause of an accident (Figure 1.4).

Each of these multicauses is equivalent to the third domino in the Heinrich Theory and can represent an unsafe act or condition or situation. Each of these can itself have multicauses and the process during accident investigation of following each branch back to its root is known as fault tree analysis.

The theory of multicausation is that contributing causes combine together in a random fashion to result in an accident. During accident investigations, there is a need to identify as many of these causes as possible. For example, a maintenance employee notices that the eaves gutter to a single-storey building is overflowing during periods of rain, suggesting that the outlet to the rainwater pipe is blocked with debris. He shins up the rainwater pipe and reaches into the outlet to remove

Table 1.5 Heinrich's Domino Theory

1. **Ancestry and social environment**	Recklessness, stubbornness, avariciousness and other undesirable traits of character that may be passed along through inheritance. Environment may develop undesirable traits of character or may interfere with education. Both inheritance and environment may cause faults of person.
2. **Fault of person**	Inherited or acquired faults of person, such as recklessness, violent temper, nervousness, excitability. These constitute reasons for committing unsafe acts or for the existence of mechanical or physical hazards.
3. **Unsafe act and/or mechanical/physical hazard**	Unsafe performance of persons, such as standing under danger areas, careless starting of machines, removal of safeguards and horseplay; mechanical or physical hazards, such as unguarded gears or points of operation, insufficient light that result in accidents.
4. **Accident**	Events such as falls of persons, striking of persons by flying objects and entanglement in moving machinery parts are typical accidents that cause injury.
5. **Injury**	Fractures, amputations, loss of eyes, cuts and lacerations are injuries arising directly from accidents.

Table 1.6 Extended Domino Theory

1. Lack of control by management, permitting
2. the basic causes (i.e. personal and job factors) leading to
3. the immediate causes (such as substandard practices, conditions or errors), which are the direct cause of
4. the accident, which results in
5. a loss (which may be categorized as negligible, minor, serious or catastrophic.

the debris. However, his foot slips on the brickwork surface to the wall and he falls backwards, fracturing his wrist in the fall. The single causation approach adopted by Heinrich would analyse this accident thus:

1. Unsafe condition – overflowing eaves gutter;

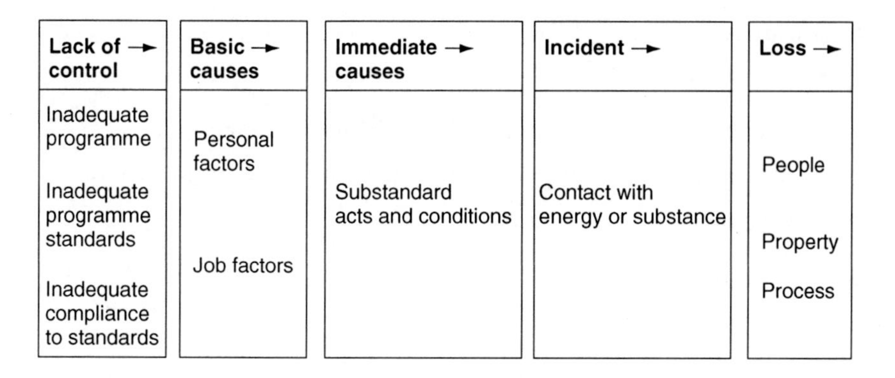

Figure 1.3 A modified accident sequence (Bird and Loftus, 1984)

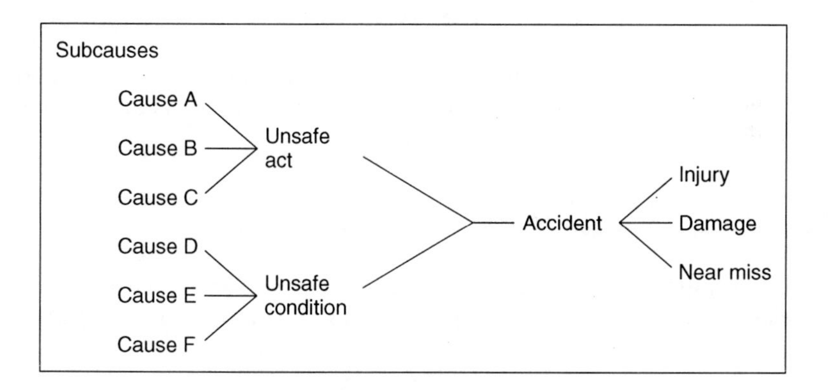

Figure 1.4 Multiple causation

2. Unsafe act – failing to use a step ladder;
3. Cause – foot slipped on brickwork.

The multiple-causation approach would view this situation differently.

1. Unsafe condition – overflowing gutter which should have been noted by maintenance employees during previous rainfall.
2. Unsafe act – the maintenance employee attempted to take a 'quick fix' solution by shinning up the rainwater pipe instead of using a step ladder.
3. Cause – unsafe working practice due to the absence of a formal procedure for work at height. Under normal circumstances, the defect should have been noted and reported to the maintenance department. This action would have initiated a specific procedure for this task including the provision of correct access equipment and a safe system of work, perhaps with a second employee securing the ladder during removal of debris by the first employee.

The chances of being hurt are related to the various multicauses identified which may be both direct or indirect multicauses.

Unsafe acts

Unsafe acts include, for example,

- taking a shortcut across a designated vehicle movement area;
- standing on a swivel chair to reach something out of normal reach; or
- working on a roof without a crawl board.

Unsafe conditions

Typical unsafe conditions include:

- inadequately guarded machinery;
- excessively high temperature levels in a working area; and
- a defective floor finish in a storage area.

Multiple causation theory endeavours to identify all unsafe acts and conditions which contribute to injury, damage or loss with a view to designing safe systems of work and safety procedures.

The ILCI Loss Causation Model

The International Loss Control Institute Loss Causation Model views the causes of accidents leading to loss as taking place in a series of stages (Figure 1.5).

Principal behavioural causes of accidents

From a consideration of the above theories and analysis of accident reports, it is possible to specify the principal behavioural causes of accidents and the actions necessary by management (see Table 1.7).

Fault tree analysis

Most accidents are the result of a sequence of events. This sequence of events, which may have been established after accident investigation, will indicate in most cases only one series of events that could have resulted in the accident. This narrow approach may lack details of some of the events which took place. An accidental chain of events is a complex relationship between human, machine and environment. The structure of a desirable accident analysis taking into account all the events which could have led to it is shown in Figure 1.6.

An ideal technique for accident analysis should focus on the possibility of occurrence of one critical event at the time and indicate the complex relationship that can cause it. Fault tree analysis is used to predict that combination of events and circumstances which could cause the critical event.

ILCI LOSS CAUSATION MODEL

1. LACK OF CONTROL

Failure to maintain compliance with adequate standards for:

- Leadership and administration
- Management training
- Planned inspections
- Job task analysis and procedures
- Job and task observations
- Job task observations
- Emergency preparedness
- Organizational rules
- Accident and incident investigations
- Accident and incident analysis
- Personal protective equipment
- Health control and services
- Programme evaluations systems
- Purchasing and engineering systems
- Personal communications
- Group meetings
- General promotion
- Hiring and placement
- Records and reports
- Off-the-job safety

CREATES:

2. BASIC CAUSES

Personal factors

- Inadequate capability
  – physical/physiological
  – mental/psychological
- Lack of knowledge
- Lack of skill
- Stress
  – physical/physiological
  – mental/psychological

Job factors

- Inadequate leadership or supervision
- Inadequate engineering
- Inadequate purchasing
- Inadequate maintenance
- Inadequate tools, equipment, materials
- Inadequate work standards
- Abuse and misuse
- Wear and tear

LEADING TO:

3. IMMEDIATE CAUSES

Substandard practices

- Operating equipment without authority
- Failure to warn
- Failure to secure
- Operating at improper speed
- Making safety devices inoperable
- Removing safety devices
- Using defective equipment
- Failing to use personal protective equipment properly
- Improper loading
- Improper placement
- Improper lifting
- Improper position for task
- Servicing equipment in operation
- Horseplay
- Under influence of alcohol/drugs

Substandard conditions

- Inadequate guards or barriers
- Inadequate or improper protective equipment
- Defective tools, equipment, materials
- Congestion or restricted action
- Inadequate warning system
- Fire and explosion hazards
- Poor housekeeping, disorder
- Noise exposure
- Radiation exposure
- Temperature extremes
- Inadequate or excess illumination
- Inadequate ventilation

Figure 1.5 The International Loss Control Institute Loss Causation Model

CAUSING:

4. INCIDENT

Contacts

- Struck against
- Fall to lower level
- Caught in
- Caught between
- Overstress, overexertion, overload
- Struck by
- Fall on same level
- Caught on
- Contact with

with the resulting:

5. LOSS

Personal harm

- Major injury or illness
- Serious injury or illness
- Minor injury or illness

Property Damage

- Catastrophic
- Major
- Serious
- Minor

Process loss

- Catastrophic
- Major
- Serious
- Minor

Figure 1.5 Continued.

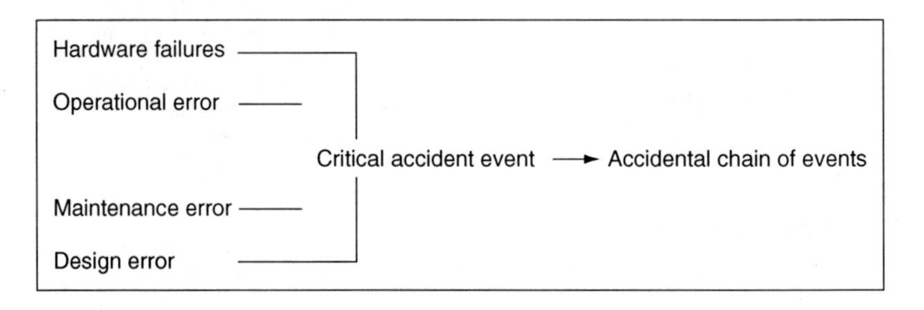

Figure 1.6 Fault tree analysis

The fault tree analysis system

Fault tree analysis is a graphical system which provides a systematic description of the combination of all possible occurrences in a system which can result in a fault or undesired event. It is a failure-orientated technique and can help determine where problems are likely to arise; information which can be used not only to improve existing products, but also to aid in initial design and development stages of new products.

The graphical construction of a fault tree starts from the 'top event' or accident, resulting in injury, release of toxic gases, fire, explosion, etc., and works its way

Table 1.7 Principal behavioural causes of accidents

Behavioural causes	Corrective action	Control
Incorrect or wrong attitude		
e.g. personal risk-taking, ignoring instructions, absent-mindedness	Line management	Line management
	Disciplinary procedure	HR manager
Lack of knowledge or skill		
Inexperience, immaturity, lack of skill and experience, ignorance of job safety aspects	Job safety analysis	Manager, line manager
	Job safety instructions Job safety training	
Job suitability		
Physical/mental incapability, e.g. deafness, restricted vision, restricted intelligence	Pre-employment health screening	Occupational health nurse, HR manager
	Regular health examinations Job placement	

down or back to the basic cause. The various events or steps, which can be represented by standard symbols are:

1. Analyse the product or system carefully. Determine the sequence of events for normal operation, normal and abnormal operating environments, and their safety implications.
2. Define the top event of the fault tree or the starting point for analysis. This may be the total failure of the system or human injury.
3. Initiate actual construction of the tree. Determine in a logical manner the events that can cause the top event and work out at each level of the tree all possible combinations of faults and events that lead to each level. Expand each branch until you reach basic causes or undeveloped events. Display this graphically with the standard fault tree symbolism.

The usefulness of the fault tree technique can be greatly enhanced by the addition of quantitative data. In this manner, not only can the fault paths be identified, but their probability of occurrence may be established.

Pre-accident and post-accident strategies

What is an accident?

The term 'accident' has been defined in a number of ways.

- Oxford Dictionary: An unforeseeable event often resulting in injury.
- British Safety Council: A management error; the result of errors or omissions on the part of management.
- Royal Society for the Prevention of Accidents (RoSPA): Any deviation from the normal, the expected or the planned, usually resulting in injury.
- Frank Bird (1974), American Exponent of 'Total Loss Control': An unintended or unplanned happening that may or may not result in personal injury, property damage, work process stoppage or interference, or any combination of these conditions under such circumstances that personal injury might have resulted.
- Health and Safety Unit, University of Aston: An unexpected, unplanned event in a sequence of events that occurs through a combination of causes. It results in physical harm (injury or disease) to an individual, damage to property, business interruption or any combination of these effects.

The pre-accident situation

In any situation prior to an accident taking place, two important factors must be considered, namely:

1. **The objective danger**. This is the objective danger associated with a particular practice, system of work, item of equipment or part of a workplace, e.g. dangerous stairs, defective lifting appliance, etc., at a particular point in time.
2. **The subjective perception of risk on the part of the individual**. People perceive risks differently according to a number of behavioural factors, such as attitude, motivation, training, visual perception, personality, level of arousal and memory. People also make mistakes.

Objectives of accident prevention

The principal objectives of any accident prevention programme should be:

- reducing the objective danger present through, for instance, ensuring good standards of structural safety; and
- bringing about an increase in people's perception of risk, through instruction and training, supervision, the operation of safe systems of work.

Pre-accident strategies

These can be classified as 'Safe Place' and 'Safe Person' strategies.

The principal objective of a 'safe place' strategy is to ensure a reduction in the objective danger to people at work. These strategies feature in much of the occupational health and safety legislation that has been enacted over the last century, in particular, the Health and Safety at Work Act 1974. They may be classified as follows:

- **Safe premises**. This relates to the general structural requirements of buildings, such as the soundness of floors and staircases, and the specific safety features of windows, glazed doors and partitions. Environmental working conditions, such as the levels of lighting, ventilation, temperature and humidity, feature in this classification.
- **Safe work equipment**. A wide range of appliances and other forms of work equipment, their power sources, location and use are relevant in this case. The safety aspects of work equipment, procedures for checking and testing new appliances and equipment together with systems for their maintenance and cleaning must be considered.
- **Safe working practices**. All factors contributing to the operation of a specific working practice must be considered, e.g. the safety of electrical appliances, procedures for handling disabled people, control of hazardous substances used at work, and the operation of chairs with lifting and tilting mechanisms.
- **Safe materials**. Significant in this case are the health and safety aspects of potentially hazardous chemical substances, such as laundry detergents, disinfectants and the specific hazards associated with the handling of these materials. Adequate and suitable information on their correct use, storage and disposal must be provided by manufacturers and suppliers. Where substances are classed as Toxic, Corrosive, Harmful or Irritant, there may be a need to assess health risks for particular circumstances of use in accordance with the Control of Substances Hazardous to Health (COSHH) Regulations.
- **Safe systems of work**. A safe system of work is defined as 'the integration of people, machinery and materials in a correct working environment to provide the safest possible working conditions'. The design and implementation of safe systems of work is, perhaps, the most important 'safe place' strategy. It incorporates planning, involvement of staff, together with training and designing out hazards which may have existed with previous systems of work. In certain cases, a safe system of work may need to be documented.
- **Safe access to and egress from work**. This refers to access to, and egress from, both the workplace from the road outside and to and from the working area. Consideration must be given, therefore, to the state of approach roads, yards, operations at high level, such as window cleaning, and the use of portable and fixed access equipment, e.g. ladders and lifts.

- **Adequate supervision**. The Health and Safety at Work Act requires that in all organizations there must be adequate safety supervision directed by senior management through supervisory management to staff.
- **Competent and trained staff**. The general duty to train staff and others is also laid down in the Health and Safety at Work Act and, more specifically, in the Management of Health and Safety at Work Regulations. All employees need some form of health and safety training and certain people may need to be competent in high risk situations, e.g. electricians. Health and safety training should be provided at the induction stage and where staff may be exposed to new risks through a change of responsibilities, the introduction of new equipment, the introduction of new technology or new systems of work. Managers must appreciate that well-trained staff are essential and, in organizations which regularly provide health and safety training and supervision, accident and ill-health costs tend to be lower.

Generally, 'safe place' strategies provide better protection than 'safe person' strategies. However, where it may not be possible to operate a 'safe place' strategy, then a 'safe person' strategy must be used. In certain cases, a combination of 'safe place' and 'safe person' strategies may be appropriate.

The main aim of a 'safe person' strategy is to increase people's perception of risk. One of the principal problems with such strategies is they that depend upon the individual conforming to certain prescribed standards and practices, such as the use of certain items of personal protective equipment. Control of the risk is, therefore, placed in the hands of the person whose appreciation of the risk may be lacking or even non-existent.

'Safe person' strategies may be classified as follows:

- **Care of the vulnerable**. In any work situation there will be some people who are more vulnerable to certain risks than others. Typical examples of 'vulnerable' groups are young persons who, through their lack of experience, may be unaware of hazards, pregnant female staff, where there may be a specific risk to the unborn child, and physically and mentally disabled clients, whose capacity to undertake certain tasks may be limited. In a number of cases there may be a need for continuing medical and/or health surveillance of such people.
- **Personal hygiene**. The risk of occupational skin conditions caused by contact with hazardous substances, such as detergents, fuels, glues, adhesives and a wide range of chemical skin sensitizers, needs consideration. There may also be the risk of ingestion of hazardous substances as a result of contamination of food and drink and their containers. Personal hygiene is very much a matter of individual upbringing. In order to promote good standards of personal hygiene, therefore, it is vital that the organization provides adequate washing facilities for use by staff, particularly prior to the consumption of food and drink and on returning home at the end of the work period.

- **Personal protective equipment**. Generally, the provision and use of any item of personal protective equipment (PPE) must be seen either as a last resort when all other methods of protection have failed or an interim method of protection until some form of 'safe place' strategy can be put into operation. It is by no means a perfect form of protection in that it requires the person at risk to use or wear the equipment all the time they are exposed to a particular hazard.
- **Safe behaviour**. Staff must not be allowed to indulge in unsafe behaviour or 'horseplay'. Examples of unsafe behaviour include smoking in designated 'No Smoking' areas, the dangerous driving of vehicles and the failure to wear or use certain items of PPE.
- **Caution towards danger**. All staff and management should appreciate the risks at work and these risks should be clearly identified in the local Statement of Health and Safety Policy, together with the precautions required to be taken by staff to protect themselves from such risks.

Post-accident (reactive) strategies

Whilst principal efforts must go into the implementation of proactive strategies, it is generally accepted that there will always be a need for reactive or 'post-accident' strategies, particularly as a result of failure of the various 'safe person' strategies. The problem with people is that they forget, they take shortcuts to save time and effort, they sometimes do not pay attention or they may consider themselves too experienced and skilled to bother about taking basic precautions.

Post-accident strategies can be classified as follows:

- **Disaster/contingency/emergency planning**. Here, there is a need for managers to ask themselves this question. 'What is the very worst possible type of incident or event that could arise in our operations?' For most types of organization, this could be a major escalating fire, but other types of major incident should be considered, such as an explosion, collapse of a scaffold used by contractors, flood or major traffic accident. The need for some form of emergency plan should be considered.
- **Feedback strategies**. Accident and ill-health reporting, recording and investigation provides feedback as to the indirect and direct causes of accidents. The study of past accident causes provides information for the development of future proactive strategies. The limitations of accident data, e.g. accident incidence rates, as a measure of safety performance should be appreciated.
- **Improvement strategies**. These strategies are concerned with minimizing the effects of injuries as quickly as possible following an accident. They will include the provision and maintenance of first aid arrangements, including the training and retraining of first aid personnel and procedures for the rapid hospitalization of seriously injured persons.

Human behaviour and pre-accident strategies

Both 'safe place' and 'safe person' strategies are directly connected with the way people behave at work. 'Safe person' strategies, in particular, are directly concerned with how people perceive risk and, again, these strategies should be considered in the risk assessment process.

Accident proneness

A person is said to be 'accident prone' when he has more accidents than other people. The concept of accident proneness is very central to the psychology of accidents and has been the subject of considerable attention in the study of accident causation. As such, the term is a truism. One theory in particular, the Unequal Initial Liability Theory, has been the subject of much discussion in the history of accident causation research. This theory postulates that there exists a certain subgroup within the general population that is more liable to incur accidents. This is based on the fact that there may be some innate personality characteristics that cause accident-prone individuals to have more accidents than non-accident-prone people.

Fundamentally, accident proneness does exist to an infinitesimally small extent and in a very small number of cases, but not to such an extent as to warrant great consideration. Generally, evidence with traffic accidents indicates that some degree of accident proneness exists among certain drivers, whereas with occupational accidents this is not the case.

Moreover, people may go through periods of accident proneness from which they eventually recover.

Unsafe behaviour

Years ago many people held the view that, if organizations complied with the law, there would be no accidents. The problem with the law is that it does not take into account, in general, the question of human behaviour and, in some cases, unsafe behaviour.

People are different in terms of skills, knowledge, experience, attitude, motivation and other psychological elements. To this extent, behavioural safety endeavours to examine those elements of human behaviour which are a contributory factor in accidents and ill-health. A substantial number of workplace accidents are instigated through unsafe behaviour and employers need to be aware that reducing accidents can only be achieved by identifying, examining and focussing upon such behaviour.

We all see examples of unsafe behaviour in the workplace, typically

- reading whilst walking;
- failure to use the handrail on a staircase;

- failure to use or wear personal protective equipment or not wearing the equipment correctly;
- over-reaching when using a ladder;
- parking vehicles in designated traffic routes.

Many more examples could be added to the list.

Conclusions

Managers need to place greater emphasis on human behaviour at work and would benefit greatly from training in areas of behaviour which can contribute to unsafe situations and accidents. They should be able to appreciate why people do the things they do, why they do not follow the safety procedures and the human characteristics that contribute to accidents.

Employees, similarly, would benefit from training based on behavioural safety aspects and the factors that can affect their behaviour adversely whilst at work.

Key points

- Human behaviour has a direct effect on the safety performance of organizations in a variety of ways.
- Occupational psychology is concerned with the study of human behaviour at work.
- Human behaviour is associated with psychological factors, such as attitude, motivation, perception, personality and memory.
- Experience, intelligence, education and training have a direct effect on people's potential for accidents.
- Changing attitudes to health and safety is one of the more difficult tasks that employers must embrace.
- Employers need to take a human factors-related approach to health and safety in line with their duty under the Management of Health and Safety at Work Regulations to consider 'human capability' when allocating tasks.
- People at work are directly influenced by the organization, the job and personal factors.
- Behavioural safety is the systematic application of psychological research on human behaviour to the problems of safety in the workplace.
- The main objective of a behavioural safety programme is to change or improve behaviour with respect to safety at work.

- Most people, irrespective of circumstances, are prepared to accept a certain level of risk.
- Consideration must be given to 'safe place' and 'safe person' strategies in the design of safe systems of work.
- Accident proneness applies to a very small proportion of the population and is not significant in accident causation and prevention.

2 Human sensory and perceptual processes

People are continually receiving information and transmitting information to others. This takes place through a number of sensory and perceptual processes, such as sight and hearing. The speed at which people process information, or inadequate, incorrect or poor information processing, is a feature of many accidents.

Perceptual processes are also associated with the skills possessed by individuals and there may be some correlation between these skills, or lack of same, and the potential for accidents.

This chapter examines the information processing function, the process of perception of danger, sensory defects, perceptual errors and how perception affects human performance, all of which are significant in accident causation.

Human sensory receptors

As stated in Chapter 1, people learn through inputs or messages to the senses which, in turn, carry information to the brain. Figure 2.1 gives percentages used in the learning process.

Sensory inputs
Sight 75%
Hearing 13%
Touch 6%
Taste 3%
Smell 3%
From what we learn, we remember and believe
10% from what we read
20% from what we hear
30% form what we see
50% from what we hear and see
80% from what we say
90% from what we say as we do

Figure 2.1 The learning process

Sensory perception and messages

Sensory inputs have a number of characteristics. The information available for the brain is all derived from the sensory input, both for events external to and inside the human body. Senses have specialized functions, for example, the eyes do not perceive sound, but this specialization is incomplete. The senses are associated, in particular, with the activity of nerve endings and nerve fibres, all of which convey messages to the brain, in some cases seeking to protect the individual from pain or serious damage to the body.

Moreover, the senses do not give continual and constant information. A constant stimulus does not arouse a continual constant stream of impulses. The nerve endings of the senses adapt and the speed of adaptation varies with the particular sense and with the stimulus. For instance, a sensation is experienced when an object is touched initially, but if the contact is maintained with a constant pressure the nerve ending will eventually cease to respond. Pain endings adapt less rapidly than touch, but again with a constant painful stimulus the sensation gradually diminishes. This adaptation is peripheral in the nerve ending itself and indicates another essential characteristic of the sensory system, that is, its high sensitivity to change. For example, people may scarcely be aware of a constant level of background light, but are immediately conscious of a change in its intensity.

Nerve fibres are also important organs of sensation, but are not specialized. Impulses travelling up these fibres are basically similar in the case of auditory or optic nerve fibres and where a sensory nerve passes from the leg or arm to the spinal cord. However, specialization of a different type takes place in the next stage, the specific areas of the brain where information is interpreted or processed, and where, ultimately, there is conscious sensation.

Unconscious sensations also take place as experienced with reflex actions which follow the initiation of sensory impulses. When, for example, a bright light is shone directly into the eye, the pupil constricts due to the contraction of the circular muscles of the iris. Similarly, if a hand touches a very hot surface, the reflex action is to withdraw the hand immediately before any pain is experienced. More complex reflex actions can take place and there are a large number of extensive reflex responses which form the basis for conscious direct actions. Reflex activity is, thus, one of the body's main protective mechanisms.

Generally, people are unable to act on all the information that is presented. They tend to filter out the insignificant, the uninteresting or redundant information and act on a small proportion of the incoming information which they feel is relevant in the situation.

The various forms of sensory perception are considered below.

Visual perception

The sense of sight is the most commonly used technique for the transmission of messages to the brain. People with standard sight have stereoscopic vision enabling them to perceive objects and views in perspective. They are able to judge the distance of an object and whether the object is stationary or moving.

Vision is based on the reflection of light off objects into the eye. This light is received by the retina and refracted, being brought to a focus by the cornea, the transparent window at the front of the eye, and the lens behind the pupil, which is flexible and can be adjusted.

With many people, vision deteriorates as part of the ageing process (prebyopia). A common cause of poor visual acuity (sharpness of vision) is a slight asymmetry of the eye (astigmatism) which is corrected with lenses. The common defects of focussing arise from discrepancies between the focal length of the lens system and the actual length of the eye, causing short sightedness or long sightedness.

Short sightedness (myopia) is due to too strong a lens, or too long an eye and distant objects become focussed at a point in front of the retina. Near objects, which need a stronger lens, can be focussed on the retina and seen sharply.

Long sightedness (hypermetropia) is caused in the opposite way to myopia. With the eye at rest, even distant objects are focussed behind the retina. The muscles of the lens can shorten its focal length, i.e. strengthen it, and if the discrepancy is not too great distant objects can be focussed sharply. However, the lens is not so adaptable that near objects can be focussed clearly on the retina.

All the above defects are corrected by suitable spectacles, in the case of short sightedness by concave lenses to weaken the refraction of the eye, and long sightedness by convex lenses to strengthen it. People with presbyopia may need separate spectacles for distant and near vision.

'Seeing' and 'perceiving'

Perception is the process of interpreting a stimulus, or a series of stimuli, by the brain through one of the sensory mechanisms. The perception process is selective in that, despite the mass of information or stimuli presented to a person at a given time, that person will select certain stimuli, reject others and organize the selected stimuli so as to give meaning. The type of stimuli that people select are affected by past experience, training and individual personality.

Reference to the diagrams in Figure 2.2 explains the difference between 'seeing' and 'believing'. People all receive the same stimuli, in this case, the four diagrams. However, some people may interpret one diagram as one thing and the rest may confirm that they see something totally different. A number of questions need to be asked.

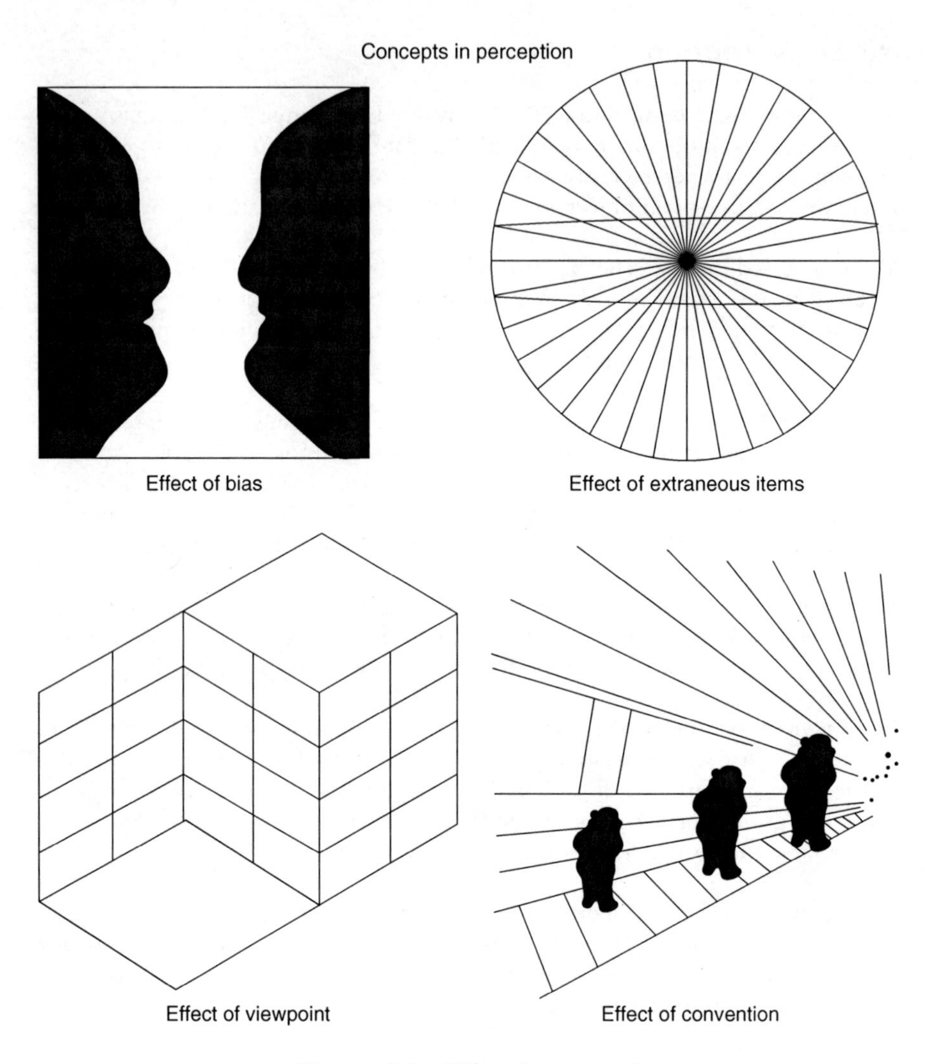

Figure 2.2 What do you see?

- **Diagram A**. Is this a picture of two men face-to-face or a Grecian urn? Here the effect of *bias* between black and white has a direct effect on individual perception.
- **Diagram B**. This diagram shows a wheel with spokes. The question to be asked is whether the two lines are parallel or slightly cambered. The fact is the two lines are parallel, but the effect of the *extraneous items*, that is, the spokes, has a distorting effect.
- **Diagram C**. Here we would have what appears to be some block work. Is the section on the left-hand side recessed back or is it projected at an angle from the block work on the right of the diagram? The effect of *viewpoint* is significant here and close study of the diagram can result in the part on the left moving backwards and forwards in the eye of the person studying it.

- **Diagram D**. This is a diagram of three men travelling on an escalator. The first question is whether the three men are of the same height or not? The second question relates to the escalator. Are the men travelling up or down the escalator? The three men are actually of the same height, but the effect of *convention*, where the surrounding lines converge at a point, makes the first man appear the tallest. Similarly, depending on how a person studies the diagram, the escalator could be going upwards or downwards.

Auditory perception

This is the process of perceiving sound and, in many cases, noise. Noise is defined as 'unwanted sound'. Exposure to noise from machinery and equipment in the workplace can result in several forms of hearing impairment. 'Sound', on the other hand, is defined as any pressure variation in air, water or some other medium that the human ear can detect.

Sound comprises a series of pressure waves impinging on the ear (sound waves). The sound waves of everyday life are composed of a mixture of many simple sound waves. Sound is generated by the vibration of surfaces or turbulence in an air stream which sets up rapid pressure variations in the surrounding air. The rate at which variations occur (frequency) is expressed in cycles per second or Hertz (Hz).

The human ear incorporates:

- the outer ear, incorporating the external pinna and the auditory canal terminating at the ear drum;
- the middle ear, which comprises a chamber containing ossicles, i.e. three linked bones, the malleus, incus and stapes (hammer, anvil and stirrup bones); and
- the inner ear, which houses the cochlea, the important organ of hearing and which is, fundamentally, a coiled, fluid-filled tube incorporating the Organ of Corti.

Sound is conveyed via the auditory canal to the ear drum and ossicles to the cochlea, and from the cochlea via the auditory nerve to the brain where the sensation of sound is perceived. Hearing loss takes place in the Organ of Corti and this may be measured by audiometry.

The human ear is sensitive to frequencies between 20 and 20 000 Hz, being particularly sensitive in the range 1000 to 4000 Hz (the frequencies of interest and progressively less sensitive at higher and lower frequencies). This fact is very important when measuring sound since two sounds of equal intensity, but of different frequency, may appear subjectively to be of different loudness.

Sound may be *pure tone*, i.e. of one frequency only, such as that produced by a tuning fork. Some industrial noise may be pure tone, but most is highly complex with components being distributed over a wide range of frequencies. Noise of this type is referred to as *broad band* noise, common examples being that produced by looms, air jets and printing presses.

Industrial noise is commonly produced by impact between metal parts. Where this features many impacts per second, as in riveting machines, the noise is generally treated as broad band noise. However, if the noise is produced by widely spaced impacts, as from a drop hammer or cartridge-operated tool, the noise is referred to as *impulse noise*. This presents special difficulties in measurement and in assessing the risk to hearing.

Hearing impairment

Exposure to excessive noise can result in hearing impairment, the condition known as 'noise-induced hearing loss' or 'occupational deafness'. Where the intensity and duration of exposure are sufficient, even 'wanted sound', such as loud music, can lead to hearing impairment.

Occupational deafness is a prescribed occupational disease which is described thus:

> substantial sensorineural hearing loss amounting to at least 50 dB in each ear, being due in the case of at least one ear to occupational noise, and being the average of pure tone loss measured by audiometry over the 1, 2 and 3 kHz frequencies.

Under the Social Security (Industrial Injuries) (Prescribed Diseases) Regulations, this definition goes on to state a wide range of activities and occupations associated with exposure to noise whereby industrial injuries benefit may be payable.

For most steady types of industrial noise, intensity and duration of exposure, i.e. the dose of noise, are the principal factors in the degree of noise-induced hearing loss (sociocusis). Hearing ability also deteriorates with age (presbycusis) and it is sometimes difficult to distinguish between the effects of noise exposure and normal age-related deterioration in hearing.

The risk of noise-induced hearing loss can be related to the total amount of noise energy taken in by the ears over a working lifetime.

The effects of noise exposure

Exposure to noise can affect hearing in three ways:

1. **Temporary threshold shift**. This is the short-term effect, that is, a temporary reduction in hearing acuity, which may follow exposure to noise. The condition is reversible and the effect depends, to some extent, on an individual's susceptibility to noise.

2. **Permanent threshold shift**. This takes place when the limit of tolerance is exceeded in terms of time, the level of noise and individual susceptibility to noise. Recovery from permanent threshold shift will not proceed to completion, but will effectively cease at some particular point in time after the end of the exposure.
3. **Acoustic trauma**. This condition involves ear damage from short-term intense exposure or even from one single exposure. Explosive pressure rises are often responsible, such as that from gun fire, major explosions or even fireworks.

Hearing loss

Symptoms of hearing loss vary according to whether the loss is mild or severe.

In the mild form of hearing loss, typical symptoms include a difficulty in conversing with people and the wrong answers may be given occasionally due to the individual missing certain key elements of the question. Speech on television and radio seems indistinct. There may also be difficulty in hearing normal domestic sounds, such as a clock ticking.

In the severe form of hearing loss, there is difficulty in discussion, even when face-to-face with people, as well as hearing what is said at public meetings, unless sitting right at the front. Generally, people seem to be speaking indistinctly and there is an inability to hear the normal sounds of the home and street.

What is important is that it is often impossible for someone with this level of hearing loss to tell the actual direction from which a source of noise is coming and to assess the actual distance from that noise. This can be a contributory factor in accidents and, in particular, pedestrian-related road accidents.

In the most severe cases, there is the sensation of whistling or ringing in the ear (tinnitus).

Tactile perception

The skin is an important sense organ of the body. Sensory nerve endings are specialized and different endings respond to touch, changes of temperature and air movement, humidity levels and to pain.

The sense of touch and pain are served by different sets of nerve endings which are not equally distributed over the body's surface. Touch endings are denser in the skin of the fingers than in the limbs or trunk. The tips of the fingers are the most sensitive to touch and it is from the way the touch endings are stimulated that the quality of materials is determined, e.g. roughness, softness, smoothness.

Nasal perception

The nose detects smells. However, the projection on the front of the face incorporating the nostrils is only a small part of the nasal system which incorporates outer and inner sections. The bridge of the outer nose, between the eyes is supported by the nasal bones. The cavity, extending to the back of the hard palate, is completely divided into left and right halves by the nasal septum. This comprises a thin sheet of bone behind and cartilage in front. The cartilaginous septum supports the outer nose below the nasal bones.

At the back of the hard palate the two halves of the nose open into a single cavity, the nasal part of the pharynx. At each side the maxilla forms the wall of the nose, the floor is the palate and the roof is the base of the skull. Three small flaps of bone, each with its lower edge curled under, protrude from the side-wall into the cavity of the nose. These conchae, the turbinate bones, increase the surface area of the cavity and shield the openings of the sinuses.

The bones surrounding the nose are hollow. Their cavities, the (paranasal) sinuses, communicate with the nose by small openings in the side-wall. Two other openings are important, that of the nasolacrimal duct, by which tears flow down, and the Eustachian tube, connecting the middle ear with the pharynx immediately below the nose proper.

The mucous membrane of the nose also lines the various extensions of the nasal cavity. It is a lax membrane, easily distended by blood and tissue fluid. When inflamed by infection, as with a common cold, or allergy, such as hay fever, it can quickly swell until the nose is blocked. In the healthy condition it is kept moist by mucus secreted by its many tiny glands. Its surface is covered with cilia, microscopic hair-like structures which undulate and keep the mucus moving towards the back of the nose. When the membrane is inflamed, the cilia can no longer cope with the excess of mucus which is formed, and in any case the sinuses could not be cleared because their openings are blocked by the swollen membranes. This gives a simple mechanical explanation of the watering eyes, headache, face ache and earache which may accompany a cold in the nose.

Nerve endings for perceiving smells are confined to a small area at the top of the nasal cavity. The nerves pass through small holes in the ethmoid bone to join the olfactory nerves. Although carried in quite different paths, the senses of smell and taste are closely related.

Taste perception

The mucous membranes of the tongue and palate contain numerous minute sensory organs, the taste buds. Each bud is a round cluster of spindle-shaped cells, from which nerve fibres lead to the facial and glossopharyngeal nerves. Four types of sensation are received from the taste buds – salt, sweet, sour (acid) and bitter. Of these, only sour is linked to the chemical structure of the substance tasted.

Some areas of the tongue are more sensitive to particular tastes than others, the tip to sweet and salt, the edges to sour and salt and the back to bitter.

Perceptual processes

Perception is the complex mental function giving meaning and significance to sensations.

An individual is constantly responding in some way to incoming stimuli. These stimuli can be accepted, rejected, ignored or distorted. It all depends upon whether the stimulus supports or contradicts the individual's beliefs, values and attitudes. This process forms the basis of perception, i.e. the way the individual interprets incoming information. Moreover, perception is developed to satisfy an individual's needs in order that he or she can cope with reality. Each individual is different, due to factors such as heredity and environment. This is what makes each individual unique and each individual's 'reality' unique.

The process of perception of danger

How people perceive danger varies from one person to another and is based, to some extent on their individual knowledge, skills and experience of similar situations. Most people would actually see the danger associated with, for example, unsafe stacking operations using a fork lift truck or the unsafe use of a flammable substance in close proximity to an ignition source. On the other hand, the fact that a machine is producing abnormal sounds during operation may not necessarily be an indication of danger to everyone.

Errors in perception caused by physical stressors

Physical stressors, such as extremes of lighting intensity and excessive noise, may be contributory factors in incorrect perception. In the first case, the overbrightness of a light source or, alternatively, poor illuminance levels, may conceal hazards that are evident under correct lighting conditions. Similarly, background noise, or noise emission from specific machinery, may mask warning signals from other plant and machinery. The wearing of ear protection may further mask these warning signals.

Factors that influence the effectiveness of hazard perception

- A boring, repetitive job may produce 'day-dreaming' which may result in the lowering of the impact of a stimulus.
- Warnings (or threats) may not be strong enough to get through an individual's perceptual set.

- Patterns of behaviour and habits can be carried from one situation to another where they are no longer appropriate or safe. For instance, people tend to drive too fast after leaving a motorway.
- The individual can get used to a particular stimulus and, if it is not reinforced, it ceases to command his attention and is ignored (habituation).
- Intense concentration on one task may make the attention to another stimulus difficult or impossible.

Information processing

The relative speed with which people process information is a precondition of many accidents, or the errors and omissions which can produce accidents. A stimulus will, in most cases, produce a response. Welford showed that the human mechanism for processing information has only a finite capacity. This means that second and subsequent stimuli have to wait until a first stimulus has been dealt with by the individual (single channel theory).

Each stimulus will produce a reaction or response and the response can be divided into two elements:-

1. Specific reaction time – the actual time it takes to perceive and process the response; and
2. Movement time – the time it takes to actually execute the response.

If a second stimulus arrives during the movement time, it has to wait. The actual movement in response is being monitored, the single channel process ensuring the execution of the original response was accurate. The particularly crucial parts of the movement are the beginning and the end, the middle often being partially or totally neglected. Thus, people generally cannot do more than one thing at a time, the speed and sequence of response varying from person to person. This factor can be significant in accident causation.

With well-known and practised tasks, e.g. driving, the monitoring action of the brain can be reduced, much depending on the speed with which a person can respond to stimuli, and not monitor specific movements. Results are achieved through continual practice or the speed–accuracy trade-off, whereby the monitoring is voluntarily removed.

The *feedback* that people receive is an important aspect of monitoring a task. However, where someone is highly skilled in that task, it can sometimes be a hindrance and destroy or aggravate performance. For example, in the teaching of people to undertake a specific practical task, such as operating a lift truck, the instructor has to put the monitoring aspect back into the task. This can affect, or even ruin, his own particular performance at that task.

Sensory defects

Defects in the ability to see, hear, touch, taste and smell are associated with many injuries and ill-health arising from workplace activities.

Obvious conditions include short-sightedness, long-sightedness and deafness, but a poor sense of touch, taste or smell can be significant in some cases where, for instance, people are unable to feel the heat from an overheating fan or the taste or smell from a hazardous airborne substance.

Basic screening techniques

A number of basic screening techniques are available for identifying varying levels of sensory defect. Sensory ability in terms of touch, taste and smell commonly form part of routine health screening of individuals by means of health questionnaires and individual examination by occupational health professionals.

- **Audiometry**. Audiometry, the assessment of an individual's hearing capability, may well form part of health surveillance programmes required under the Control of Noise at Work Regulations.
- **Vision screening**. Similarly, vision screening may be desirable for all drivers of an organization's vehicles to ensure continuing levels of road safety and for display screen equipment 'users' under the Health and Safety (Display Screen Equipment) Regulations.

Perception and the limitations of human performance

Perception

Human performance is directly related to a person's ability to perceive, understand, interpret and process information presented at a specific point in time. As stated earlier, defects in perceptual systems, together with failing perception as part of the ageing process, are significant contributory factors in accidents

Misperceptions

People suffer misperceptions which are an important feature of the potential for human error. Misperception tends to occur when a person's limited capacity to give attention to competing information, perhaps under stress, produces 'tunnel vision', or when a preconceived diagnosis, notion or explanation conceals the actual sources of inconsistent information. In other words, there is a strong tendency for a person to assume that an established pattern holds good so long as most of the indications are to that effect, even if there is an unexpected indication to the contrary.

One potent source of error in such situations is the inability to analyse and reconcile conflicting evidence or information. This may derive from an imperfect understanding of a process itself or the actual meaning conveyed by instruments incorporated with a process. Any analysis of the control measures required involves the need for operators to understand the process itself along with the ergonomic and technical features of the instrumentation.

This effect was identified in the official report into the incident at the Three Mile Island nuclear power station in the United States (1979), which identified 'human factors' as the main causes of the incident. In this case, misleading and badly presented operating procedures, poor control room design and layout, poorly designed display systems together with inadequate training all, in one way or another, resulted in operators receiving misleading or incomplete information. The subsequent official enquiry, however, emphasized how failures in human factors design, inadequate training and procedures, together with inadequate management organization led to a series of relatively minor technical faults being magnified to a near disaster with potentially significant human and economic consequences.

The limitations in human performance

Human performance is concerned with the mental and physical capabilities of people and is affected by many factors mentioned earlier.

In the assessment of tasks, not only should the requirements for such a task be identified, but also factors which may limit performance of the task. Manual handling work is a simple example. People involved in regular manual handling operations need to have a good level of physical capability and fitness. Where handlers suffer from, for example, ligamental or muscular strains, their ability to undertake the task is reduced in terms of speed and efficiency of handling. On this basis, mechanical handling aids should be provided and used wherever the risk of back injury is identified.

The mental capability requirements of tasks are more complicated and limitations in human performance in these tasks may contribute to accidents. Mental tasks include, for example, the addition of figures, operation of a word processing programme, reading and understanding data or activities such as map reading. Each of these tasks requires specific mental skills which have to be developed through instruction and training, incorporating repeated practice. These tasks require the development of memory and cognitive skills and the rate at which people acquire, develop and retain these skills is an important limitation in human performance.

Conclusions

People are continually receiving information and transmitting information to others. This takes place through various inputs or messages to the senses which, in turn, carry information to the brain. Sensory perception takes place through the various senses of sight, hearing, touch, taste and smell. All these senses may incorporate defects, such as visual or hearing impairment, which can be a contributory factor in accidents. There are also many factors that influence the effectiveness of hazard perception and the human mechanism for processing information has only finite capacity. There are also a number of factors which limit human performance, such as mental and physical capability, both of which vary considerably from one person to another.

Key points

- The principal sensory inputs are sight and hearing; secondary sensory inputs are touch, taste and smell.
- Perception is a complex mental process giving meaning and significance to sensations.
- Errors in perception are a contributory factor in many accidents.
- Perception is directly related to the ability of an individual to process information.
- People frequently suffer misperceptions through tunnel vision or preconceptions of situations.
- The mental capability requirements of tasks should be specified with a view to preventing accidents arising through misunderstanding, misperception, mistaken actions and other forms of human error.

3 Organizations and groups

Most people belong to some form of organization on a work-related, social, professional or community basis. Membership of the group entails complying with the group norms or standards for behaviour. Failure to comply with these norms can lead to rejection from the group or, in some cases, the need to demonstrate future compliance with group norms.

Organization theory

An organization can be defined as 'groups of people united by a common goal'. That goal may be

- the manufacture of products in order to make a profit;
- the provision of a service or services, whether for profit or not;
- the practice of standards of performance set by a profession.

Organisation Theory can be traced back hundreds of years and is characterized by

- centralized authority;
- clear demarcation of responsibility with clear lines of authority;
- marked division of labour;
- specialization and the use of particular expertise;
- working rules and procedures; and
- the specific separation of the staff function from the line function in management.

Features of organizations

Organizations have goals, environments and systems. Organizations also have clients or customers who purchase their products and/or use their services. Organizations operate on the basis of a hierarchy which is fundamentally based on power. Orders pass down the organization and information passes back up the system. The flow is one-way in both cases. Promotion within the hierarchy is based on the premises of merit and hard work. Organizations operate principally on a formal basis. They are deliberately impersonal and are based on ideal relationships at all levels and between all levels of the organization.

Organizations comprise as a rule:

- the *functional or line organization*, which is based on the type of work being done; and
- the *staff organization*, which is in direct contrast to the line organization and represented by those people who have advisory, service or control functions.

Organizations comprise individuals who belong to groups, e.g. production, research, engineering, marketing, sales, distribution, etc., within departments or divisions forming the overall organization. In turn, the organization belongs to a wider population, for example, an industry, existing alongside other organizations in a wider context. They may also participate more widely, within a continent or on a global basis.

Conflict can arise between the line and staff organizations for a number of reasons, such as differing motivations, misunderstanding of individual roles, differing cultures and objectives, differing priorities and levels of commitment.

Forms of organization

Organizations take many forms, such as

- the rigidly organized bureaucracy;
- the bureaucracy run by a senior executives' group;
- the bureaucracy that has created cross-departmental teams and task forces;
- the matrix organization;
- the project-based organization; or
- the loosely-coupled organic network.

Weaknesses of formal organizations

Formal organizations suffer from communications failures and are frequently perceived by their employees as uncaring and lacking in interest and commitment to ensuring appropriate levels of health, safety and welfare. In particular, they ignore emotional factors in human behaviour.

The organizational environment

For organizations to succeed and survive, they must operate within an appropriate organizational environment. These include:

- the role and function of management;
- the human factors issues – human resources and their effective management;

- external and internal influences on the organization;
- potential sources of conflict;
- the significance of individual behaviour at all levels;
- decision-making;
- communications;
- the organizational culture and style;
- strategies for dealing with change with the organization;
- strategies for dealing with conflict;
- strategies for survival.

The *survival of the organization*

One of the main objectives of any organization is to survive in the current climate of rapid change, instability, uncertainty and an ever-changing world market. Organizations that survive are those that learn and innovate. Managers must look, listen, learn and operate as a learning organization. They must operate as team members all the time.

A major challenge confronting managers is how to design organizations that will survive during this period of instability and chaos. To survive, organizations must shift towards structures that are more horizontal than vertical, with greater employee involvement and participation. At the pinnacle of the shift is that type of organization known as 'the learning organization'.

Learning organizations

A learning organization is one in which everyone is engaged in identifying and solving problems, enabling the organization to continuously experiment, improve and increase its capacity to learn, grow and achieve its purpose.

It is designed to solve problems in a unique way so as to satisfy the needs of clients. The differences between the traditional focus and the mindset of managers in a learning organization are shown in Table 3.1.

Important factors for survival

- **Leadership**. To what extent do you create and inspire a clear, shared vision, shape the organizational culture and help to achieve it?
- **Organization**. To what extent have you developed a flat organization, based on teams who have authority to implement changes?

Table 3.1 Comparison between traditional and learning organizations

	Learning organization	**Traditional organization**
Leadership focus	Effectiveness	Efficiency
Structure	Horizontal teams, task forces, project management	Hierarchical with few (if any) teams, task forces or project management
Empowerment	Empowered employees, shared responsibility	Staff perform specialized tasks for which they are responsible
Strategy	Decentralized decision-making, participative strategy	Strategic decision-making is centralized
Culture	Strong adaptive culture, love change	Rigid culture, resist change
Communications and information sharing	Open information, horizontal communication, face-to-face	Vertical communication and reporting systems, predominantly paper-based, information limited on a 'need to know' basis

- **Empowerment**. To what extent do your employees have the power, training and freedom to make decisions and perform effectively? Do they share responsibilities for outcomes?
- **Participation**. To what extent is strategy generated from all levels of staff, based on deliberate systematic efforts to encourage staff to do so?
- **Culture**. To what extent do staff feel responsible for the whole organization, have a sense of caring and compassion for each other, welcome change, continuously strive to improve and innovate?
- **Information**. To what extent do you follow an 'open book' approach where everyone understands the whole organization as well as their part in it?

Basic organizational structures

As a rule, organizations operate on the basis of a hierarchy. Orders pass down the system and information passes back up the system. Organizational structures are affected by a number of factors, such as:

- the overall aims, policies and objectives of the organization;
- the size of the organization in terms of the number of employees;
- the range of products manufactured and/or services provided;

- the geographical location of business units, e.g. locally, nationally or worldwide; and
- the legal requirements affecting the organization's operations.

Objectives and policies

Objectives are statements of what an organization aims to achieve. *Policies*, on the other hand, are statements concerning how these objectives will be achieved.

Organizations state their objectives with respect to, for instance, the quality of a product or service and the health and safety of their employees and other persons affected by their operations. These objectives are then reinforced by means of a Quality Policy Statement and a Statement of Health and Safety Policy.

Formal and informal organizations

Organizations may be classed as operating on a formal or informal basis.

Formal organizations are established to achieve identified goals, aims and objectives. They have clearly defined rules, structures and channels of communication. Typical examples are national and local government, businesses and international organizations.

Informal organizations, such as groups of people working together on specific tasks, are common features of formal organizations. In these cases, the members of the group set their own rules and members are expected to comply with these rules.

Organizational structures

There is no standard format for an organization's structure, but a typical structure might be as depicted in Figure 3.1. Here, there is a clearly identifiable line of authority from the top downwards. The responsibilities and functions of people at various levels of the organization are commonly specified by way of job descriptions which identify the key objectives of the job and the job performance characteristics.

Formal and informal groups within an organization

Organizations incorporate groups of people who communicate on both a formal and informal basis.

Formal groups include the Board of directors, members of the management team, specific committees or working groups who meet on a regular basis to plan future

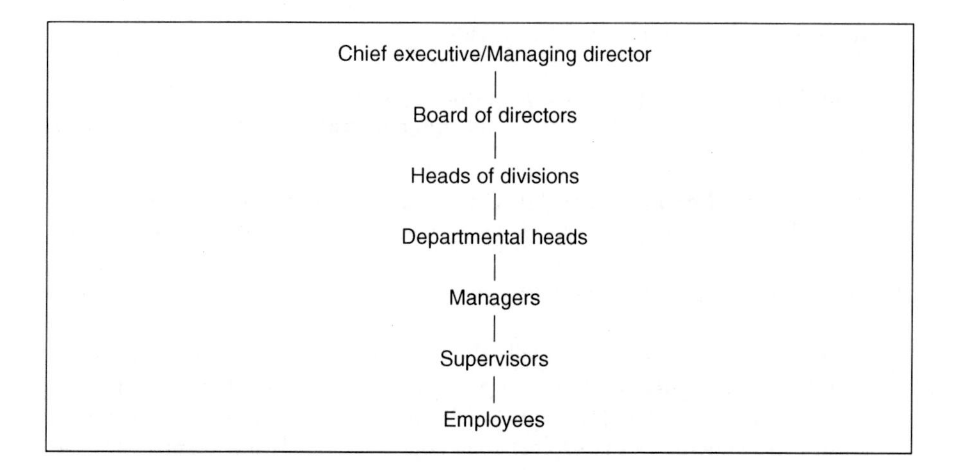

Figure 3.1 A typical organizational structure or hierarchy

activities and who report back within the organization on a regular basis. Health and Safety Committees come within this classification.

Informal groups are groups of employees who work together, who may form part of a team for undertaking a particular task and who communicate on a regular basis. Typically, they may be incorporated in sections of a department, reporting to a line manager.

Work groups

Work groups are defined as a group or collection of individuals interacting with each other in the pursuance of a common work-related task or objective and who, for this purpose, are dependent upon each other in the achievement of the task or objective. Organizations include groups of people involved in a range of activities. Each member of the group brings with him differing levels of energy. For the group to work together, these energies must be harnessed in order to achieve the group's objectives.

In the attainment of objectives there must be co-operation between members of the group as some of the individual goals of members are dependent on the actions of others. Inevitably an element of competition will exist within a group in that some members will endeavour to perform better than others, thereby affecting the unity of the group as a whole. For there to be unity in a group, therefore, the individuals in the group must have interdependent goals, each relying on the other to achieve the stated goals.

The benefits of belonging to a group are associated with the experience of working together and existing socially, being specifically identified by the organization as a group, the sharing of goals and from the social approval displayed by members of the group towards those people who comply with the group norms.

Groups can be divided into two categories:

- **Sociological groups**. These groups exist when individuals pursue interdependent goals which assist the other members of the group to attain their own particular goals.
- **Psychological groups**. This form of group exists when individuals perceive themselves as pursuing interdependent goals, but on a more competitive basis.

A distinction can also be drawn between:

- **Primary groups**. These are small subgroups of the secondary group in which relationships are informal and on a personal basis. The number of people in a primary group may vary but, commonly, members may subdivide into smaller groups to maintain and satisfy personal and social needs and communication.
- **Secondary groups**. The secondary group is the larger group, such as members of a section of a department. It has a formal structure with members of the group having defined roles and responsibilities.

Characteristics of groups

Social groups can differ in terms of their characteristic features. Much will depend upon

- the number of members, or size of the group, and fluctuations in the size of the group;
- the extent of their organization and operation in a formal manner;
- the degree to which they are stratified, that is, the extent to which group members are related to one another in some form of hierarchy;
- the degree to which they exercise, or attempt to exercise, control over the behaviour of their members;
- the degree of participation which is permitted, expected or demanded of members;
- the ease of access to membership in the group and the ease with which a member can leave or be rejected from the group;
- the degree of stability of the group over time and the continuity of its membership over time;
- the degree to which group members relate to each other intimately, on a personal basis and with respect to a wide range of interests and activities, rather than in a formal way and only with respect to a narrowly defined set of activities; and
- the degree to which the group is subdivided into smaller groups or cliques, and the extent to which such cliques are in conflict with each other.

Membership

Groups have particular standards of behaviour (the group norms) and members are expected to comply with the group norms. The norms are informally binding on members. Failure to comply with these norms can result in expulsion from the group or some form of sanction imposed on the individual by the group. In addition, pressure for conformity by members can be exerted by the rest of the group. This can have the effect of changing attitudes and beliefs, motivating members to better levels of performance in some cases.

In many cases, the actual size of the job, in terms of the scale of work, the duration of the task and the skills required can determine the size of the work group. Some groups may exist only on a temporary basis as in the case of construction activities where, once the project is completed, the group disbands and members may never meet again as a group.

Consultation with specific work groups is increasingly being seen as a significant means of raising standards of performance, agreeing work patterns and safety procedures. In many cases, the consultation process will identify the need for training as a group.

Group dynamics

Group dynamics is the study of the nature of groups. Group dynamics seeks to investigate and learn more about the development and functioning of groups, the factors which affect relationships between groups and individuals and between that group and other groups.

Group behaviour

Membership of a group has a direct effect on behaviour. Groups play a significant role in determining and developing attitudes, decisions, beliefs, actions and behaviour towards other people. For groups to survive and operate effectively there must be:

Conformity

'Conformity' implies 'agreeing with' or 'going along with' the behaviours, attitudes and beliefs of the group. This may entail personal or private acceptance, where people make the behaviours, beliefs and values of the group their own, or mere compliance, where members go along with the group as a result of group pressure without necessarily changing their private attitudes or beliefs.

There are a number of reasons why people conform. An individual may

- go along with the opinions of his group because they appear to be right (*social comparison*);
- wish to avoid the social disapproval and censure that may result from contravening the group norms (*avoidance of social disapproval*); or
- suffer some form of internal conflict as a result of what he perceives and what other people tell him (*dissonance reduction*).

Other factors affecting conformity include:

- the degree of an individual's orientation towards the group;
- the need to be liked and accepted;
- the perceived consensus within the group; and
- the means by which a new member of a group is gradually induced into the norms and standards of the group.

Consultation and bargaining

People need to reach agreement with each other if they are to live and work together in social groups. There must be formally established rules and procedures covering the organization of work and how the results of work are to be shared in terms of individual reward. These rules, or norms, arise in the process of socialization and as a result of internal bargaining and consultation.

The consultation process uses established norms as a focal point in discussion and resolution of differences between members of the group. In this process, a certain amount of bargaining can take place, where people can make offers, seek a compromise and arrive at a mutually agreeable solution.

People attempt to maximize rewards and minimize costs in social situations, that is, to seek pleasure and avoid pain. *Rewards* are events that produce some form of gratification, including material gain, enhanced self-esteem and bring social approval. *Costs*, on the other hand, are the reverse of rewards. They result in loss of material gain, lowered self-esteem, increased tension and anxiety.

Vulnerable groups

Certain groups of people, by virtue of their age or physical condition, are considered to be more vulnerable to accidents and ill-health at work. These groups include new or expectant mothers, young persons and disabled persons. The law requires employers to make extra provision for these groups (Management of Health and Safety at Work Regulations: Approved Code of Practice).

New and expectant mothers

A risk assessment should take account of the risks to new and expectant mothers. Where the risk assessment identifies risks to new and expectant mothers and these risks cannot be avoided by the preventive and protective measures taken by an employer, the employer will need to:

- alter her working conditions or hours of work if it is reasonable to do so and would avoid the risks or, if these conditions cannot be met;
- identify and offer her suitable alternative work that is available, and if that is not feasible suspend her from work. (The Employment Rights Act requires that this suspension should be on full pay. Employment rights are enforced through tribunals.)

Young persons

The employer needs to carry out a risk assessment before young persons start work and to see where risk remains, taking account of control measures in place. For young workers, the risk assessment needs to pay attention to areas of risk described in regulation 19(2) (i.e. beyond physical or psychological capacity, involving exposure to harmful agents, involving exposure to radiation, etc.). For several of these areas, the employer will need to assess the risks with the control measures in place under other statutory requirements.

When control measures have been taken against these risks and if a significant risk still remains, no child (young worker under compulsory school age) can be employed to do this work. A young worker, above the minimum school leaving age, cannot do this work unless:

- it is necessary for his or her training;
- she or he is supervised by a competent person; and
- the risk will be reduced to the lowest level reasonably practicable.

Disabled workers

The Disability Discrimination Act 1995 makes it unlawful for an employer to discriminate against a disabled job applicant or worker with respect to selection for jobs, terms and conditions of employment, promotion or transfer, training, employment benefits and dismissal or other detrimental treatment. A person has a *disability* for the purposes of the Act if he has a physical or mental impairment which has a substantial and long-term adverse effect on his ability to carry out normal day-to-day activities. A *disabled person* is a person who has a disability. An employer must make reasonable *adjustments* where working arrangements and/or the physical features of a workplace cause substantial disadvantage for a disabled person in comparison with those who are not disabled.

Reasonable steps which an employer may need to take and which may need to be covered in the risk assessment process involving a disabled person, include

- altering working hours;
- allowing time off for rehabilitation or treatment;
- allocating some of the disabled person's duties to someone else;
- transferring the disabled person to another vacancy or another place of work;
- giving or arranging training;
- providing a reader or interpreter;
- acquiring or modifying equipment or reference manuals;
- adjusting the premises.

Physical features of the premises which may require adjustment are

- those arising from the design or construction of a building;
- exits or access to buildings;
- fixtures, fittings, furnishings, equipment or materials;
- any other physical element or quality of land or the premises.

Typical examples of physical features that might cause substantial disadvantage to a disabled person include the absence of ramps for wheelchair users, inadequate lighting for someone with restricted vision, doors that are too narrow for wheelchair users and a chair which is unsuitable.

Leadership as a feature of organizations

Without sound and effective leadership, organizations will fail. Leadership is the process of successfully influencing the activities of a group of people towards the achievement of an objective or goal. What are the essential characteristics of a good leader? Such a person has the ability to influence others by his expertise in his field, command of language, presence, charisma and who commands respect in order to achieve the group objectives ethically.

Leadership entails a range of skills such as decision-making, delegating appropriately, the use of persuasive and clear communication, motivating others and being empowered through knowledge.

In his book *The Human Side of Enterprise*, the late Douglas McGregor drew attention to a number of factors that are significant as far as organizations are concerned. Firstly, people know how to use physical science and technology to benefit themselves. The major problem is how to use the social sciences to make human organizations effective. Given proper conditions, vast resources of creative energy can be made available. McGregor introduced the twin concepts of *Theory X* and *Theory Y* with respect to the way organizations operate.

Theory X

The conventional understanding of management's task is that of harnessing human energy to organizational requirements. This is based on the belief that without active intervention by management, people would be passive, even resistant, to organizational needs and that they are, by nature, indolent, lacking in ambition, self-centred and resistant to change. It follows from this belief that to achieve the best results with people, management must make direct efforts and motivate and control the actions of subordinates in order to modify their behaviour to fit the needs of the organization. To achieve this objective, management's actions must fall between two extremes, the *Hard Approach* and the *Soft Approach*.

The *Hard Approach* is 'strong management', that is, directing the individual by coercion and threats, close supervision and tight controls. The problem with this approach is that 'force breeds counterforce', that is, tough management creates antagonism and disinterest on the part of employees. In times of full employment, it also creates a high rate of labour turnover.

The *Soft Approach* is 'weak management', a permissive approach aimed at achieving a tractable labour force prepared to accept management's directions. In practice, people take advantage of this approach, taking more and giving less.

McGregor believed that Theory X was based on correct observation of human behaviour, but that it is incorrect to believe that this behaviour is innate. Rather, these attitudes to work have been brought about by people's experience of the conditions that exist in organizations, in which case they may be modified by changing the conditions.

Theory X can be summarized as follows:

- the average human being has an inherent dislike of work and will avoid it if he can;
- because of this human characteristic of dislike of work, most people must be coerced, controlled, directed and treated with punishment to get them to put forth adequate effort towards the achievement of organizational objectives;
- the average human being prefers to be directed, wishes to avoid responsibility, has little ambition and wants security above all.

Managers who subscribe to Theory X will place great emphasis on pay and job security. They will not understand, for instance, that most of their employees are really trying to satisfy other motives, such as those relating to self-esteem and the realization of potential, or self-fulfilment. Their policies will be directed towards the manipulation of subordinates through the use of various incentives, such as money, in particular, that are largely ineffective in improving worker performance and subsequent production figures.

Theory Y

Modern organizations provide good pay, a fair level of security, good working conditions and adequate welfare facilities. The problem is that management does not get the co-operation and high levels of productivity that it expects because higher needs are not satisfied and employees experience frustration as a result.

Theory Y is McGregor's own theory based on new assumptions about human motivation and behaviour. It is based on the belief that people are not passive or resistant to organizational needs by nature, but have been made so by their experience of working for organizations. The motivation, the capacity for assuming responsibility and directing behaviour towards organizational goals are all present in people. Management does not put them there.

It follows, then, that management's task is to make it possible for people to recognize and develop these characteristics for themselves. The essential task in this is to arrange organizational conditions and methods of operation so that people can best achieve their own goals, that is, the satisfaction of their needs, by directing their own efforts towards the organizational objectives.

Certain concepts in accord with Theory Y are

- **Decentralization and delegation**. This removes people from overly close control and satisfies their egoistic needs by giving them responsibility.
- **Job enlargement**. This provides greater responsibility and status at the bottom of the organization.
- **Participation and consultative management**. This involves people in decisions and group situations and makes it possible for management to tap the knowledge of subordinates. It also satisfies the subordinates' social and egoistic needs.
- **Performance appraisal**. The Management of Performance Standards approach involves the agreeing of 'targets', or objectives, with the individual so that the self-evaluation is possible.

Theory Y can be summarized thus:

- the expenditure of physical and mental effort in work is as natural as play or rest;
- external control and the threat of punishment are not the only means for bringing about effort towards organizational objectives as people will exercise self-direction and self-control in the service of objectives to which they are committed;

- commitment to objectives is a function of the rewards associated with their achievement;
- the average human being learns, under proper conditions, not only to accept but to seek responsibility;
- the capacity to exercise a relatively high degree of imagination, ingenuity and creativity in the solution of organizational problems is widely, not narrowly, distributed in the population; and
- under the conditions of modern industrial life, the intellectual potentialities of the average human being are only partially utilized.

The manager who works to Theory Y attempts to unite the goals of the individual and the organization, so that the employee does not need to be coerced into work. This does not imply that the Theory Y approach is a 'soft' approach, however. It is one based on human motivation.

Classical organization theory

Classical organization design is based on Theory X. It can be summarized in eight precepts (Bass, 1965):

1. Some one person should be responsible for each essential activity.
2. Responsibility for each activity should not be duplicated and should not overlap.
3. Each position should have a number of clearly stated duties.
4. Every person should know exactly what their duties are.
5. Authority for making decisions should be commensurate with responsibility for them.
6. Authority should be delegated to persons close to the point of action.
7. Managers should have a limited number of subordinates, say, four to seven.
8. Every manager should know to whom they report and who reports to them. The chain of command should be clearly defined.

While the above precepts may be appropriate in many cases generally and specifically in the field of health and safety, the classical organization theory has a number of deficiencies. For instance, it fails to develop each individual's potential. Essential activities can be neglected at times, leading to major effort or even disasters. Furthermore, it typically creates a vertical structure with the resulting communications problems.

On the other hand, this theory may be appropriate in certain areas of health and safety, for instance in the identification of individual responsibilities, as outlined in an organization's Statement of Health and Safety Policy under the Health and Safety at Work Act.

Characteristics of highly productive managers and organizations

The following characteristics are classed as important for managers and organizations:

- a preponderance of favourable attitudes on the part of each member towards other members, superiors, the work and the organization as a whole;
- identification with the objectives and a sense of involvement in achieving them;
- all motivational forces:
  - the *Ego*, that is, the desire to achieve and maintain a sense of personal worth and importance;
  - desire for security;
  - curiosity, creativity and the desire for new experiences; and
  - economic forces,

  are harnessed;
- the organization consists of a tightly knit, effectively functioning social system comprised of interlocking work groups and high levels of skill in personal interaction permit effective participation in group decision-making;
- measurements of organizational performance are used primarily for self-guidance rather than to superimpose control.

Management style

Which style of management is more effective?

The answer to this question depends whether or not certain assumptions about management behaviour can be confirmed. For example,

- employee-centred supervisors are more productive than job-centred supervisors;
- supervisors with high goals and a contagious enthusiasm as to the importance of achieving these goals achieve the best results;
- there is a marked inverse relationship between the average amount of 'unreasonable' pressure the employees in the department feel and the productivity of that department;
- general, rather than close, supervision is conducive to high productivity;
- highly productive managers make clear to their subordinates what their objectives are and what needs to be accomplished, then give them the freedom to get on with the job;

- the highly productive foreperson has a helpful, nonpunitive attitude to errors;
- there is a clear correlation between productivity and the workers' perception of the supervisor rather than between productivity and the general attitude towards the company;
- high productivity is associated with skill in using group methods of supervision;
- work groups can have goals that will influence productivity and cost either favourably or otherwise;
- there is a clear relationship between group solidarity and productivity;
- to function effectively, a supervisor must have sufficient influence with his own superior to be able to affect his superior's decisions.

Management style is, therefore, an important feature of an organization. While styles vary considerably, many of the above factors are significant to the success of the organization as a whole.

The autocrat and the democrat

In order to investigate an aspect of group functioning under different types of leadership and different types of group 'atmosphere', experiments with groups of children were carried out in the 1930s by R. Lippitt and R.K. White. The aim was to establish small groups of children led by adult experimenters who adopted two totally different leadership styles, that is, the 'authoritarian' and the 'democrat'. The groups were involved in a range of tasks and meetings were held regularly over a period of weeks.

The two different approaches incorporated the following features.

Authoritarian style

- All policies were determined by the appointed leader.
- Techniques and steps for attaining the goals were dictated by the leader one at a time; future direction was uncertain to a large degree.
- The leader usually dictated the work task and work companions of each member.
- The leader was 'personal' in praise and criticism without giving objective reasons.
- The leader remained aloof from active group participation, except when demonstrating.

Democratic style

- Policies were determined as a result of group decisions.
- An explanation of the overall process was given at the first meeting.
- When technical advice was needed, the leader suggested several alternatives from which choices could be made.
- Members were free to choose work companions and the division of tasks.
- The leader was 'objective' or 'fact-minded' in praise and criticism and tried to be a regular group member in spirit without doing much of the actual work.

The outcome

The outcome of this experiment indicated that differences in the behaviour of the authoritarian-driven group and democratically led group were significant. The following are some of the more interesting differences:

- The authoritarian group tended to be either more aggressive or more apathetic than the democratic group. When aggression was expressed, it tended to be directed at the other group members rather than towards the leader. Two scapegoats were the targets of such concentrated hostility that they left the club. In the apathetic authoritarian group, it seemed that the lack of aggression was due merely to the repressive influence of the leader as when the leader left the group for a short period, aggressive outbursts occurred.
- In the authoritarian groups, there were more submissive approaches to the leader and more attention-seeking approaches. The approaches to the democratic leader were more friendly and task-related.
- In the authoritarian groups, the relations between group members tended to be more aggressive and domineering than in the democratic groups.
- Group units appeared higher in the democratic groups and subgroups tended to be more stable than in the authoritarian atmosphere, where they tended to disintegrate.
- Constructiveness of work decreased sharply when the authoritarian leader was temporarily absent, whereas it dropped only slightly when the democratic leader was absent.
- When frustrations were induced in the work situation, the democratic groups responded by making organized attacks on the difficulty, whereas the authoritarian group tended to be disrupted by the recrimination and personal blame that occurred as a result.

Peer group pressures and norms

As stated earlier in this chapter, everyone belongs to groups at all stages of their lives. Groups may be formed on a social and work basis. Social groups include, for example, families, children in a class at school, members of a club, football team or dramatic society. Work groups include people working together on a production line in a factory, in an office or shop. In these groups, members or 'peers' are perceived as having equal status.

These groups may be formed at the early stages by children in a playground, by apprentices at various stages of their training, by supervisors in a department or by groups of employees working together. These 'peer groups' put pressure on their members whereby membership of the group entails conforming to the group norms or standards. Similarly, membership of a professional body entails meeting certain levels of competence, conforming to stated levels of performance and not doing anything which brings that professional body into disrepute. In all cases, failure to meet the group norms entails either rejection from the group or the need to prove continuing commitment to achieving the group norms.

Types of organizational communication

Communication is the transfer of information, ideas and emotions between one individual or group of individuals and another. The basic function of communication is to convey meaning. A communication system can transmit information upwards, downwards and sideways within an organization on a one-way or two-way basis. When communication is purely one-way, as with an order, there is no opportunity to elicit a reaction or response from the recipient of the message. In two-way communication, however, the recipient can provide a response and is commonly encouraged to do so. Two-way communication is essential (see Chapter 10, Principles of communication).

Attitudes to safety and risk management

Both management and employee attitudes vary dramatically with respect to safety and risk management. Numerous theories are put forward from the notion that all accidents are Acts of God over which we have no control to the view that 'Safety is management's responsibility and nothing to do with me!'

If levels of safety are to be improved in an organization, it is essential that attitudes to safety are assessed. This can be achieved by

- undertaking a selection of activities including the development of interest groups with employees drawn from each sector within the organization;
- individual interviews with key people, such as departmental managers and supervisors; and
- undertaking attitude surveys through the use of questionnaires.

Attitude surveys

Attitude surveys should incorporate questions aimed at understanding the vision of safety and risk management held by the organization and the principles of safety it would like upheld by employees. Surveys can be seen as the first step to understanding an organization's safety needs in terms of obtaining a measurement of employee perceptions, concerns, attitudes and areas of satisfaction across a range of key cultural, safety and performance dimensions. They provide information about the extent to which there is alignment of employee perceptions to the organization's vision, values and safety culture, as well as identifying the extent to which employee opinions vary across differing groups.

Safety attitude surveys enable the establishment of baseline measures on key organizational issues and factors, providing opportunities for on-going measurement of the success of change and safety initiatives.

As with any survey process, the outcome of a safety attitude survey should be brought to the attention of all employees. In many cases, this outcome forms the starting point for exercises directed at improving attitudes, such as the introduction of safety groups to discuss certain issues, the establishment of a health and safety committee and the provision of health and safety information, instruction and training.

Criteria for risk acceptability (tolerability)

'Risk' is the chance that something adverse will happen (see Table 3.2). More specifically, it can be defined as 'the likelihood of a specified undesired event occurring within a specified period or in specified circumstances'. (HSE, 1988).

Issues for consideration

- The magnitude of the risk
- The benefits to individuals and society resulting from a risk
- Voluntary risks compared with involuntary (imposed) risks
- The immediate effects compared with the delayed effects
- Known risks compared with unknown risks, or risks which have an uncertain magnitude
- The 'dread' or fear associated with particular risks
- Familiar and unfamiliar risks
- Those risks which are inevitable compared with risks over which an individual can exercise choice
- Those risks where the individual is in control

Table 3.2 Major events occurring or estimated per year in Great Britain

Event	Approximate chance per annum	Basis
A fire killing 10 or more people	1	Experience
A railway accident killing or seriously injuring 100 or more people	1 in 15 or 20	Experience of last 40 years
An aircraft accident killing 500 people	1 in 1000	Very limited world experience, scaled
A tidal surge too large for the Thames Barrier to control	1 in 1000	GLC design specification
Event at Canvey Island causing 1500+ deaths or serious injuries	1 in 5000	Expert estimates of risk following improvements
'Conservative', i.e. similar event causing 18 000+ deaths or serious injuries	1 in 100 000	Likely to overestimate rather than underestimate
Aeroplane crashing into any one of London's many football stadia whilst empty	1 in a million	
Aeroplane crashing into full football stadium	1 in a hundred million	Based on pattern of actual crashes in home counties

- The feasibility of risk reduction
- Frequency of reminders of the risk
- Occupational or non-occupational exposure to risk
- The reversible and irreversible consequences of risk
- The fact that everyone must die in the end, a fact often overlooked in risk tolerability studies.

Problems in the assessment of risk

Risk assessment has to take rational account of

- Individual and societal risks
- Death, injuries and fates worse than death
- Risks to employees

- Risks to members of the public
- The value of life
- Multiple fatality situations
- Engineering feasibility for risk removal
- Resource allocation.

Methods available for assessing risks include:

- Professional judgement
- 'Bootstrapping'
- Formal analysis.

The objective is to develop criteria for judging risks which are

- usable;
- comprehensible;
- capable of being used to compare different types of risk; and
- accepted by experts, promoters and third parties as being valid and fair.

Traditional criteria

Professional judgement

Experts make decisions on the basis of 'what seems right' in terms of their experience and the conventions of their profession.

'Bootstrapping'

Bootstrapping is a technique of accepting new risks of the same or less magnitude of risks which have been tolerated historically, e.g. the use of the Fatal Accident Rate.

Fatal accidents frequency rate

One way of measuring risk is the Fatal Accident Rate (FAR). This was introduced in the early 1970s and is used extensively in the chemical industry. The criterion in FAR is usually based on the expected number of fatalities occurring in 100 000 000 working hours. It is the number of deaths resulting from industrial injury in a group of 1000 men during their working lives (40 years).

Formal methods

Formal methods are based on cost–benefit analysis (CBA). CBA provides a framework for identifying and quantifying in monetary terms all the desirable and undesirable consequences of a given activity. Costs and benefits are valued from society's point of view, in terms of real resources used or saved as a result. A typical option might involve some form of informal weighting. For instance,

Project A. The capital cost of £0.4m reduces damage by £0.45m and fatal injuries by one.

Project B. The capital cost of £0.4m reduces damage by £0.15m and fatal injuries by four.

Selecting project A instead of project B means damage is reduced by an extra £0.3m, but three less fatalities are prevented. If project A is selected, life is valued at less than £0.1m. If project B is selected, life is valued at least £0.1m.

The full cost–benefit approach

- **Cost aspect**

  What does the proposal require with existing efforts?

  What are the additional equipment requirements, labour, etc. charges, losses of production?

- **Benefit aspect**

  What is the present scale of the problem in terms of actual and potential risks?

  What is the contribution of the proposed action to reducing or eliminating those risks?

Public perceptions of risk

Why do members of the public not accept the risks judged as tolerable as a result of expert study, bootstrapping techniques or cost–benefit analysis?

- Errors by members of the public as to the 'true' risk.
- Lack of confidence in 'experts'.
- Whilst benefits may accrue to society, the risks bear on groups for whom little benefit exists.
- Tolerance is affected by the unknown and the 'dread' element of risk.
- Attitudes to risk are affected by 'non-verifiable beliefs' (superstitions) and by attitudes to other related matters. Changing views on the risk itself may require changing views on a range of other issues.

Organizational safety culture

Organizations endeavour to promote an appropriate culture with respect to, for example, quality management, customer service and sales management. This implies the establishment and implementation of a range of objectives and means for achieving these objectives, together with frequent monitoring of the performance of employees to ensure working practices devised to meet the objectives are being maintained (see Chapter 16, Health and safety culture).

The effective development and promotion of health and safety is an important feature of the health and safety culture of any organization.

Health and safety organization

The organization of health and safety at work hinges around the Statement of Health and Safety Policy. Such a statement should incorporate the 'organization and arrangements' for implementing this policy. Nearly all organizations are hierarchical in structure, i.e. they incorporate different levels of authority and responsibility, and this hierarchy can be depicted in an organization chart. In many organizations, the health and safety adviser reports within the human resources function.

An organization's structure is a means for attaining the established goals and objectives set down by the organization. In order to attain these goals, an organization must utilize the following concepts:

- **Objectives**. Objectives should be set at the highest level and determine the objectives to be set for each group and subgroup.
- **Authority and direction**. To be given responsibility requires being given the necessary power and authority to implement that responsibility. 'Direction' implies the planning and giving of orders.
- **Delegation of responsibility**. Senior management is responsible for achieving the overall objectives, but each group leader is then delegated responsibility for ensuring his group achieves its objectives.
- **Accountability**. Every person delegated responsibility is accountable for his actions to the person who delegated the responsibility to him.
- **Division of activities**. Various activities within the whole are divided into specialist groups in order to improve efficiency and to give greater control to higher management.

Generally, within the strata of management, authority and responsibility are passed downwards, whereas accountability is passed upwards.

Performance monitoring

Monitoring performance

Monitoring implies the continuing assessment of the performance of the organization, of the tasks that people undertake and of operators on an individual basis against agreed standards and objectives. Performance monitoring is a common feature of most organizations in terms of financial performance, production performance and sales performance. It implies clear identification of the organization's objectives, perhaps through a written 'mission statement' and the setting of policies and objectives which are measurable and achievable by all parts of the organization, supported by adequate resources. Performance monitoring is generally accepted as a standard feature of good management practice.

The Management of Health and Safety at Work Regulations emphasize the need to monitor performance. This may take place following the setting of health and safety policy, organizational development, risk assessment, establishment of the role of competent persons, or following the actual development of techniques of planning, measuring and reviewing performance.

At the organizational level, employers should be aware of their strengths and weaknesses in health and safety performance. They may be identified through various active forms of safety monitoring, such as safety inspections, sampling exercises and audits, and through reactive monitoring systems, such as the investigation of accidents and occupational ill-health, together with reactive analysis of accident and sickness absence returns. At task level, the implementation or otherwise of formally established safe systems of work, permit to work systems, in-company codes of practice and method statements are an important indicator of performance. Reactive monitoring through feedback from training exercises, in particular, which are those which are aimed at increasing people's perception of risk, improving attitudes to safe working and generally raising the level of knowledge of hazards, will indicate whether there has been an improvement in performance or not.

The concept of 'competent persons', outlined in the Management of Health and Safety at Work Regulations, raises a number of important issues. The general concept of 'competence' is based on skill, knowledge and experience, linked with the ability to discover defects and determine the consequences of such defects (Brazier v. Skipton Rock Company Ltd. [1962] 1 AER 955). In the designation of competent persons, however, not only do we need to consider the above factors, but also the system for monitoring and measuring their performance against agreed objectives.

Performance monitoring should not be used solely for designated competent persons, however. Most organizations practising Management by Objectives or operating performance-related pay systems have a standard form of job and career review or appraisal for staff, but how many of these systems take into account the health and safety performance of the various levels of management?

On-the-job performance monitoring should take into account the human decision-making components of a job, in particular the potential for human error. Is there a need for job safety analysis leading to the formulation of job safety instructions, a review of current job design and/or an examination of the environmental factors surrounding the job?

Auditing performance

An audit is a procedure which subjects each area of an organization's activities to a critical examination with the principal objective of minimizing loss. A safety audit could be defined as 'the systematic measurement and validation of an organization's management of its health and safety programme against a series of specific and attainable standards' (RoSPA).

Safety audits, as a form of safety monitoring, should identify strengths and weaknesses in health and safety performance. Every component of the total system is included, in particular management policy and systems, group and individual attitudes, training arrangements, features of processes, emergency procedures, documentation, etc. What is important about any auditing system is that the results should be measurable against agreed standards. Many safety audits, however, simply fail to achieve this objective, being, in many cases, a series of questions which require a simple 'Yes' or 'No' to a series of specific questions.

On this basis, many organizations tend now to use a form of health and safety review which identifies key components of performance ranked according to significance. Key components may be decided on an annual basis taking into account such factors as past accident experience, impending legislation, the identified risks and potential areas of loss. In many respects, they can be compared to a safety sampling exercise, but on a much broader scale and directed at identifying strengths and weaknesses in health and safety management systems.

A typical example of a health and safety review for transport maintenance workshops is shown in Figure 3.2. It is similar in format to a safety sampling exercise.

Health and safety reviews may be undertaken by the organization's health and safety practitioners on a 6-monthly or annual basis. Fundamentally, they are directed at identifying improvement or deterioration in health and safety performance against nationally agreed standards. In many cases, they may be linked to the organization's reward structure at all levels of management and be incorporated as a feature of quality management. As such they have much to offer as a means of motivating management to ensure effective levels of health and safety management are maintained.

TRANSPORT MAINTENANCE WORKSHOPS HEALTH AND SAFETY REVIEW			
Workplace		**Procedures/Systems**	
Cleaning and housekeeping	10	Safe systems of work	10
Structural safety	10	Accident reporting	5
Fire protection	10	Sickness reporting	5
Internal storage	5	Hazard reporting	5
Machinery and equipment	10	Safety monitoring	10
Electrical safety	10	Emergency procedure	5
Servicing operations	10	Cleaning schedule	5
Hazardous substances	5	Maintenance schedule	5
Access equipment	5	Notices and abstracts	5
Hand tools	5	Publicity/information	5
Max	**80**	**Max**	**60**
People		**Environmental factors**	
Personal protection	10	Internal	10
Manual handling	10	External	10
Safe behaviour	10	Welfare facilities	10
Max	**30**	**Max**	**30**

Total score = (Max 200)
% Improvement / Deterioration =

Action

1. Immediate
2. Short term (28 days)
3. Medium term (6 months)
4. Long term

Reviewer _____ Date _____

Figure 3.2 Health and safety review

Corporate responsibility for safety

The concept of corporate liability is well-established in section 37 (Offences by bodies corporate) of the HSWA thus:

1. Where an offence under any of the relevant statutory provisions committed by a body corporate is proved to have been committed with the consent or connivance of, or to have been attributable to any neglect on the part of, *any director, manager, secretary or other officer of the body corporate, or a person who was purporting to act in any such capacity*, he as well as the body corporate shall be guilty of that offence and shall be liable to be proceeded against and punished accordingly.

2. Where the affairs of a body corporate are managed by its members, the preceding subsection shall apply in relation to the acts and defaults of a member in connection with his functions of management as if he were a director of the body corporate.

To ensure directors, company secretaries, partners and other members of the body corporate appreciate their duties under criminal law, their role, function and accountabilities for health and safety should be incorporated in an organization's Statement of Health and Safety Policy under the HSWA. From a civil liability viewpoint, such persons constitute the *mens rea* (the ruling mind) of an organization and can be jointly and individually sued where there may be evidence of negligence.

They should be trained in these duties and kept abreast of current health and safety legislation.

The role of the supervisor

The supervisor has a significant role in the management in health and safety. The most important features of an effective supervisor can be summarized thus:

1. **Introduction**. Getting to know the employees in his charge and to be known by them.
2. **Instruction**. Passing on information and theory in a clear manner with regard to safe systems of work, the correct use of personal protective equipment, accident reporting procedures, etc.
3. **Demonstration**. Actually showing how a job is done safely.
4. **Practice**. Making reasonable allowance for employees to become proficient in tasks, including any precautions necessary to ensure safe working.
5. **Monitoring**. Observing and measuring employees' extent of proficiency, including compliance with formal safety procedures.
6. **Reporting**. Making a fair evaluation of employees' performance for management.
7. **Correcting and encouraging**. Correcting and encouraging employees as necessary.

Supervisor training

All the above factors should be considered in the training of a supervisor and particularly in terms of his responsibilities and duties for ensuring sound levels of health and safety performance in his section.

Conclusions

People generally work in groups and these groups form the main structure of an organization. One of the principal objectives of an organization is to survive and, in order to survive, an organization must be structured in such a way to ensure that its people are continually learning and being involved. Leadership is a crucial factor, supported by an effective communication system.

Organizations are exposed to many risks, some of which may be acceptable at a particular point in time, but there is a need to regularly reassess these risks through various forms of performance monitoring, such as safety audits.

Fundamentally, health and safety must be driven from the top of the organization. There must be clear identification and specification of the duties of those people forming the body corporate and efforts should be made to ensure appropriate attitudes on the part of senior management and other levels of management towards safety and risk management.

Key points

- Organizations have goals, environments and systems, together with customers who purchase their products and/or services.
- Organizations tend to operate on a formal basis, comprising a functional, or line, organization and a staff organization.
- The survival of an organization depends, to a great extent, on whether it is a learning organization or not.
- The success of an organization depends upon the quality of leadership.
- Without sound communication processes, an organization will fail.
- To ensure continuing improvements in health and safety, attitudes to health and safety at all levels need to be addressed.
- The supervisor, as the link between senior management and employees, has a significant role in management and, in particular, safety management.

4 People factors

People are the most important feature of any organization. Without the skills, commitment and loyalty of its people, an organization will simply not function. Employers need to consider the people or human factors issues in health and safety management.

What are human factors?

According to the Health and Safety Executive (HSE), 'human factors' refers to environmental, organizational and job factors, together with human and individual characteristics which influence behaviour at work in a way which can affect health and safety.

The term is often used to cover a range of issues, including

- the perceptual, mental and physical capabilities of people and the interactions of individuals with their jobs and working environments;
- the influence of equipment and system design on human performance; and
- those organizational characteristics which influence safety-related behaviour at work.

These issues are directly affected by:

- the system for communication within the organization; and
- the training systems and procedures in operation,

all of which are directed at preventing human error (Reducing error and influencing behaviour: HS(G)48: HSE).

These factors are considered below.

The Organization

Those organizational characteristics which influence safety-related behaviour include

- the need to promote a positive climate in which health and safety is seen by both management and employees as being fundamental to the organization's day-to-day operations, i.e. managers must create a positive safety culture;

- the need to ensure that policies and systems which are devised for the control of risk from the operations take proper account of human capabilities and fallibilities;
- commitment to the achievement of progressively higher standards which is shown by senior managers and cascaded through successive levels of the organization's management structure;
- demonstration by senior management of their active involvement, thereby galvanizing managers into action; and
- leadership, whereby an environment is created which encourages safe behaviour.

The job

Successful management of human factors and the control of risk involves the development of systems of work designed to take account of human capabilities and fallibilities. Using techniques such as job safety analysis, jobs should be designed in accordance with ergonomic principles so as to take into account limitations in human performance.

Major considerations in job design include

- identification and comprehensive analysis of critical tasks expected of individuals and appraisal of likely errors;
- evaluation of required operator decision-making and the optimum balance between the human and automatic contributions to safety actions;
- application of ergonomic principles to the design of man–machine interfaces, including displays of plant and process information, control devices and panel layouts, as in the case of display screen equipment workstations;
- design and presentation of procedures and operating instructions;
- organization and control of the working environment, including workspace, access for maintenance, noise, lighting and thermal conditions;
- provision of correct tools and equipment;
- scheduling of work patterns, including shift organization, control of fatigue and stress, and arrangements for emergency operations; and
- efficient communications, both immediate and over periods of time.

Personal factors

This aspect is concerned with how personal factors, such as attitude, motivation, training, human error and the perceptual, physical and mental capabilities of people, can interact with health and safety issues.

Attitudes are directly connected with an individual's self-image, the influence of groups and the need to comply with group norms or standards and, to some extent, opinions, including superstitions, like 'All accidents are Acts of God'.

Changing attitudes is difficult. Attitudes may be formed as a result of past experience, by the level of intelligence of the individual, specific motivation, financial gain and the skills available to an individual. There is no doubt that management example is the strongest of all motivators to bring about attitude change on the part of employees.

Important factors in motivating people to work safely include joint consultation in planning the work organization, the use of working parties or committees to define objectives, the attitudes currently held, the system for communication within the organization and the quality of leadership at all levels.

The appropriate safety climate

The HSE publication *Reducing Error and Influencing Behaviour* (2001) identifies the need on the part of an organization to create a climate that promotes the commitment of all employees to health and safety and which emphasizes that deviation from the corporate health and safety goals, at whatever level, is not acceptable. This assumes, of course, that the organization has actually established corporate health and safety goals and that these goals are appropriate in the circumstances.

Producing the appropriate safety climate requires clear, visible management commitment to health and safety from the most senior levels in the organization. This commitment should not be merely a formal statement of intent but be evident in the day-to-day activities of the organization, so that it is readily known and understood by employees. Individuals may be reluctant to err on the side of caution in matters that have health and safety implications if their decision to do so is likely to be subject to unwarranted criticism from their superiors or their fellow employees.

The attitude of a strong personality at senior management or board level within the organization may have either a beneficial or an adverse effect on the safety climate. Inevitably, the more junior employees will be influenced by this person's example.

Safety procedures soon fall into disuse if there is no system for ensuring that they are followed. Too often, procedures lapse because of management neglect or operators are discouraged from working to them by peer group or other group pressures, such as the need to meet production targets. Where managers become aware of deficiencies in safety procedures but take no action to remedy them, the workforce readily perceives that such actions are condoned.

Individuals may not understand the relevance of procedures nor appreciate their significance in controlling risk. Sometimes procedures are faulty, irrelevant or lack credibility. When accidents arise, management cannot blame individuals for taking shortcuts that appeared safe and were permitted to become routine, if they have not explained the importance of, or monitored the procedures, that they originally laid down.

To promote a proper working climate, it is essential to have an effective system for monitoring health and safety that identifies, investigates and corrects deviations. This could be through the operation of safety audits, inspections or surveys. There should be clearly defined standards and goals that are capable of being monitored and effective systems for reporting and investigating the causes of incidents, near misses and unsafe situations.

The introduction and operation of these systems requires considerable effort by management and only by allocating adequate resources can they be confident that failures will be prevented or controlled. In short, the organization needs to provide:

- clear and evident commitment from the most senior management downwards, which promotes a climate for safety in which management's objectives and the need for appropriate standards are communicated and in which constructive exchange of information and ideas at all levels is positively encouraged;
- an analytical and imaginative approach, identifying possible routes to human factors; this may well require access to specialized advice;
- procedures and standards for all aspects of critical work and mechanisms for reviewing them;
- effective monitoring systems to check the implementation of the procedures and standards;
- incident investigations and the effective use of information drawn from such investigations; and
- adequate and effective supervision with the power to remedy deficiencies when identified.

Management involvement

Management commitment is, perhaps, the most important of the points listed above. This commitment can be demonstrated as part of the process of encouraging a positive safety culture by

- the board clearly stating its intentions, expectations and beliefs in relation to health and safety;
- appropriate resources being made available in order to translate plans into positive achievements;

- managers being accountable for their performance as far as health and safety is concerned and senior managers being seen to take an active interest in the whole health and safety programme;
- the reward of positive achievements; and
- actively involving lower levels of management who must, in turn, ensure that health and safety has a high profile within their areas of responsibility.

Getting results

Clearly, if management has no intention of becoming involved in health and safety improvement measures, then any expressed commitment to it is unlikely to be taken seriously, and employees will not be motivated to use safe working procedures.

It is appropriate at this stage to state the *Ten Principles of Safety* of the Du Pont Corporation, a world leader in the field of health and safety at work.

1. All injuries and occupational illnesses can be prevented.
2. Management is directly responsible for preventing injuries and illness, with each level accountable to the one above and responsible for the level below. The chairman undertakes the role of chief safety officer.
3. Safety is a condition of employment; each employee must assume responsibility for working safely. Safety is as important as production, quality and cost control.
4. Training is an essential element for safe workplaces. Safety awareness does not come naturally – management must teach, motivate and sustain employee safety knowledge to eliminate injuries.
5. Safety audits must be conducted. Management must audit performance in the workplace.
6. All deficiencies must be corrected promptly either through modifying facilities, changing procedures, better employee training or disciplining constructively and consistently. Follow-up audits are used to verify effectiveness.
7. It is essential to investigate all unsafe practices and incidents with injury potential, as well as injuries.
8. Safety off the job is as important as safety on the job.
9. It is good business to prevent illnesses and injuries. They involve tremendous costs – direct and indirect. The highest cost is human suffering.
10. People are the most critical element in the success of a safety and health programme. Management responsibility must be complemented by employees' suggestions and their active involvement.

Organizations would do well to emulate the example given by the Du Pont Corporation and its chief executive who personally undertakes safety audits.

The individual differences in people

Everyone is different. These differences influence patterns of work behaviour and may limit the effectiveness with which an individual carries out a task. Such differences will also influence how safely tasks are undertaken.

The individual differences in people are associated with a range of factors.

Genes

Physical features of people, such as the colour of the eyes and hair, height and body dimensions, are strongly influenced by genetic factors. Factors, such as personality and intelligence, are genetically inherited and are significant when endeavouring to change behaviour in the case of, for example, the implementation of safe systems of work. In the majority of cases, these two features cannot be changed.

Environmental factors

The environment is defined as 'that which surrounds us; the surroundings'. From early childhood, people are influenced by a number of environments, such as the home environment, the school environment and the work environment. In most cases, peer groups are established which have a great influence on attitudes held and personality.

Attitudes held and personality

Each individual has a unique set of attitudes to a range of circumstances and situations. Personality characteristics vary significantly (see 'Personal factors' earlier in this chapter). Both these aspects of behaviour affect a person's perception of the world in general and how they react to specific situations.

Perception of situations

People are influenced by their perception of situations which can influence their behaviour at that point in time. For example, an enforcement officer, on arriving at a workplace, may be influenced by evidence of poor levels of external housekeeping, dangerous driving by fork lift truck operators or unsafe work activities by contractors, before he has announced his arrival at the reception point. He may well, therefore, form the opinion that this is an unsafe workplace and this will affect his attitude and approach during a subsequent formal inspection of that workplace.

Incorrect perception of situations, where people misread the information that is presented at that point in time, can be a contributory factor in accidents. This is particularly the case with overtaking situations leading to road accidents.

These differences arise from an interaction between the inherited characteristics, which are passed on from parents, and the various life and learning experiences through which an individual passes from the moment of conception. Many of the factors below mould the individual into a unique person different from all other individuals.

Psychological, sociological and anthropological factors leading to individual differences

From birth, individuals are continuously developing and changing their own particular set of behavioural characteristics with respect to, for example, attitudes held, personality, intelligence, motivation and the way they think. This development is part of the evolutionary process resulting in people acquiring particular behaviour patterns and different tendencies for behaviour. Some of these behaviour patterns are inherited, whereas others will be developed as a result of a range of experiences, particularly in early life (Figure 4.1).

- Genes and chromosomes
- Experiences in the womb
- Birth trauma
- Family influences
  - cultural patterns of child rearing
  - parents (personality, strictness, style, affection, etc.)
  - brothers and sisters
  - position in the family
  - socioeconomic group
  - background
- Geographical location
- Preschool influences
- Education – opportunities, quality, support
- Occupational factors
  - training and retraining
  - opportunities
  - status (of the occupation and position held)
  - group membership
- Hobbies and interests
- Own family influences – marriage, children
- Ageing

Figure 4.1 The individual differences in people

Sociology is concerned with the effects of group membership upon the behaviour, attitudes and beliefs of people. Most people operate within groups, at work, at school, etc. Membership of the group entails complying with the group norms with respect to behaviour, attitudes and beliefs shared by the group. These sociological factors have a significant effect on the development of the individual influencing attitude in particular and making that person different from people who are not members of the group.

Anthropology is broadly defined as the science of the human race. Anthropological factors, such as body dimensions, strength and stamina, contribute significantly to the individual differences in people. Anthropometry, the measurement and analysis of Man's physical, mechanical and functional characteristics, endeavours to design environments to accommodate these characteristics. Typical examples include the arrangements for driving positions in vehicles with respect to leg length, arm length and front, side and rear vision. Similarly, the analysis of display screen equipment workstations considers a number of anthropometric factors, such as the arrangement of work desks and the design of chairs based on the variability of people's body dimensions.

The ways in which people differ are many and varied. It is important to bear this fact in mind from the point of view of work effectiveness and safety. It is vital to know what a particular job entails (job description) and to specify the characteristics required to enable an individual to perform that job effectively (personnel specification).

The seven-point plan

Alec Rodger (1952) classified these special characteristics into seven categories (Table 4.1).

Other points can be specified as needed. It will be clear that some of the above factors are easier to specify and measure than others and there is a tendency to place more emphasis and importance on certain factors simply because they can be easily assessed.

However, there are other aspects of individual differences that are equally important in the working situation, in particular, personality, intelligence, aptitude, perception, motivation and commitment and these are even more difficult to specify and assess.

Human limitations

All people are different in terms of their limitations and capabilities which feature in their performance at work. In particular, the following features of human performance can be a contributory factor in accidents.

Table 4.1 The individual characteristics of people

1. Physique	Height; Build; Hearing; Eyesight; General health; Looks; Grooming; Dress; Voice
2. Attainments	General education; Job training; Job experience
3. General intelligence	Tests; General reasoning ability
4. Special aptitudes	Mechanical; Manual dexterity; Skill with words; Skill with figures; Artistic ability
5. Interests	Intellectual; Practical/constructional; Physically active; Social; Aesthetic
6. Disposition	Acceptability; Leadership; Stability; Self-reliance
7. Circumstances	Age; Marital status; Dependants; Mobility; Domicile

Physical

Factors such as reach, lifting ability and capacity, skeletal features, sensory features (visual, aural) and energy level are significant in many tasks.

Physiological

People may be affected by illness, the effects of drugs and medication, fatigue, the oxygen supply, environmental contaminants, such as fumes, gases, etc., the effects of alcohol, time zone adjustment, ageing and circadian (diurnal) rhythm.

Psychological

Variations in individual ability, aptitude, knowledge, interests, personality, memory and motivation may affect an individual's potential for accidents.

Psychosocial

Factors such as cultural context, group pressures, individual risk-taking and situational influences can affect performance at work.

Degradation of human performance

Human performance is directly affected by the working environment and sound levels of working environment promote optimum levels of performance. Many factors influence the human system and performance can degrade over a wide range of environmental conditions, breaking down in a limited number of cases, as follows.

Fatigue

People who work excessive hours without rest breaks and periods of sleep may suffer from tiredness and general fatigue.

Diurnal (circadian) rhythm

Body rhythms follow a cyclical variation linked to the 24-hour light–dark cycle and sleeping–waking cycle. This is known as diurnal or circadian rhythm. Interruptions in this rhythm, as experienced by shift workers and night workers, place considerable stress on operators resulting in reduced performance as high as 10 per cent below average.

Adjustment may take place after 2–3 days and goes on increasing up to a period of approximately 14 days, provided that the individual continues both to live and work on a night-time schedule and does not return to normal day-time living at weekends.

Night workers and shift workers are also subject to noise created by routine daytime activities and suffer from disturbed sleep. They also suffer major disruptions to social life as most social events take place in the evening. This can result in conflicts within families, and increased gastrointestinal disturbances, such as ulcers and nervous disorders.

Rotating shift patterns, e.g. a week on nights followed by a week on days, or 12-hour shifts rotating from, for instance, 6 a.m. to 6 p.m., noon to midnight, and 6 p.m. to 6 a.m. on different weeks, can result in high levels of stress on operators and their families.

Loss of motivation

Where there is no stimulation from management in terms of performance targets and reward for achievement of those targets, for instance, workers rapidly become demotivated and their performance deteriorates.

Lack of stimulation

Many jobs are boring, repetitive and demotivating resulting in a lowered level of arousal. Stimulation of performance can be achieved by job rotation, productivity bonus schemes (provided the rewards are seen to be fair to all concerned), working in small teams and counselling in certain cases to endeavour to reduce stress.

The relevance of human factors within the sociotechnical system

What is a sociotechnical system?

The term originally arose from work by Eric Trist and Fred Emery, founders of the Tavistock Institute, London. Fundamentally, in the development of organizations, 'sociotechnical systems' is an approach to complex organizational work design that recognizes the interaction between people and technology in workplaces. The term also refers to the interaction between society's complex infrastructures and human behaviour. Within this concept, society itself and most of its substructures, are complex sociotechnical systems.

Sociotechnical systems theory is an approach to the design and management of systems that aims to satisfy four closely related objectives:

- user satisfaction;
- system efficiency;
- successful system implementation; and
- effective change management.

The approach is commonly used in managing organizational change or in certain projects where business process re-engineering may be involved.

Human factors and sociotechnical systems

Human factors has an important role in the sociotechnical system, involving areas such as job design, task analysis, job enrichment, job enlargement, job rotation, motivation, satisfaction, autonomy, process improvement and the move towards self-managing teams in the workplace.

Sociotechnical systems theory has two important objectives:

1. to design systems that improve the welfare and quality of users' lives; and
2. to improve the performance of the organization by adding shareholder value, increasing productivity and competitiveness.

The principal message from such systems is that new systems or technologies should never be designed or implemented without considering the 'softer' issues. This means that if the context within which the technology will be used is ignored, overlooked or not properly understood, the technology may fail to be useful to its users. It may be disregarded or actively rejected by operators.

Individual decision-making processes

Decision-making is the cognitive process of selecting a course of action from among multiple alternatives. It is a reasoning process which can be either rational or irrational and based on explicit assumptions or tacit assumptions.

People are continually making decisions – when to get up, what to wear, what to eat, what to do, when to go out, etc. Every decision-making process produces a final choice, which can be some form of action, with a commitment to that action, or an opinion on a particular matter.

Structured rational decision-making is an important feature of the way people perform at work and of the management process. In the latter case, specialists apply their knowledge in a given area to making informed decisions or, conversely, decision-making may take place in groups, where a board or committee decides on a particular course of action.

Biases in the decision-making process

Many forms of bias can confuse the decision-making process. These biases may query or call into question the correctness of a decision. A number of these biases are outlined below.

- **Conservatism and inertia**. Unwillingness to change thought patterns that people have used in the past in the face of new circumstances.
- **Experiential limitations**. An unwillingness to look beyond the scope of past experiences and rejecting that which is unfamiliar or not experienced.
- **Wishful thinking or optimism**. Wanting to view a situation in a positive light, which distorts a person's thinking and perception of the situation.
- **Repetition**. Willingness on the part of a person to believe what he has been told most often and by the greatest number of different sources.
- **Source credibility**. People may reject a view or opinion where there is a bias against the person or organization expressing the view or opinion. People are inclined to accept a statement by someone they like or admire. Political parties frequently suffer this form of bias.

- **Group thinking**. A member of a group may be put under pressure to conform to the opinions held by that group and to conform to the group norms.
- **Inconsistency**. An unwillingness to apply the same decision-making criteria in similar situations.

Making decisions

Fundamentally, people make decisions in different ways based on factors such as their level of knowledge, skills and experience of similar situations. The decision-making process entails considering options, reasoning and deciding on the best solution.

However, the decisions that people make may be affected by a range of biases which can distort or confuse the final outcome, resulting in incorrect or unwise decisions. Incorrect decision-making is a contributory factor in many accidents, particularly road accidents.

Individual change

During the course of their lives, people undergo many changes in terms of motivation, attitude, personality, knowledge, skills and outlook. These changes may be related to a range of experiences during infancy, childhood, puberty, school days and eventual adulthood which affect human characteristics and the limitations of people. The process of growing up is one of continuing change with the setting of new objectives, the rejection of old ideas and planning for the future.

People are continually adjusting to change associated with the social and work environment, new technology, new information, legal requirements, changes in job objectives and activities and in the acquisition of new skills. In many cases, the changes in behaviour necessary arising from these factors can create problems and concern for the individual due to, in many cases, an in-built resistance to change.

Human reliability and human reliability assessment

The Management of Health and Safety at Work Regulations require employers to consider the capabilities of employees as regards health and safety when entrusting tasks to their employees. Thus, when allocating work to employees, they must ensure that the demands of the job do not exceed the employees' ability to carry out the work and the level of their training, knowledge and experience. This requirement to consider human reliability is a relatively new aspect of health and safety legislation.

Increasingly, attention is being paid to human error, incorrect human action or the failure to act by people as a causative factor in accidents of varying severity. In 1987, the Advisory Committee on the Safety of Nuclear Installations published the work of a study group which was to report to that committee on the part played by human factors in the incidence of risk in the nuclear industry and the reduction of that risk. The Second Report of the Study Group *Human Reliability Assessment: A critical overview* (1991), provides a comprehensive insight into the various classes of human error, the factors which contribute to human error and the technique of human reliability assessment.

Human reliability assessment

Human reliability assessment (HRA) includes the identification of all points in a sequence of operations at which incorrect human action or the failure to act (sins of omission) may lead to adverse consequences for workplaces, plant and people.

HRA techniques assign a degree of probability on a numerical scale to each event in a chain and then by aggregating these, arrive at an overall figure for probability of human error for the whole chain of events. Such an assessment may point to the steps that need to be taken to reduce the likelihood of error at certain points by introducing organizational, procedural, ergonomic or other changes in systems of work.

A systematic approach

A systematic approach to analysing and optimizing human reliability in a particular task takes place in a series of stages:

- identifying the human performance requirements, both mental and physical, and setting objectives;
- performing task analysis;
- performing human error analysis;
- identifying:
  - root causes;
  - consequences;
  - recovery points; and
  - error reduction strategies;
- quantifying to enable cost–benefit analysis of the error reduction strategies proposed;
- choosing and implementing a selected error reduction strategy;
- monitoring the effectiveness of that error reduction strategy; and
- assessing whether the human performance objectives have been achieved.

Where performance objectives have not been achieved, it may be necessary to select an alternative error reduction strategy.

The human causes of accidents

Charles Darwin, author of *The Origin of the Species*, suggested that man increased his likelihood of having accidents at the point where he decided to stop walking on all four limbs and stand erect, a position he was not originally intended to take. Whether this is the case is a matter for conjecture. However, there are a number of aspects of human life and activity which increase the potential for accidents.

Age

At the various stages of life, infancy, childhood, youth, manhood and old age, the potential for accidents varies dramatically. The baby going through the early stages of walking will have many accidents, some resulting in minor injuries. As the baby progresses to the child stage, his confidence increases but he may still be prone to falls and other accidents associated with a lack of fear common in young children.

Teenagers are at the stage in their lives where experience leads them to take certain actions to protect themselves. However, physiological and psychological changes are taking place as part of their development and they need to adjust themselves quickly to different control patterns. They may be prone to experimentation in a range of risk-related situations.

At adulthood, the lessons have been learnt, dangerous situations have been experienced and people are able to distinguish between what is dangerous and what is safe. However, with this group, behavioural factors such as overconfidence, inadequate skills, lack of awareness, lapses of attention and carelessness may be contributory factors in accidents.

As old age approaches, physical and mental agility reduces. Failing eye sight and hearing may further contribute to an increased potential for accidents as with people who may be suffering presbycusis (age-induced deafness). In this case, many road accidents involving older people are associated with their not being able to identify the source from which a sound is coming, such as a vehicle, together with their distance from the source.

Work activities

People working in factories, in particular on production processes, may experience boredom and fatigue due to the repetitive nature of the work. Some people, however, are quite happy to undertake repetitive tasks, such as the manufacture of components, the preparation of sandwiches or the packing of products. This can be associated with their temperament, the fact that they do not have to make decisions, accept responsibility and have some degree of interest in the task.

Other people find this sort of work boring, demotivating and degrading, which raises the question of the relative levels of intelligence of people involved in this type of work. It could be argued that the more intelligent person finds this sort of work boring because there is no challenge to their intelligence or creative ability. As a result, the potential for accidents may be increased due to the more intelligent worker not having their mind on the job, thereby increasing the potential for human error.

The cause–accident–result sequence

People are directly or indirectly involved in accidents. The cause–accident–result sequence explains how the indirect causes of accidents (personal factors and source causes) contribute to the direct causes of accidents (unsafe acts and unsafe conditions) which result in an accident. Each accident has direct results and indirect results (for the injured person and the organization). This sequence is shown in Figure 4.2.

Human capability and safety – The legal situation

The Management of Health and Safety at Work Regulations place an absolute duty on every employer, when entrusting tasks to his employees, to take into account their capabilities as regards health and safety (Regulation 13(1)).

The Approved Code of Practice to the regulations explains this requirement to some extent thus:

> When allocating work to employees, employers should ensure that the demands of the job do not exceed the employees' ability to carry out the work without risk to themselves and others. Employers should take account of the employees' capabilities and the level of their training, knowledge and experience. Managers should be aware of relevant legislation and should be competent to manage health and safety effectively. Employers should review their employees' capabilities to carry out their work, as necessary. If additional training, including refresher training, is needed, it should be provided.

Very little guidance, if any, is provided in the HSE guidance accompanying the regulations on the question of human capability, however. Emphasis is placed on the need for determining the level of training and competence required for each type of work through the risk assessment process.

INDIRECT CAUSES	DIRECT CAUSES	ACCIDENTS	DIRECT RESULTS	INDIRECT RESULTS
Personal factor *Definition:* Any condition or characteristic of a man that causes or influences him to act unsafely. 1. Knowledge and skill deficiencies: (a) Lack of hazards awareness (b) Lack of job knowledge (c) Lack of job skill. 2. Conflicting motivations: (a) Saving time and effort (b) Avoiding discomfort (c) Attracting attention (d) Asserting independence (e) Seeking group approval (f) Expressing resentment. 3. Physical and mental incapacities.	Unsafe act *Definition:* Any act that deviates from a generally recognized safe way of doing a job and increases the likelihood of an accident. Basic types 1. Operating without authority 2. Failure to make secure 3. Operating at unsafe speed 4. Failure to warn or signal 5. Nullifying safety devices 6. Using defective equipment 7. Using equipment unsafely 8. Taking unsafe position 9. Repairing or servicing moving or energized equipment 10. Riding hazardous equipment 11. Horseplay 12. Failure to use protection.	The accident *Definition:* An unexpected occurrence that interrupts work and usually takes this form of an abrupt contact. Basic types 1. Struck by 2. Contact by 3. Struck against 4. Contact with 5. Caught in 6. Caught on 7. Caught between 8. Fall to different level 9. Fall on same level 10. Exposure 11. Overexertion/strain.	Direct results *Definition:* The immediate results of an accident. Basic types 1. 'No results' or near miss 2. Minor injury 3. Major injury 4. Property damage.	Indirect results *Definition:* The consequences for all concerned that flow from the direct result of accidents. For the injured 1. Loss of earnings 2. Disrupted family life 3. Disrupted personal life 4. And other consequences. For the company 1. Injury costs 2. Production low costs 3. Property damage costs 4. Lowered employee morale 5. Poor reputation 6. Poor customer relations 7. Lost supervisor time 8. Product damage costs.
Source causes *Definition:* Any circumstances that may cause or contribute to the development of an unsafe condition. Major source 1. Production employees 2. Maintenance employees 3. Design and engineering 4. Purchasing practices 5. Normal wear through use 6. Abnormal wear and tear 7. Lack of preventive maintenance 8. Outside contractors.	Unsafe conditions *Definition:* Any environmental condition that may cause or contribute to an accident. Basic types 1. Inadequate guards and safety devices 2. Inadequate warning systems 3. Fire and explosion hazards 4. Unexpected movement hazards 5. Poor housekeeping 6. Protruding hazards 7. Congestion, close clearance 8. Hazardous atmospheric conditions 9. Hazardous placement or storage 10. Unsafe equipment defects 11. Inadequate illumination, noise 12. Hazardous personal attire.			

Figure 4.2 The cause–accident–result sequence

Human capability and risk assessment

Risk assessment is the cornerstone of all modern protective legislation, such as that relating to health and safety and food safety. However, only the Manual Handling Operations Regulations and the Health and Safety (Display Screen Equipment) Regulations provide any form of guidance on the issue of human capability and this is extremely limited.

Manual Handling Operations Regulations (1992)

HSE Guidance to the Manual Handling Operations Regulations makes the following points:

Individual capability – reducing the risk of injury

Personal considerations

Particular consideration should be given to employees who are or have recently been pregnant, or who are known to have a history of back trouble, hernia or other health problems, which could affect their manual handling capability. However, beyond such specific pointers to increased risk of injury the scope for preventive action on an individual basis is limited.

Clearly an individual's state of health, fitness and strength can significantly affect the ability to perform a task safely. But even though these characteristics vary enormously, studies have shown no close correlation between any of them and injury incidence. There is therefore insufficient evidence for reliable selection of individuals for safe manual handling on the basis of such criteria. It is recommended, however, that there is often a degree of self-selection for work that is physically demanding.

It is also recognized that motivation and self-confidence in the ability to handle loads are important factors in reducing the risk of injury. These are linked with fitness and familiarity. Unaccustomed exertion – whether in a new task or on return from holiday or sickness absence – can carry a significant risk of injury and requires particular care.

Health and Safety (Display Screen Equipment) Regulations 1992

Fatigue and stress

Many symptoms described by display screen workers reflect stresses arising from their task. They may be secondary to upper limb or visual problems but they are more likely to be caused by poor job design or work organization, particularly lack of sufficient control of the work by the user, under-utilization of skills, high-speed repetitive working or social isolation. All these have been linked with stress in display screen work, although

> clearly they are not unique to it; but attributing particular symptoms to particular aspects of a job or workplace can be difficult. The risks of display screen workers experiencing physical fatigue and stress can be minimized, however, by following the principles underlying the Display Screen Equipment Regulations 1992 and guidance, i.e. careful design, selection and disposition of display screen equipment; good design of the user's workplace, environment and task; and training, consultation and involvement of the user.

Clearly, in both cases, there is a need to assess human capability with respect to physical and mental capability, respectively. Health surveillance by an occupational health nurse may go some way in identifying factors significant to the tasks undertaken, such as physical strength and fitness for manual handling work, visual acuity and the potential for fatigue with display screen work. In the latter case, much will depend upon the type of work undertaken, such as data processing and other forms of repetitive tasks. Consideration should be given to software ergonomics in this case (see Chapter 8, Ergonomic principles).

Conclusions

Human factors refers to the environmental, organizational and job factors, together with those human and individual characteristics which influence behaviour. Fundamentally, all people are different in terms of genes, their attitudes, personalities and perception of situations. People also have limitations which are associated with their individual physical, physiological, psychological and psychosocial characteristics. In recent years, considerable emphasis has been placed on this relationship between human factors and the sociotechnical systems operating within organizations.

The individual decision-making processes of people also need consideration together with the biases in decision-making which can be contributory factors in accidents. Human error, incorrect human action and the failure to act by people also contribute to accidents with varying degrees of severity. In some cases, there may be a need for assessment of human reliability.

Fundamentally, people are still the one significant cause of accidents in terms of their attitudes to work and individual factors, such as age, inherent skills and knowledge.

Key points

- Human factors in the field of health and safety is concerned with the capabilities of people and their interaction with the job and work environment, the influence of equipment and system design on performance (the ergonomic considerations) and organizational characteristics which influence safety-related behaviour.

- People are different as a result of psychological, sociological and anthropological variations.
- The limitations of people vary considerably.
- Personality and personality traits are important contributory factors in accidents.
- Human factors considerations are an important feature in the development of sociotechnical systems.
- Decision-making is a cognitive process of selecting a course of action from among multiple alternatives.
- People are continually undergoing change during the course of their lives, the process of adjustment of such change varying significantly from person to person.
- Human reliability assessment endeavours to reduce the likelihood of human error which can result in adverse consequences for people, machinery and workplaces.

5 Perception of risk and human error

Perception of risk

In Chapter 1, Human behaviour and Safety, reference was made to the concept of 'safe person' strategies and the need to increase people's perception of risk. But what is 'risk'? 'Risk' is defined as 'the probability or likelihood of the harm from a particular hazard being realized'. Probability theory is based on the outcomes or consequences of a set of events arising from a particular situation.

People perceive risk through the normal sensory inputs of sight, hearing, touch, taste and smell. The principal sensory input is sight, although hearing is an important input. Perception relies to a great extent on learning and past experience and is heavily influenced by motivational factors, the context in which a stimulus is produced, for example, the context in which people hear something, the level of arousal of the individual, ergonomic factors, such as the layout and design of controls and displays on work equipment, and the level of training received. Generally, no two people perceive risk in the same way. Much depends on an individual's perceptual set, however.

Perceptual set

Everyone has their own perceptual set, a mechanism which selects, rejects, modifies, ignores and interprets that which is perceived. Perceptual set is affected by

- context (background environment);
- expectations held;
- instructions;
- motivation (needs);
- emotions;
- past experience;
- individual differences in people;
- culture; and
- the potential for reward or punishment.

All these factors combine to produce a particular perceptual set in a particular situation and produce what is actually perceived by the individual.

Perceptual sensitization and defence

There are two mechanisms that are of specific importance when considering practical applications of perception theory.

1. **Perceptual sensitization**. An individual can be or become 'sensitized' to certain stimuli if they are relevant, important or meaningful to that person. Sensitization can be permanent or temporary and may or may not be accepted consciously by the individual. The stimuli to which the individual is sensitized have a much greater impact than would normally be expected.
2. **Perceptual defence**. No one likes to feel threatened or anxious. The mechanism of 'perceptual defence' tends to protect the individual from such unpleasant situations by obliterating these threats and making them difficult to perceive at a conscious level. The individual does not wish to register the problem or difficulty, so it is pushed into the unconscious.

Perception of hazards

Both these two mechanisms can have an important influence on the perception of hazards and risks in the workplace. For instance, a health and safety practitioner may well be 'sensitized' to observing hazards, whilst those actually exposed to the danger are exhibiting perceptual defence and may not accept that any danger exists.

Fundamentally,

- People do not see what is there. They see what they expect to be there.
- People do not see what they do not expect to be there.
- People do not see what they do not want to be there.

Perception and sensory inputs

Perception involves a range of sensory inputs, as outlined in Chapter 2, Human sensory and perceptual processes. How people perceive risk entails a complex process based on many factors, such as past experience of similar situations, their level of knowledge and training, motivation and various innate abilities. Incorrect perception can be associated with failing or defective sensory processes, as with failing eyesight, incorrect assumptions as to the circumstances prevailing at the time, stress and overfamiliarity with a work process. No two people necessarily perceive the same situation in the same way. This fact can commonly be identified, for example, in the review of witness statements following an accident.

There is a need, therefore, to consider these various aspects of human performance with respect to the causation of accidents.

Human performance and accident causation

The model of human performance in relation to accident causation developed by Hale and Hale (1970) is one of the most well-known simulations of the accident

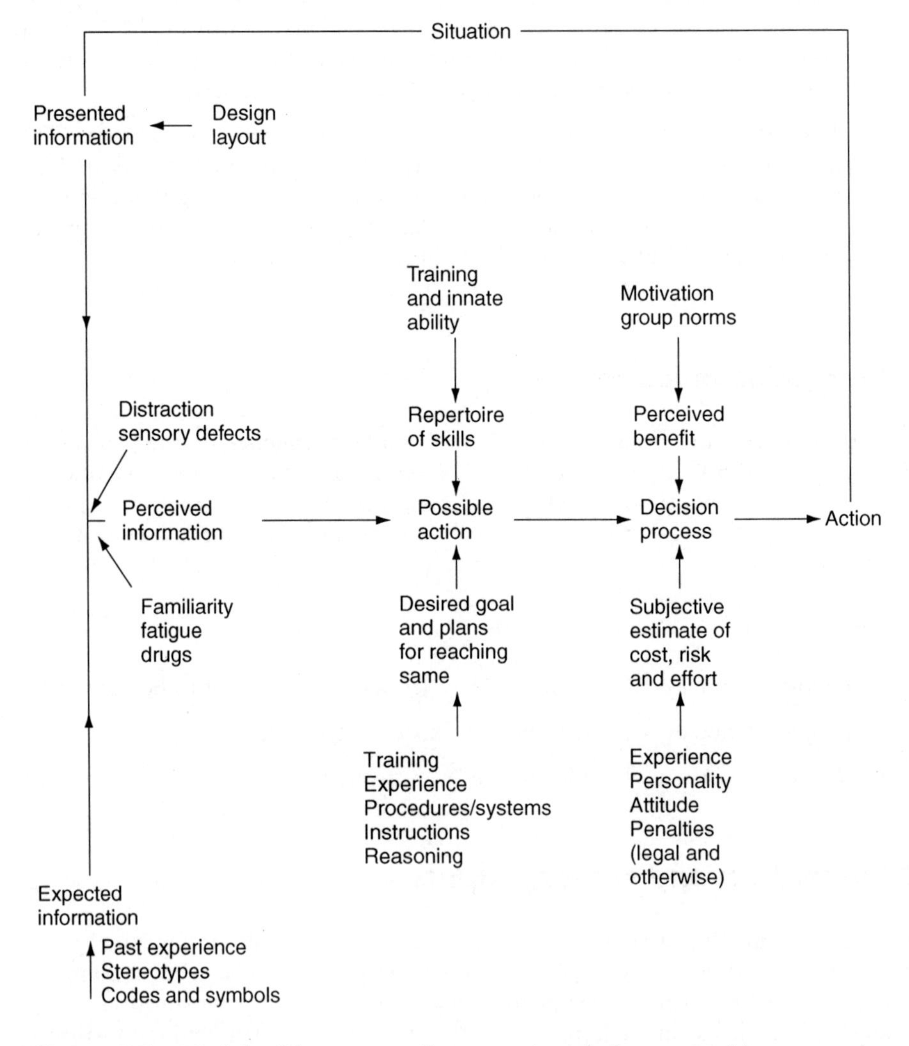

Figure 5.1 Model of human performance in relation to accident causation (Hale, A.R. and Hale M., (1970): Accidents in Perspective: *Occupational Psychology* 44: 115–122)

causation process. In this model, use is made of a closed loop system, which considers the major factors of

- the original situation of danger;
- presented, expected and perceived information;
- the action;
- the decision-making process;
- the action required;
- the feedback to the original situation.

Secondary features, such as past experience, stress, sensory defects and motivation contribute to the major factors in accident causation.

A typical situation involves the potential accident victim receiving *presented information*, which may arise through the design or layout of the working area. *Expected information* includes that person's past experience and the presence of codes and symbols.

The *presented information* and *expected information* combine to produce the *perceived information* which may be affected by sensory defects, stress and familiarity with the situation. From this initial perception of the situation, the individual decides on a series of *possible actions*, which may be based on his *repertoire of skills* and a *desired goal and plans for reaching that goal.* The level of training, in particular, greatly influences these two factors.

Following this stage, the individual moves into the *decision process* which entails a consideration of the *perceived benefit* and a *subjective estimate of the cost, risk and effort* required. Factors such as the quality of supervision, experience and personality affect these last two factors.

From the decision process, the individual must decide on the *action* necessary, which can be affected by his physical state and health. As a result of the action taken and its outcome, the individual receives feedback which he can apply to future situations of a similar nature.

Filtering and selectivity of perception

In the process of perception, the brain classifies information on the basis of its type and source. It can further select those stimuli that are relevant through a filtering mechanism which reduces the significance of other less relevant stimuli.

For example, in the driving situation the driver is continuously receiving information as a result of the visual perception process, such as the distance from the vehicle in front, road and weather conditions, road signs and the information as to vehicle speed from the odometer. These are all significant pieces of information in terms of safe driving. Less relevant stimuli, such as people walking on the pavement, signs over shops and parked vehicles tend to be filtered out.

Similarly, the process of hearing can be directed in such a way as to be selective in terms of identifying, for example, irregular noises coming from a running engine, the normal running sounds being filtered with a view to detecting the fault.

Optimum bias

Perception may be affected by 'optimum bias', whereby the individual demonstrates a tendency to be over-optimistic about the outcome of a risky situation. This is the case with both experienced and inexperienced operators.

Experienced operators may consider that their knowledge of the system automatically protects them from danger. Conversely, inexperienced operators may consider that the design of a process and the management systems in operation will protect them and tend to be less attentive to the precautions necessary on their part.

Risk compensation (homeostasis)

This term, sometimes referred to as 'risk homeostasis', 'danger compensation', 'risk-offsetting behaviour' or 'perverse compensation' is an effect whereby individual people and animals may tend to adjust their behaviour in response to perceived changes in risk. It is seen as self-evident that individuals will tend to behave in a more cautious manner if their perception of risk or danger increases. Similarly, people may behave less cautiously in situations where they feel safer or more protected.

Such an effect is seen with people and is associated with the use of safety features, such as car seat belts, antilock braking systems and cycle helmets. Where people perceive that they are safer when using these features, they tend to drive or cycle faster, thereby compensating for the increased feeling of safety.

Risk compensation is a theory which endeavours to understand the behaviour of people in potentially hazardous situations. The basic theory suggests that a person will accept a given level of risk in a specific activity. If their perceived level of risk alters, their behaviour will compensate to place them back at the accepted level of risk.

Risk homeostasis is a controversial theory that suggests that people have an implicit, preferred or target level of risk. One outcome of this theory is that any safety improvements that reduce the risk exposure of an individual will potentially lower the perceived threat below the target level. This creates the opportunity for people to alter their behaviour. In particular, they may trade performance objectives for a slightly increased level of risk. For example, drivers of cars fitted with advanced braking systems may drive faster and brake later than drivers of cars not fitted with such systems.

Cultural aspects of risk

People operate within a range of cultures in terms of the family culture, the organizational culture at work, the safety culture which the organization endeavours to promote and a wider socioeconomic culture. They may also operate within a particular professional culture, such as lawyers, or trade culture, such as bricklayers.

'Culture' implies a state or set of manners and beliefs held by a particular individual, group or organization. In recent years, much attention has been paid to the 'macho' culture within organizations, the culture which adopts the approach that, in order to be accepted into a group, a member must display a series of aggressive attitudes and approaches to the way work is carried out. This type of culture can be seen in sales-orientated organizations in particular.

In many cases, conforming to the culture implies acceptance of a range of risks. For example, delivery drivers may be expected to drive faster than other drivers in order fulfil their organization's promises to get the goods to the destination promptly. Many tasks incorporate an element of risk, such as working on a roof or servicing of electrical equipment. The people who undertake these tasks may consider their skill, knowledge and experience automatically protect them from danger and may operate within a risk-taking culture typified by failure to follow safety procedures and adopting their own 'shortcuts' when carrying out the work.

Social amplification of risk

Social and individual factors can act to amplify perceptions of risk and thereby create secondary effects, such as economic losses or regulatory impacts. Many factors and processes contribute to the formation of public opinion on risk, such as events and issues reported and discussed in the media. Moreover, members of the public are not passive recipients of expert risk knowledge, but draw upon multiple information sources and understanding to rationalize risk, such as their own personal experience. What this implies is that there are critical points where the orientation, tempo or strength of the social image of a hazard changes significantly.

The Social Amplification of Risk Framework (SARF) was developed in the United States in the 1980s. The framework described the various processes by which some hazards and events become a form of social and political concern and activity (amplification) despite experts believing that they present a relatively low statistical risk, while other more serious events receive comparatively little public attention (attenuation). It must be appreciated that both amplification and attenuation hold equal places in the framework.

The tabloid media are extensively involved in the amplification of risk, in promoting 'scares' on a range of issues, such as the risk of AIDS, much depending upon the particular audience they hope to reach.

Communication of risk

Risk communication as a discipline takes a sociological approach with respect to basic risk perception and the formulation of policies, legislation dealing with hazards, the key issues of public involvement, the risk and environmental management.

Risk communication is based on two specific concepts. Firstly, risk communication is a subject of study in which researchers endeavour to understand social, psychological and cultural aspects of risk as they apply to the creation, transmission and effect of risk information. Conversely, risk communication can be an instrument to control information by people with vested interests, such as organizations and governments.

The discipline derives from risk perception studies, based on a psychological approach, which endeavour to ascertain how members of the public are influenced by certain variables in perceiving risk as acceptable or not.

In order to improve the effectiveness of risk communication, a number of factors need consideration, in particular

- the variables which affect public perception of risk;
- statistic content analysis of the media, such as national and local newspapers;
- qualitative content analysis of appropriate public documents and reports;
- collection and analysis of background information on the community context.

Risk perception and communication (HSE)

The research into individual and public perception of risk has been extensive over the last 20 years, with references to many risk-related topics, such as those from radon, anaesthetics, smoking, hazardous wastes and radiation.

The Health and Safety Laboratory's publication *Review of the Public Perception of Risk, and Stakeholder Engagement (HSL/2005/16)* provides considerable guidance on how members of the public perceive a range of risks.

A number of variables influence the perception of risk, such as

- perceived control;
- psychological time and risk;
- familiarity;
- trust and distrust;
- the framing of risk;
- risk communication and numerical representations of risk;
- perception of hazardous substances; and
- the context or the individual?

Several important points emerge from the main findings of this publication with respect to risk perception as follows.

- At the level of the individual, risk perception is a result of many factors, as opposed to rational judgements based on the likelihood of harm. There are a number of explanations for why the perception of risk is not based on these rational judgements. These reasons include systematic biasing of risk information, the use of mental shortcuts and the way that risk information can be presented.
- Greater scope exists for these biases when the risk is complex or the effect of the harmful consequence is delayed. These biases do not occur in isolation, but are influenced by the situation in which the individual perceives the hazard. The relationship between the situation and the bias is unclear, but is not thought to be insignificant.
- Longstanding evidence from the psychometric approach to risk perception indicates that acceptance of a hazard is related to the qualitative characteristics of that hazard. The accepted range of characteristics include:
    - The nature of the hazard – familiarity and experience of the risk; understanding of the cause–effect mechanism; uncertainty; voluntary exposure to the risk; artificiality of the hazard; violation of equity of benefits arising from the hazard.
    - The risk's consequences – ubiquity of the consequences of the risk (geographically and across time); fear of the risk's consequences (catastrophic potential); delay effect (e.g. the salience of the risk is a function of delay in deleterious consequences); reversibility (potential to restore to original state); negative impact on individual; social and cultural values.
    - Management of the risk – personal control over the risk; trust and distrust in perceived institutional control of risk.

  There is indication that hazards should be judged on a case-by-case basis to account for the separate contexts of each hazard. There is little indication of the relative weights that should be afforded each characteristic and the extent this varies with contextual factors.

- There is wide support for the idea that risk perception is influenced by social relations and trust in risk management institutions, and increasing concern of the limitation of approaches that do not account for social explanations. Though there is little understanding of how these social factors interact with the range of risk characteristics identified by the psychometric tradition.
- It is acknowledged that risk has a 'dual nature' that relates to the extent it is understood as existing 'objectively', or is a product of mental processes. Integrating these two understandings is problematic, and is one of the central difficulties in incorporating public perceptions of risk into risk policy decision-making.
- Research into risk perception is increasingly rejecting single theoretic perspectives, especially when perceptions of risk are not through direct experience, but are mediated, e.g. via the mass media. There is indication that the public responds to media coverage of hazards in a more rational and active way than might be presumed. Neither is the relationship between media coverage and risk perception as undirectional or directly proportional as might be thought. There can be differentiation at national and local level between risk amplification and attenuation over the same risk issue.
- It is important to consider the extent that a person has knowledge of, or is familiar with a hazard when investigating factors that influence perception of risk. For example, there is evidence that the media play a more important role in people's interpretation of hazards when they have less experience or knowledge of those hazards. Similarly, evidence suggests that when the public is less familiar or directly involved with a hazard, trust in risk management institutions can act as a shortcut for mediating judgements of risk acceptability.
- A number of criticisms are raised regarding the methodologies used to investigate risk perception that are relevant to high hazard industries, principally:
  - It is problematic to measure perception of risk without acknowledging the context in which it is experienced.
  - It is also problematic to measure what someone has not experienced or considered previously.
  - It is important that data collection methods do not impose conceptualizations on the participants, nor researchers frame the problem according to their own values.
  - When the public is asked to compare risks, care should be taken on how the information is presented. There are also problems in comparing risks, as 'real world' risks are multifaceted and not identical.
  - Care should be taken when aggregating individual judgements of risk as indicative of group responses, as the process of individual decision-making and group decision-making are not synonymous.

These recommendations arising from the main findings of the HSL publication should be taken into account by organizations when making judgements or decisions on the public's perception of risk-related outcomes arising from their activities and processes.

Improving risk communication

In order to improve the communication of risk to the public, a number of aspects need consideration, in particular:

- the significance of monitoring public perceptions of risk on a continuing basis;
- the use of hazard templates, namely frameworks that people hold for making sense of risk information;
- ensuring the views of minority groups are understood and taken into account; and
- understanding how decisions are made by the media with respect to the reporting of risky events or situations and the potential for risk amplification by the media.

Perceptual expectancy and stereotyping

Bottom-up and top-down processing

Perception proceeds in two particular ways. In bottom-up processing, information is analysed commencing at the bottom with small units or features, building upward into a complete perception of a situation. Conversely, top-down processing utilizes a person's pre-existing knowledge of the world to rapidly organize features into a meaningful whole. Top-down processing is a feature of perceptual expectancy. For example, a sprinter waiting in the starting blocks at a race is set to respond in a particular way when the starting pistol is fired. Should a similar sound occur, such as a lorry back-firing, the sprinter may 'jump the gun'.

Similarly, past experiences, knowledge, motives, context or suggestion may create a perceptual expectancy that sets an individual to perceive in a certain way. One of the problems is that people frequently 'jump the gun' in the process of perception. Perceptual sets often lead people to see what they expect to see as opposed to what is actually there.

Stereotyping

Stereotyping can be defined as 'any theory or doctrine indicating that the actions of an individual reflects on his own culture, sex, age, race, class, occupation or nationality. Fundamentally, stereotyping is a tool which people use to label others. People can be stereotyped in numerous ways, for example, by their speech, in particular, local accents, their age and even where they live.

In the case of risk and risk-taking, certain occupations are stereotyped. They are perceived by others as being high, medium or low risk. These perceptions are based on past experience of that occupation. Traditionally, construction work, work on off-shore installations and work at height have been perceived as high risk, whereas other occupations, such as working in an office, are perceived as low risk.

Information processing

For an individual to act in a certain situation, he must be capable of interpreting and processing the information which, in most cases, represents a range of stimuli presented at that point in time. In fact, the relative speed with which people process information is a precondition of many accidents. Generally, people can only process one stimulus at a time and subsequent stimuli arriving have to wait until the first stimulus has been processed (Single Channel Theory). This aspect of human performance is, therefore, very important in the design of tasks where people need to make quick decisions as to the next course of action.

This is particularly the case in driving situations where the driver is continually receiving information in the form of stimuli from the road ahead, the road behind (through the driving mirror and wing mirrors), road signs with respect to speed restrictions, traffic direction and lane control, the condition of the road, e.g. dry, wet, slippery, and the activities of cyclists and pedestrians which can cause distraction.

Each stimulus produces a reaction or response and the response can be divided into two elements:

1. *specific reaction time*, the actual time it takes to perceive and process the response; and
2. *movement time*, the time taken to actually execute the response.

If a second stimulus arrives during the movement time, it has to wait. The actual movement in response is being monitored, the single channel process ensuring the execution of the original response was accurate. The particularly crucial parts of the movement are the beginning and the end, the middle often being partially or totally neglected. Thus people generally cannot do more than one thing at a time, the speed and sequence of response varying from person to person. This factor can be significant in accident causation.

With well-known and practised tasks, e.g. driving, the monitoring action of the brain can be reduced, much depending on the speed with which a person can respond to stimuli, and not monitor specific movements. Results are achieved through continual practice or the speed–accuracy trade-off, whereby the monitoring is voluntarily removed.

The feedback that people receive is an important aspect of monitoring a task. Where someone is highly skilled, it can be a hindrance and destroy performance.

In the teaching of trainees to undertake a specific practical task, such as driving a fork-lift truck, the instructor has to put the monitoring aspect back into the task. This can ruin that person's own particular performance at that task.

In the information processing operation, people ascribe different values to the various outcomes of their decisions. These values are commonly influenced by extraneous factors, such as the financial benefit that could accrue, or the possibility of saving effort and time. These decisions are subjective and may be influenced by individual personality and past experience of similar situations.

The level of brain arousal can also affect the efficiency and rate of mental processing. Arousal is defined as an increase in alertness and muscular tension. However, individual levels of arousal vary significantly. Generally, at low arousal levels, performance tends to be poor. As arousal increases to an optimum, performance rises accordingly, but then drops as further arousal takes place. Frequent changes in arousal levels take place during the typical working cycle.

Capacity to act following the making of a decision will vary according to the physical and mental limitations of people, such as their strength, speed, intelligence, perception of events and expectations of the outcome of their decision. No two people, therefore, react to a particular stimulus in the same way or at the same speed. Speed of reaction is particularly important in the case of vehicle and machinery-related accidents.

With information processing, a distinction can be drawn between on-line processing and off-line processing.

On-line processing

This is the spur-of-the-moment decision-making that an individual regularly has to take in order to survive and is associated with the concept of single channel theory outlined above. On-line processing is of limited capacity, however, and can best be used by grouping actions together as 'habits' or 'packages' that can be put into operation as a group sooner than they could as a series of separate actions.

This process is commonly associated with tasks that involve repetitive skills, such as bakery employees feeding whole loaves through bread-slicing and wrapping machines, certain aspects of driving, such as starting, stopping and changing gear, and operating bottle-filling machines. These habits, which are developed over a long period of time, can be difficult to break due to the fact that the brain's monitoring element is greatly reduced or completely absent. This can be a contributory factor in accidents, particularly where the machine develops a fault and the operator continues the process without realising the fault has arisen.

Induction and on-the-job training programmes endeavour to inculcate the correct habits, including dealing with emergencies, right from the start, with a view to preventing human error-related accidents.

Off-line processing

Off-line processing involves the process whereby people actually simulate in their minds the outcomes of different courses of action prior to making any final decision as to which course of action to take. This planning process, and the individual planning skills of the people concerned, are important features in the prediction of accidents or other adverse consequences that could arise.

Off-line processing is a skill based on factors such as individual knowledge, intelligence, experience of similar situations and the amount of practice received in using the skill concerned. It has its limitations because, for example, there may be attempts to simplify decisions, perhaps by ignoring the less important aspects of the situation, in order to make it more easily manageable in the mind. These limitations, some of which are unconsciously brought in, can result in incorrect, inaccurate or unsafe decisions being made by an individual.

Skill-, rule- and knowledge-based behaviour

The behaviour of people is greatly affected by their individual repertoires of skills, the rules they are required to follow and their level of knowledge at a particular point in time. As they progress through life they adapt their behaviour to these three factors. Various techniques, in particular human reliability assessment (HRA), endeavour to predict incorrect human action, or a failure to act, arising from these facets of behaviour which could lead to adverse consequences in major installations, such as nuclear installations.

A skill-based behaviour is a type of behaviour that requires either very little or no conscious control to undertake a task or implement an intention once that intention is formed. With this behaviour, performance is smooth, uninterrupted and consists of well-developed and integrated patterns of behaviour.

Skill-based errors can be predicted in some cases. They may be identified through laboratory-based experiments or from an individual's experience in other skilled tasks. In the case of keyboard skills, for example, the probability of an operator striking the wrong key can depend upon the form, nature and complexity of the material being produced. With this type of work there is a definite speed/accuracy trade-off, the greater the typing speed, the greater the potential for typing errors.

A rule-based behaviour is typified by the establishment and use of rules, systems and procedures in order to identify a course of action in a well-established work situation. Rules can take many forms, for example, instructions for correct use of chemicals issued by a supervisor, or safe procedures acquired as part of a training activity. Generally, operators are not required to understand the underlying principles of a procedure or system to perform a rule-based control.

Rule-based errors depend very much on time, a person's frame of mind and his mental, physical or emotional state with respect to compliance with established working rules and procedures. This form of error is difficult to predict in most cases and may need to be assessed through a process of observation. There is a very fine dividing line between what is an actual error as opposed to a 'violation', a deliberate action contrary to some rule required to be implemented by the organization.

A knowledge-based behaviour involves a greater level of reasoning. People judge many situations on the basis of the knowledge they possess. Operators of machinery, for example, are required to know the principles and legal requirements relating to the safeguarding of that machinery.

Knowledge-based errors arise from the detailed knowledge of the system or process possessed by the individual. The resulting 'mental model' of the system, however, may not be correct. It may be that one particular facet of knowledge is incorrect and this could contribute to this form of error. As with rule-based errors, this form of error is difficult to predict.

Rasmussen's model

Rasmussen distinguished between the 'knowledge-based', 'rule-based' and 'skill-based' levels of operator behaviour. These implicate different aspects of human functioning and are subject to different types and degrees of disturbance or error on account of external signals or internal processing failures.

He postulated that good design needs to support all three levels of operation, not just one level.

The knowledge-based level

This level of behaviour occurs where there are no rules or instructions to inform or guide the operator. In many cases, it entails conscious and time-consuming

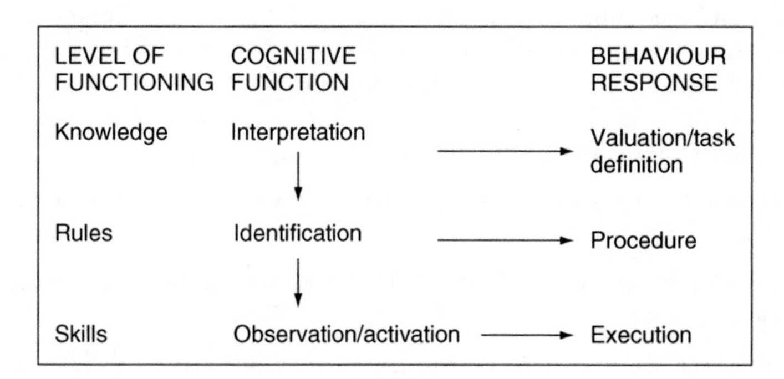

Figure 5.2 Rasmussen's model of the levels of functioning (Rasmussen, J. (1983): Skills, rules and knowledge: Signals, signs and symbols and other distinctions in human performance models: *IEEE Transactions on Systems, Man and Cybernetics* SMC 13(3))

problem-solving in unusual or novel situations. These conditions, commonly associated with emergencies, need a degree of interpretation, thought, reasoning and consideration about, for example, the state of the plant and equipment, based on his current level of knowledge, that is, the mental model. This level implies a certain degree of planning in most situations.

The rule-based level

At this level, behaviour becomes a conscious activity in terms of identifying an appropriate course of action. It is based on what is familiar in terms of rules, requirements and procedures and matching these learned rules to a correct interpretation or diagnosis of the current situation. Generally, this behaviour occurs less frequently and is reserved for situations that are rarer, but which can have serious consequences when they occur.

The skills-based level

This level entails virtually automatic and unconscious responses to routine situations and features as part of the normal workplace environment. Much of this is not available to conscious thought.

Individual behaviour in the face of danger

The Hale and Glendon model (1987)

This model demonstrates how the common control failures vary from one level of behaviour to another and, in the same way, the various types of accidents and the safety strategies used in preventing or controlling exposure to hazardous situations.

The Hale and Glendon model demonstrates initially how danger arises in a workplace or work system. Whilst some form of danger may be ever present, it is controlled by a range of control measures associated with, for example:

- the design and layout of the workplace and equipment;
- management procedures, such as the operation of a planned preventive maintenance system;
- the people, in terms of their level of training, skills and the quality of supervision; and
- organizational factors, such as the restriction of safety critical tasks to designated and trained operators.

On this basis, providing all risks have been identified, assessed and controlled, the potential for accidents or damage is greatly reduced. Where, however, there is a deviation from this state, then the accident process commences.

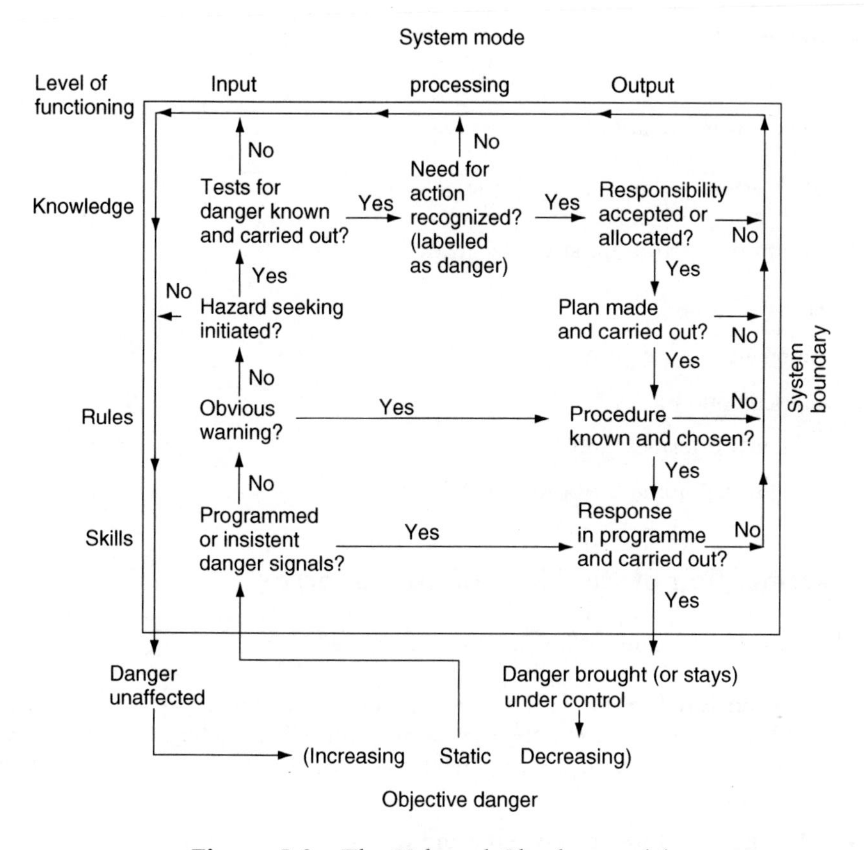

Figure 5.3 The Hale and Glendon model

People have a crucial role in this system. They must ensure correct implementation of the accident prevention measures and controls in order to prevent deviations. They may also have the task of identifying and correcting deviations which do occur, modifying the system and its controls in the event of new demands, information, hazards and processes. In the Hale and Glendon model (Figure 5.3), these actions are modelled as a range of identification and control tasks as they relate to danger.

Human error

Is human error a significant feature of accidents? How frequently we have seen the cause of an accident written down to 'carelessness' on the part of the individual. The fact is that, whilst people may occasionally be genuinely careless, in most cases the cause is more likely to be human error.

Limitations in human capacity:

- to perceive;
- to attend to;

- to remember;
- to process; and
- to act on information;

are all relevant in the context of human error.

Typical human errors are associated with

- lapses of attention;
- mistaken actions;
- misperceptions;
- mistaken priorities; and
- in a limited number of cases, wilfulness.

Classification of the kinds of human error

There are several kinds of human error. They include

- **Unintentional error**. This may arise when an individual may fail to perform a task correctly, for example, operating a control or reading a gauge. These are typical 'slips' or 'lapses', frequently associated with 'carelessness' or 'lack of attention' to the task.
- **Mistakes**. In this situation, the individual shows awareness of a problem but forms a faulty plan for solving it. He will thus carry out, intentionally but erroneously, action(s) which are wrong, and which may entail hazardous consequences. Typical examples are in the operation and maintenance of plant and machinery and in assembly work. Failure to correct individual mistakes through training and supervision can lead to disaster situations.
- **Violations**. In this case, a person deliberately carries out an action that is contrary to some rule which is organizationally required, such as an approved operating procedure. Deliberate sabotage is an extreme example of a violation. Violations involve some complex issues concerning conformity, communications, morale and discipline. In piecework systems, the removal of machinery guards or the defeating of safety mechanisms in order to increase output and, therefore, remuneration, is a common violation.
- **Skill-based errors**. These arise during the execution of a well-learned, fairly routine task. They are amenable to prediction either from laboratory experiment, or from experience in other skilled tasks, even when those tasks are performed in different industries. For instance, the probability of a skilled typist striking the wrong key depends upon the nature and complexity of the material being keyed but, on average, the error rate turns out to be much the same in an office of any kind, as in laboratory experiments. Skill-based errors commonly occur amongst the more highly skilled members of the

organization. Such people, because of their acquired skills and experience in the work process, frequently take the view that the safety precautions imposed are purely for the benefit of the underskilled and underexperienced operator. This is the classic sign of familiarity breeding contempt. Such an attitude to work is accompanied by overconfidence in the task which can lead to, particularly, machinery-related accidents.

- **Rule-based errors**. These are the types of error which occur when a set of operating instructions or similar set of rules is used to guide the sequence of actions. They are more dependent on time, on individual cast of mind and on the temporary physical, mental or emotional state of the person concerned. They are, therefore, harder to predict from past statistics or experiment.
- **Knowledge-based errors**. Also called 'errors of general intention', these are the errors which arise when a choice decision has to be made between alternative plans of action. Such errors arise from the detailed knowledge of the system possessed by the person and the resulting 'mental model' may be incorrect. Such errors, which are extremely difficult to predict, can only be forecast by an analyst who possesses the insight to predict this form of model.

Model for classification of error

Figure 5.4 classifies errors or unsafe acts as falling under intended or unintended actions. Unintended actions can result in slips (attentional failures) and/or lapses (memory failures), both of which are largely skill-based. Intended actions, on the other hand, are associated with mistakes, which can be either rule-based or knowledge-based, and violations as outlined above.

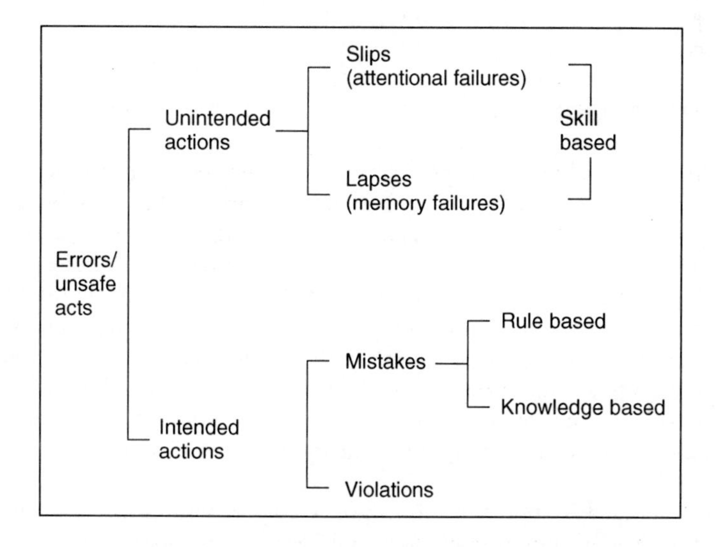

Figure 5.4 Classification of human error

The significance of human capability

Human capability and the potential for error are important factors in the selection of people for different tasks. Tasks should be designed using techniques like job safety analysis, with particular reference to the various influences on behaviour that the task may produce.

Moreover, techniques based on human reliability assessment (see Chapter 4, People factors) can predict the potential for human error and, linked with ergonomic principles in the design of work layouts, systems and machinery and vehicle controls and displays, should be used in the process of job design. (Advisory Committee on the Safety of Nuclear Installations (1991): Second Report of the Study Group 'Human Reliability Assessment': HMSO).

Classification of human error

Professor James Reason stresses that the problems of human fallibility and human error can be viewed in two ways – the person approach and the system approach. The *person approach* focusses on the errors of people, blaming them for inattention, moral weaknesses or forgetfulness. The *system approach*, on the other hand, concentrates on the conditions under which people work and endeavours to establish defences to avoid errors or to mitigate these effects of same. Each has its model of error causation and each model gives rise to quite different philosophies in the management of error.

High reliability organizations, which tend to have less accidents than other organizations, acknowledge the fact that human variability is a matter to be recognized in the avoidance of error.

The person approach

This is the traditional form of approach which focusses on the unsafe acts in which operators indulge, i.e. errors and procedural violations. They may be associated with aberrant mental processes, such as inattention, forgetfulness, carelessness and even recklessness. Any associated countermeasures are, therefore, directed at reducing unwanted variability in human behaviour, such as the use of safety posters, reminding people to do things in a particular safe way, such as correct manual handling techniques, or appealing to people's sense of fear, as with early safe driving posters. Other methods include the preparation and implementation of formal safe systems of work, such as permit-to-work systems, retraining of operators, together with naming, blaming and shaming techniques.

Managers taking this approach look upon errors as some form of moral issue, namely that 'bad things happen to bad people'. This approach is commonly known as the 'Just World Hypothesis'.

The person approach still remains the dominant approach with many organizations. Employees are perceived by many managers as 'free agents' who are capable of selecting safe and unsafe methods of work. Pointing the finger and blaming individuals is a 'quick fix' solution that requires no further action by managers. Health and safety practitioners who examine and analyse accident reports produced by managers see the causes of accidents commonly written down as being due to carelessness, lack of attention and human error on the part of the accident victim. Ticking the box on the accident report form that indicates 'failure in preventative maintenance' or 'inadequate management supervision' can imply a failure in the management system with all the subsequent enquiries as to how the system failed.

A further weakness with the person approach is that, by concentrating on the individual origins of human error, this isolates unsafe acts from their system context. As a result, two important aspects of human error may be overlooked. Firstly, it is often the well-trained and time-served 'expert', such as an engineer or designer, who makes the worst mistakes. The notion that errors are only associated with untrained and inexperienced people is a myth. Secondly, far from being random, errors tend to fall into recurrent patterns, namely that a similar set of circumstances, such as in maintenance of electrical equipment, can provoke similar errors, regardless of the operators involved.

Effective risk management relies heavily on establishing and promoting the right safety culture (see Chapter 16, Health and safety culture). This includes detailed analysis of all events resulting in loss to the organization – accidents resulting in injury, occupational ill-health, scheduled dangerous occurrences, those resulting in lost time and 'near misses'. Without this analysis there is no way of identifying recurrent error traps (see below) and the measures necessary to avoid these traps.

The system approach

The system approach takes the view that people are fallible, they make mistakes and that employers must accept that fact. On this basis, errors are to be expected, even in the best organizations. The approach recognizes that organizational processes cannot totally work in unison with the human condition. Errors are seen as consequences rather than causes, having their origins not so much in the perversity of human nature as in 'upstream' systemic factors. These include recurrent 'error traps' in the workplace and the organizational processes that give rise to these error traps arising from, for example, poor ergonomic design of equipment, incorrect workplace layout and management inattention to their responsibilities for health and safety.

A central concept is that of 'system defences'. All dangerous processes and technologies incorporate barriers and safeguards. When an adverse event arises, such as a fatal accident, the important factor is not who committed an error, but how and why the system defences failed.

The 'Swiss cheese' model of system accidents

The system approach is based on defences, barriers and safeguards. High technology systems, such as those for power stations and many chemical manufacturing processes, have a series of defensive layers. These may take the form of engineering defences, such as physical barriers, alarms, automatic shut-down devices and sprinkler systems. Other defences may rely on the skill, knowledge and experience of people, such as engineers, health and safety specialists and chemists. These various aspects are supported by procedures and administrative controls and the total function is to protect people and the organization's assets from loss. The majority of systems undertake these functions effectively, but there can be weaknesses which can downgrade these systems.

Ideally, each defensive layer would be intact. However, this is not generally the case and, in reality, they are more like the slices of a Swiss cheese having many holes in its internal structure. However, compared with a static Swiss cheese, these holes are continually opening, closing and moving their location. The presence of holes in one slice of the Swiss cheese does not normally result in an adverse situation. Generally, this can happen only when the holes in many slices line up for a brief second to permit a trajectory of accident opportunity, bringing a hazard into contact with a person, thereby causing injury.

Active failures and latent conditions

Active failures are the unsafe acts committed by people who may be in direct contact with an accident victim or a particular system. Such failures take a number of forms – slips, lapses, mistakes and procedural violations which may be hard to foresee. Active failures have a direct and generally short-lived impact on the integrity of defences. This was identified in the investigation into the Chernobyl incident where operators wrongly violated plant procedures and turned off a range of safety systems, thereby creating the immediate trigger for the catastrophic explosion in the core of the plant. Those who follow the person approach above would have looked no further for the causes of such an adverse event once they had identified the unsafe acts. What the investigators at Chernobyl did not do was to examine the history behind these unsafe acts and the management systems that permitted these acts to recur.

Latent conditions, on the other hand, are perceived as the 'resident bugs' within the management system arising from the mistaken decisions of a range of people – architects, designers, people who formulate procedures and systems, construction contractors and senior level management. Strategic decisions made by these people have the potential for fouling up the system resulting in two forms of adverse effect:

1. They can result in error-provoking conditions within the workplace associated with, for example, time pressures, operator stress and fatigue, inadequate equipment and materials.

2. They can create permanent holes or weaknesses in the defences through, for example, design and construction defects, unworkable procedures, understaffing, inefficient indicators and alarms.

One problem is that these conditions may lie dormant within a high technology system for many years before they combine with active failures and local triggers to create an opportunity for some form of adverse event, such as a fire or explosion. These conditions are foreseeable, however with a system of safety management that is proactive sooner than reactive in its approach.

Techniques for predicting and analysing human error include

- checklist analysis:
  - for situational related errors; and
  - for management systems-related errors;
- workplace walk-through analysis;
- HAZOPs, job safety analysis, etc; and
- quantitative human reliability analysis.

Employee selection processes

Successful Health and Safety Management (1991) (HS(G)65) states:

Arrangements made by companies who manage health and safety well will include:

> recruitment and placement procedures which ensure that employees (including those at all levels of management) have the necessary physical and mental abilities for their jobs, or can acquire these through training and experience. This may require assessments of individual fitness by medical examination, and tests of physical fitness, or aptitudes and abilities

Selection of employees for particularly high risk tasks should take the above factors into consideration. The selection process should consider not only the skill, knowledge and past experience of applicants but a number of other human attributes, such as

- personality and aptitude;
- motivation;
- attitude to risk;
- reliability;
- commitment to safety procedures;

- the ability to work under time pressure; and
- understanding of the process, systems and procedures.

All these factors may be significant in the prevention of human error.

In certain cases, aptitude tests may identify the potential for human error on the part of individuals.

Use of system responses to help prevent human error

The HSE estimates that human error is involved in approximately 80 per cent of accidents. In its publication *Reducing Error and Influencing Behaviour* (1999) (HSG48), it states:

> Many accidents are blamed on the actions or omissions of an individual who was directly involved in operational or maintenance work. This typical but short-sighted response ignores the fundamental failures which led to the accident. These are usually rooted deeper in the organisation's design, management and decision-making functions.
>
> Organisations must recognise that they need to consider human factors as a distinct element which must be recognised, assessed and managed effectively in order to control risk.

Error management is concerned with:

- limiting the incidence of dangerous errors; and
- creating systems that are better able to tolerate the occurrence of errors and contain their damaging effects.

The systems approach requires a comprehensive management programme aimed at several different targets:

- the person;
- the team;
- the task;
- the workplace; and
- the organization as a whole.

The person

In the case of people, a high level of information, instruction, training and supervision is needed to advise them of the factors that can contribute to human error, the causes of human error, the risks arising from human error and the measures they must take personally, including high levels of vigilance at all times. They should also be advised of the system for reporting of defects and other shortcomings in the employer's protection arrangements.

The team

Teams work together and, as such, should be informed and trained together to ensure effective interaction within the team. There should be emphasis on the significance of the team as a group, the need to look after each other in the face of danger and to follow established safety procedures.

The task

Any task should be clearly analysed and defined through techniques, such as task analysis and job safety analysis, with the view to developing safe systems of work. Risk assessment of the task or work activity may need to be carried out by the employer with a view to identifying the hazards and precautions necessary to comply with current health and safety legislation. The potential for human error in tasks must be considered in a risk assessment.

The workplace

Workplace safety procedures should take into account current legal requirements, such as the Workplace (Health, Safety and Welfare) Regulations and the Electricity at Work Regulations, 1989. There may be a need to produce a workplace risk assessment to identify hazards and the precautions necessary. Risk assessments should consider the potential for human error.

The organization as a whole

The organization's commitment to ensuring the health and safety at work of employees and others should be formally established in the Statement of Health and Safety Policy under the HSWA.

A more simplistic approach to human error

Taking into account the various theories above, a more simplistic approach to the problem of human error as a causative factor in accidents can be based on three specific features of human behaviour, as follows.

Ignorance

In many cases, both employers and employees are ignorant of the legal and practical requirements for ensuring safe working and of the dangers that exist in the workplace. Others do not appreciate the consequences of their actions and how the safety of others may be affected. Workers will follow a particular work practice for a period without considering safety implications and it is only when someone is actually injured that they take notice ('Trial and error learning').

Employers should be aware of this factor in the introduction of new materials, substances and equipment and in the health and safety training of employees.

Carelessness

A small proportion of the working population simply do not care about the safety of themselves or others. This can be seen in unsafe working practices adopted by workers, thoughtless and self-motivated actions by employers in endeavouring to get the job done on time irrespective of hazards that may be present, shortcuts devised and taken by employees, driving situations and total disregard for general and fire safety procedures where people smoke in dangerous areas. Evidence of careless fork-lift truck driving, for instance, can be seen in damage to wall surfaces, doors and other structural items in workplaces.

Carelessness can be associated with a lack of self-discipline and low level of arousal in the case of employees and a failure in supervision on the part of management.

Inadequate communication

Some people, both managers and employees, fundamentally, lack basic communication skills. This inability to communicate on health and safety matters, in particular, can be a contributory factor in accidents. Typical examples of communications failure include failing to report hazards and 'looking the other way' when dangerous situations arise. In some cases, the nature of the work undertaken, such as piecework, the operation of boring and monotonous tasks and the arrangement of workstations, can result in inadequate communication.

Contribution of human error to major catastrophes

Human error has been a contributory factor in a number of catastrophic incidents. For example,

- in the Ladbroke Grove rail crash (1999), a train driver passed a red signal which resulted in the collision. In this case, statements by other drivers indicated the same red signal had been passed on eight previous occasions.

- in the Brent Cross crane collapse (1964), a number of features of human error in both design and operation of a mobile road crane were identified:
  - basic design faults;
  - the failure of inspection to remedy the design fault;
  - defects in fabrication;
  - the failure to notify the designer's limitations on use of the mobile crane to the user;
  - errors in inspection by contractors;
  - the system for calculation of residual and main stresses in the crane;
  - the fact that the mobile crane was standing on a 1:30 slope which reduced its stability.

On this basis, it is essential that designers of systems, structures and equipment take into account the potential for human error at the design stage. This may entail specific studies into the potential for human error and its significance during the operation of systems and equipment and in the erection of structures.

Conclusions

How people perceive risk is a feature of, in particular, their upbringing, training, attitudes and past experience. Faulty perception can be a contributory factor to human error.

It is essential, therefore, that in the design of work systems, employers are aware of the potential for human error and make appropriate adjustments to systems aimed at preventing human error from arising. Systems should, as far as possible, be foolproof.

Key points

- Perception of risk is a feature of training, upbringing, attitudes, culture and past experience, and is incorporated in an individual's perceptual set.
- Perceptual sensitization and perceptual defence are two important mechanisms in the perception of hazards.
- Risk compensation is a feature of people whereby behaviour may be adapted or adjusted in response to perceived changes in risk.
- A number of variables influence the perception of risk, such as familiarity with the task, perceived control over the risk and the framing of the risk.
- The speed and accuracy with which people process information is a significant precondition of many accidents.

- Skill-, rule- and knowledge-based behaviour feature prominently in the way people deal with risk.
- Human error is associated with limitations in human capacity to perceive, attend to, remember, process and act on information received.
- Human fallibility and human error can be associated with both unsafe acts and badly designed work systems.
- Error management is concerned with both limiting the incidence of dangerous errors and creating systems that are better able to tolerate the occurrence of errors and contain the damaging effects.

6 Organizational control and human reliability

Control of an organization's operations and activities relies, to a great extent, on the physical and mental capabilities, that is, the reliability of its people. People are directly affected by the organizational environment.

The organizational environment

What is the environment?

The environment can be interpreted as that which surrounds us or our surroundings. Everything exists in the context of a wider environment:

- people within organizations;
- organizations within industries;
- industries within countries;
- countries within the global context.

Organizations comprise individuals who belong to groups, within departments or divisions, forming the total organization. In turn, an organization belongs to a wider population, such as an industry, existing alongside other populations in a wider organizational ecology.

Features of the environment

The concept of an environment is created by drawing a boundary at some level within a system of relations, thereby separating a particular element from the rest of the system. For instance, in its everyday operation, an organization interacts with clients, customers, competitors, suppliers, trade unions, shareholders, government agencies and other individuals and agencies, such as enforcement agencies, that have an immediate influence on the organization's well-being. These relations constitute the organization's *task environment*. However, extending beyond this, there are forces that constitute a broader *contextual environment* involving aspects such as the economic, social, political, technological and cultural factors that shape the organization's overall operations. It is essential that organization's pay attention to both these aspects of the environment.

What is the organizational environment?

This is the set of forces and conditions that operate beyond an organization's boundaries but which affect a manager's ability to acquire and utilize resources.

Organizational environments may be affected by:

- **Barriers to entry**. Factors that make it difficult and costly for an organization to enter a particular task environment or industry.
- **Boundary spanning**. Interacting with individuals and groups outside the organization to obtain valuable information from the task and general environments.
- **Competitors**. Organizations that produce goods and services that are similar to a particular organization's goods and services.
- **Demographic forces**. The outcome of changes in, or changing attitudes toward, the characteristics of a population, such as age, gender, ethnic origin, race, sexual orientation and social class.
- **Economic forces**. Interest rates, inflation, unemployment, economic growth and other factors that affect the general health and well-being of a nation or the regional economy of an organization.
- **Environmental change**. The degree to which forces in the task and general environments change and evolve over time.
- **Global forces**. Outcomes of changes in international relationships, changes in nations' economic, political and legal systems, and changes in technology, such as falling trade barriers, the growth of representative democracies and reliable and instantaneous communication.
- **Stages of industry life cycles**. The changes that take place in an industry as it goes through the stages of birth, growth, shakeout, maturity and decline.
- **Political and legal forces**. Outcomes of changes in laws and regulations, such as the deregulation of industries, the privatization of organizations and increased emphasis on environmental protection.
- **Suppliers**. Individuals and organizations that provide an organization with the input resources that it needs to produce goods and services.
- **Technological forces**. Outcomes of changes in the technology that managers use to design, produce or distribute goods and services.

Organizational environment elements

The organizational environment is comprised of a number of elements, such as the organizational structure, processes, operations and, most importantly, people with varying management styles, attitudes, personalities and abilities. Furthermore, organizational environments are developed over time to equip the organization to meet the challenges ahead in two ways:

1. by learning from mistakes made in the past; and
2. innovation, based on research, technological development and insight into ever-changing markets and demands.

What is constantly changing is the nature of the commercial or political pressures to stay ahead. In particular, it is the ability to innovate that gears an organization to be consistently looking for improvements and ways to take significant steps forward in terms of competitive advantage, public service or cost.

The importance of innovation

An environment where the organization:

- is very hierarchical;
- where individual functions do not communicate;
- where management is autocratic; and
- where everything is done 'strictly by the book'

can spawn a culture that is unlikely to be particularly supportive of innovation. It may well survive in small protected areas, such as a research and development department, but it will survive in its own microclimate and will not take root and spread throughout the organization.

If the organization wants innovation to take hold and thrive, then unsupportive elements need to be identified and addressed. For example, an organization may wish to change behaviours. For these changes in behaviours to be sustained an organization needs to change attitudes and, sometimes, even beliefs. If attitudes do not change, then old behaviours will creep back into practice. For attitudes to change, the environment needs to change in order to reinforce the required behaviours. If the reward system is changed to recognize collaboration and the spread of good ideas, then people's attitudes will change. Over time they will come to believe that collaboration is beneficial and 'the way we do things around here'.

What needs changing?

The challenge is to understand what is fundamental to the success of the organization and what can be changed without necessarily upsetting this fundamental focus. A number of questions must be addressed.

- How can the current environment be assessed in a disciplined way?
- How can the target environment be defined?
- What can be done to address the gaps?

These questions need to be asked across four fundamental categories that reflect the major influences on the organization's environment.

Direction

- Where is the organization going?
- Is it going in the right direction?
- Is it achieving progress at the right speed?

Capability

Are managers capable of achieving the objectives set in terms of

- the proposed direction of the organization;
- the existing resources in terms of management skills, competencies, technology, plant, equipment and materials;
- current financial resources?

Is the organization up-to-date with current technology, management systems and legal requirements?

Infrastructure

Where does the organization fit within

- the industry infrastructure?
- the national infrastructure?
- the global infrastructure?

What are the particular opportunities and threats arising from each of these infrastructures and are they being handled effectively by the organization?

Behaviours and ways of working

Do people at all levels of the organization aspire to the success of the organization in terms of attitudes held and motivating factors?

Is there evidence of conflict between, for example

- senior and line management;
- individual line managers;

- line management and trade unions;
- individual employees and/or groups of employees?

Are ways of working effective and efficient?

Generally, the actual organizational environment varies dramatically from one organization to another. Management, however, must use the resources available to maximize the effectiveness of this environment in order to succeed.

Innovation and learning from mistakes made are the key elements of a successful organizational environment.

Organizational factors

Organizations operate on the basis of a hierarchy, which is one based on power held by specific individuals from the chief executive or managing director downwards. Generally, orders pass down the organization and information passes back up. The flow is one-way in both cases. Promotion within the organization is, allegedly, based on merit and hard work.

Organizations are deliberately impersonal and operate on the basis of ideal relationships between staff at all levels. More accurately, organizations tend to operate according to the 'Rabble Hypothesis of Man', namely 'Every man for himself!'

Two specific groups or sets of people tend to operate within organizations. There is the functional or line organization which operates on the basis of the type of work being carried out, such as production, administration, engineering and construction activities. The staff organization, on the other hand, operates in direct contrast to the line organization and is represented by those people who have advisory, service or control functions. This includes people such as health and safety specialists and quality assurance managers.

Weaknesses of formal organizations

The formal organization is by no means perfect. Weaknesses include, in particular, communications failures. They commonly ignore emotional factors in human behaviour and are frequently perceived as uncaring and lacking in interest or commitment to ensuring appropriate levels of health, safety and welfare for staff and others (see Features of organizations, below).

Safety organization

As with any other management function, the safety of employees and other persons needs to be properly organized. The organization of health and safety should be dealt with in the Statement of Health and Safety Policy.

When considering organizational factors relating to health and safety, it is essential to be aware of the structure and functions of the organization, e.g. company, local authority, professional body. Safety organization can operate on both a formal and informal basis.

Formal organizations

These are established to achieve set goals, aims and objectives. They have clearly defined management structures, policies, rules and channels of communication. They are frequently divided into productive and nonproductive organizations.

Objectives are statements of what an organization intends or aims to achieve, e.g. financial objectives.

Policies are statements concerning how the objectives are to be achieved, e.g. policies on quality.

Organizations feature a number of functions, such as marketing, production, engineering, personnel, finance and administration.

Line functions are those with direct responsibility for achieving the organization's objectives. They are concerned with the output of goods or services. Production is a pure line function.

Staff functions are necessary for line functions to be carried out. They assist, advise and facilitate the operation of the line function. Typical examples of staff functions are the personnel function and the health and safety advisory function.

The hierarchy

Most organizations incorporate a hierarchical structure and it is common for the hierarchical structure to be depicted as a line diagram or organizational chart as shown in Figure 6.1.

Attainment of goals

The setting and attainment of goals, which are both measurable and achievable on the part of the individual, is one of the principal functions of a hierarchy. For example, the production director may be tasked by the managing director to raise productivity by 10 per cent during a 12-month period.

Informal organizations

Many informal organizations operate within the framework of a formal organization. This can include, for example, groups of supervisors who exchange

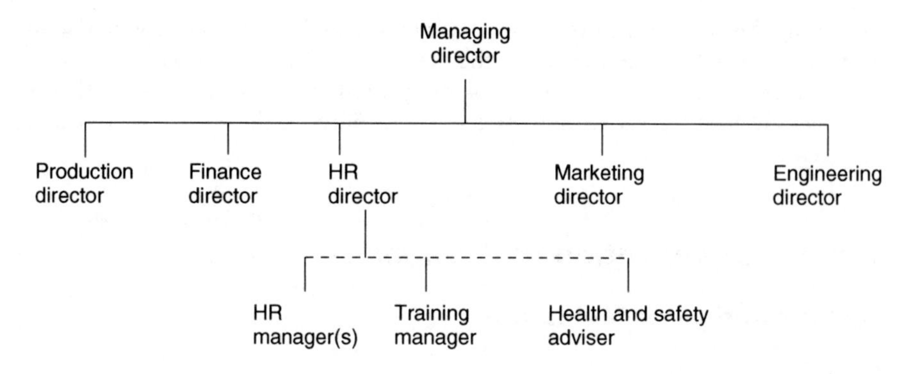

Figure 6.1 A typical hierarchy

information, seek advice from each other and endeavour to put into operation the demands of senior management. There are no formal rules in terms of membership of the organization.

Similarly, groups of workers, such as machinists, maintenance personnel and drivers, may operate as an informal organization with respect to meeting production targets, quality standards and safety standards.

Organizational and individual goals

People within organizations are motivated to work towards goals or objectives that ultimately result in increased remuneration, promotion and peer recognition. However, in many cases, these individual goals are not aligned with the organization's broader goals. The negative consequence is that, while people in the organization strive towards personal motivations, very few people, if any, are directly concerned with the organization's ultimate objective, namely that of producing and delivering products and/or services to their customers.

Integration of organizational goals with those of the individual

Most people have functional roles. For instance, they may produce an intermediate product or participate in the management of a particular section of an organization's operations. However, they never see the final outcome of the operation, that is the product going to market for sale. As such, they tend to become locked into their own particular part of the operation and cannot see 'the big picture'.

People have their own particular goals with respect to work, such as future promotion, the acceptance of their ideas and recommendations by the organization and the increased remuneration accompanying these goals.

Integrating the goals of the organization with those of the individual entails shared beliefs and attitudes with respect to the core focus of the organization, such as the production of a quality product or the provision of a service. In particular it implies authority, responsibility and accountability on the part of all levels of management.

Systems organization and reliability

Background to organization

The systems approach

Human failings can affect the safety performance of an organization, particularly where complex problems are involved. People are an important feature of the organizational system. It is necessary, therefore, to develop a systems approach to explain human behaviour.

This approach is based on the idea that change in one part of an organization cannot be viewed independently. Because the system is said to be open the system is seen to interact with the environment in which it operates. Figure 6.2 is a typical diagram of how this interaction takes place.

Systems fall into two categories:

1. **Sociotechnical**. This system implies that an organization is a combination of technology (tasks, equipment and working arrangements) and a social

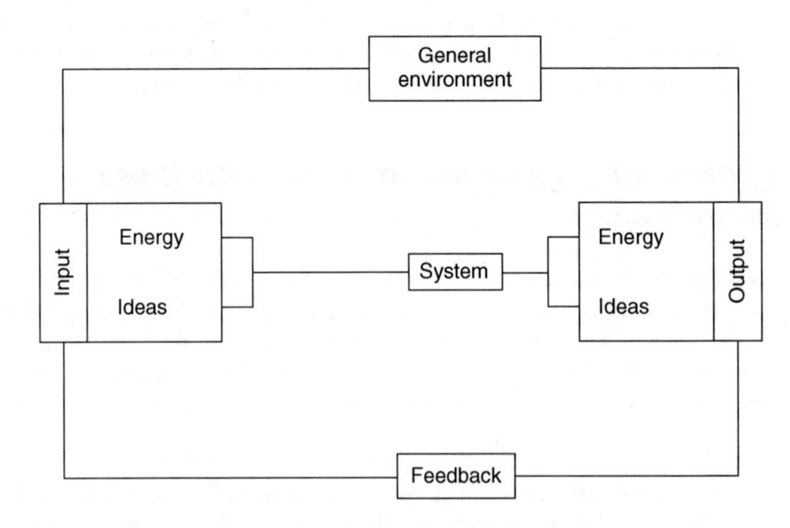

Figure 6.2 The systems approach

system (interpersonal relationships). The two systems are in constant interaction.

2. **Open-system**. In this system, the organization imports resources and information from the environment in which it operates and exports products and services to that environment.

According to Likert (1960), the total system is comprised of three levels:

1. society as a whole;
2. organizations of similar functions; and
3. subgroups within the larger system.

Therefore, society evolves a system incorporating feedback and its evolution affects its interaction with the organizations which develop within its environmental influence.

Types of organization

Self-regulating organizations

The human body is often considered to be a nearly perfect example of a system of self-regulating organization. The pattern of activity is self-fulfilling and balanced. Information is fed back and it becomes self-adjusting as a result of new information. The theory is that a self-regulated or self-controlled system with effective feedback mechanisms should be adaptive to change.

Therefore, an organization does not stay in the same condition for any length of time. Each organization undergoes a continual process of change, the rate of which is controlled by internal factors.

The organization of a system means the development of that concept into a living, active, vital working unit. The self-regulating organization is one which is progressively changing in its evolution, recognizing that its objectives may have to be altered or the order of priority of objectives may change.

Organizations are constantly changing. In fact, if they do not develop in response to what their sensors tell them about changes in their environment, they will stagnate and become extinct. (Figure 6.3).

Natural organizations

Groups of people come together because they identify common needs, aims and objectives. They have an instinct to do so and they realize that in membership of

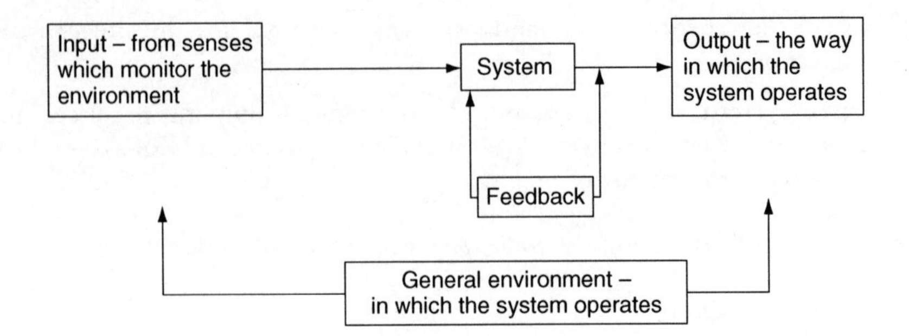

Figure 6.3 Systems organization

a group of similarly motivated people there is an element of strength and purpose which cannot be achieved by the individual on his own.

This is where the organization's own continued 'life' depends upon the group itself continually reappraising its purpose and undergoing a process of self-regulation. Thus a group can come together for one reason, but stay together for others. It is the culture and mission statement which binds the group together around shared values and goals. This social organization or co-existence in groups or companies, which comes about at least partially to satisfy our gregarious instinct, is identifiable both in working groups and non-working groups.

Cybernetics

Cybernetics studies the flow of information around a system and the way in which information is used by a system as a means of controlling itself.

Management systems

To function correctly, a system must have controls, i.e. a set of rules, regulations or instructions by which it operates. Most systems, known as management systems, such as production, planning, sales, budgetary control and health and safety, are readily recognized with cybernetics as subsystems, the system being the entire organization. The effectiveness of any subsystem can only be assessed when seen as part of the whole.

Successful subsystems do not necessarily add up to successful systems. The subsystem may not fit. They may not have the right elements. The measure of the extent of convergence of mutual or common objectives may not have been considered. Hence there is great advantage in bringing quality, efficiency and safety together into a total management system.

Feedback

Feedback is an essential part of any control process in an organizational system. Its existence marks the point of difference between planning and control. Feedback provides evidence or information for decisions which adjust the system all the time. Feedback is usually obtained by reporting variation from, for instance, objectives and also the means designed to achieve these objectives.

A thermostat provides a good example of feedback in a closed system. In open systems, people may be the instruments which measure, compare and take appropriate corrective action. Feedback can be observed as an error-correcting mechanism, a form of *negative feedback*. To give negative feedback, the controller makes a correction in the opposite direction from that in which the system is drifting. *Positive feedback* has the effect of increasing the drift and causes the system to go out of control.

Organization of work

It is essential that people at work are organized. This entails some form of organizational structure. The formal structure takes place as duties are assigned, procedures adopted, rules and regulations are circulated and processes are begun. The formal organization serves as a guide and tends to give an organization an identity of its own, independent of the people who join the workforce.

As it grows and more labour is added, management authority systems are devised. From a sociological viewpoint, social organization arises wherever people interact on a continuous basis in pursuit of common goals and the organization consists of the behaviour expectations which people have towards each other as group members.

Management

The evolution of management organization has accompanied the growth and expansion of industry and commerce. The expansion of technology and the profit motive led to large-scale production. Corporate business enterprise arises when individuals were unable to finance the expansion taking place and the stockholders were invited to group together and finance the developing enterprise.

Management hierarchies developed on the basis of individual responsibility and accountability to stockholders for running the business successfully and providing profit for the stockholders.

Workers

On the basis that 'unity is strength', people in similar occupations and/or work locations established their own organizations, the objectives being clearly agreed and stipulated by the members. In the early days, groups of workers formed craftsmen's guilds which were subsequently followed by trade unions and workers' associations.

The reliability of organizational systems

Any organization is only as good as the reliability of its management systems, for example, financial control, forecasting, planning and organizational development. However, many organizations fail due to the inability of its leaders to consider human behaviour.

Problems can arise in adopting a systems approach to human behaviour. In many cases, there is conflict between the organizational objectives and those of individuals.

Weber and the bureaucratic model

Bureaucracy grows as enterprises grow. A bureaucracy is simply a hierarchical arrangement of unit organizations, an orderly arrangement of people and work based on divisions of function and authority. The hierarchy may be viewed as several layers of authority, which form the pyramid of authority.

Weber studied and analysed bureaucratic structures and functions and drew a close relationship between such structures and functions and the general theory of power and dominance. He established that the greater the degree of power and dominance exerted by management over the workers, the less reliable was the system. This view brought in the concept of *participative organizational control* or *participative management*. As long as the communication organization is appropriate and effective, it is felt that this may be the most effective system, since it is nearest to the cybernetic concept of self-regulation.

Resolving conflict and introducing change

What is 'conflict'?

This term is variously defined as 'a trial of strength', 'to be at odds with' and 'to clash'. Conflict costs employers time, money, employee commitment and reputation. The hidden cost of unresolved conflict in organizations is enormous. Identifying effective ways to manage and resolve organizational conflict can have a significant effect on productivity and profitability.

Forms of conflict

Conflict can arise at a number of levels:

- **Intrapersonal conflict**. People can be at conflict within themselves over a range of matters – money, decisions and individual personality factors, such as lack of confidence.
- **Interpersonal conflict**. The conflict is primarily between two or more individuals, that is, at a team level within an organization or involving, for example, friends, neighbours and contemporaries.
- **Systemic conflict**. The conflict is a symptom of a wider organizational issue that needs to be addressed, such as pay rates, working hours, staff turnover and changes in working practices.
- **Industrial conflict**. An industry is at conflict with, for example, the government or the trade unions over a range of issues.
- **Global conflict**. Individual countries are in conflict with other countries over, for example, commodity prices, armaments, poverty, social norms and economic factors.

Potential sources of conflict

Conflict may arise between the line and staff organizations for a number of reasons, such as:

- differing motivations;
- misunderstanding of individual roles;
- differing cultures and objectives; and
- differing priorities and levels of commitment.

One of the principal causes of conflict, however, is the introduction of change which is too rapid for people to take in, where they are expected to change behaviour, where new concepts are being introduced and where they may feel threatened as a result of impending change. Managers need to understand the human reaction to change and introduce change through a change management programme (see Chapter 17, Change and change management).

Workplace representation

Joint consultation between employers and employees is an important feature of the safety management process. This consultation may take place with managers and trade union-appointed safety representatives, through the operation of a health and safety committee or as part of a normal employer/employee consultative process.

Consultation requirements

Consultation requirements are embodied in:

- The Safety Representatives and Safety Committees Regulations 1977
- The Health and Safety (Consultation with Employees) Regulations 1996.

Safety Representatives and Safety Committees

These regulations apply in organizations where there are safety representatives who are appointed by a recognized trade union to represent the health and safety interests of the members of that trade union.

Objective of the regulations

The main objective of the regulations is to provide a basic framework within which each undertaking can develop effective working relationships. As such, these working relationships must cover a wide range of situations, that is, different forms of workplace and work activity.

Appointment of safety representatives

A recognized trade union may appoint safety representatives from amongst the employees in all cases where one or more employees are employed by an employer by whom it is recognized. The employer must be notified by the trade union of the names of the safety representatives. Each safety representative has certain prescribed functions as follows:

- to represent employees in consultation with employers;
- to co-operate effectively in promoting and developing health and safety measures;
- to make representations to the employer on any general or specific matter affecting the health and safety of their members;
- to make representations to the employer on general matters affecting the health and safety of other people employed at the workplace;
- to carry out certain inspections;

Safety representatives are also entitled to represent their members in consultation with the HSE and to receive information from inspectors. They are entitled to attend meetings of the safety committee, if appropriate, but may not necessarily be members of a safety committee. It should be noted that none of these functions impose a duty on safety representatives.

Time off with pay

Employers must give safety representatives time off with pay for performing their functions and for any reasonable training they undergo.

Inspections of the workplace

Safety representatives are entitled to carry out workplace inspections and employers must give them reasonable assistance. The Approved Code of Practice to the regulations recommends one inspection every 3 months or in prescribed circumstances, e.g. following a major injury accident to a member of that trade union. Safety representatives can further inspect the scene of a reportable accident or dangerous occurrence.

Inspection of documents

Safety representatives can inspect any document which the employer has to maintain, other than documents relating to the health records of identifiable individuals. Typical examples of documents that a safety representative would be entitled to access to is the Statement of Health and Safety Policy, risk assessment documentation, internal health and safety codes of practice and the outcome of safety monitoring undertaken by a health and safety specialist.

Approved Code of Practice

The Approved Code of Practice accompanying the Safety Representatives and Safety Committees Regulations outlines specific arrangements for the appointment of safety representatives and the running of safety committees.

Safety representatives

1. **Qualifications**. As far as is reasonably practicable, safety representatives must have 2 years' experience with the employer or in similar employment.
2. **Functions of safety representatives**. They must
   - keep themselves informed of legal requirements;
   - encourage co-operation between employer and employees;
   - undertake health and safety inspections of their area of the workplace; and
   - inform employer of the outcome of such inspections.

Employers

Employers must provide information on

- plans and performance;
- hazards and precautions;
- occurrence of accidents, dangerous occurrences and occupational disease; and
- any other information, including the results of any measurements taken.

Safety committees

1. **Basic objectives**. To
   - promote co-operation between employers and employees; and
   - act as a focus for employee participation.
2. **Functions**. To consider
   - the circumstances of individual accidents and cases of reportable diseases; and
   - accident statistics and trends;
   - to examine safety audit reports;
   - to consider reports and information from the HSE;
   - to assist in the development of safety rules and systems;
   - to conduct periodic inspections;
   - to monitor the effectiveness of health and safety training, communications and publicity;
   - to provide a link with the inspectorate.
3. **Membership**. Safety committees should be reasonably compact but allowing for representation of management and all employees. It should be recognized that the safety representative is not appointed by the safety committee and vice versa. Neither is responsible to or for the other. Management representation should include line managers, supervisors, engineers, human resources specialists, occupational health specialists and health and safety advisers. A safety committee must have authority to take action. Specialist knowledge should be available.
4. **Conduct**. Meetings should be held as often as necessary. Agendas and minutes are to be provided.

5. **Arrangements at individual workplaces**. The following aspects must be considered

   - division of conduct of activities;
   - clear objectives/terms of reference of the committee;
   - the membership and structure of committee to be clearly defined in writing; and
   - the system for publication of matters notified by safety representatives.

Health and safety committees

In conjunction with the Safety Representatives and Safety Committees Regulations 1977, the HSE published guidance on the objectives, function, membership and conduct of health and safety committees.

Basic objectives

- The promotion of co-operation between employers and employees in instigating, developing and carrying out measures to ensure the health and safety at work of employees.
- To act as a focus for employee participation in the prevention of accidents and the avoidance of industrial diseases.

Specific functions of health and safety committees

- The study of accident and notifiable disease statistics and trends, so that reports can be made to management on unsafe and unhealthy conditions and practices, together with recommendations for corrective action.
- Examination of safety audit reports on a similar basis.
- Consideration of reports and factual information provided by inspectors of the enforcing authority appointed under the Health and Safety at Work Act.
- Assistance in the development of workplace safety rules and safe systems of work.
- A watch on the effectiveness of safety content of employee training.
- A watch on the adequacy of safety and health communication and publicity in the workplace.
- The provision of a link with the appropriate enforcing authority.

In certain cases, safety committees may consider it useful to carry out an inspection by the committee itself. However, it is the employer's responsibility to take executive action, to have adequate arrangements for regular and effective checking of health and safety precautions and for ensuring that the declared health and safety policy is being fulfilled. The work of the safety committee should supplement these arrangements; it cannot be a substitute for them.

Membership of safety committees

The membership and structure of safety committees should be settled in consultation with management and the trade union representatives concerned through the use of the normal machinery. The aim should be to keep the total size as reasonably compact as possible and compatible with the adequate representation of the interests of management and of all the employees, including safety representatives. The number of management representatives should not exceed the number of employees' representatives.

Management representatives should not only include those from line management, but such others as engineers and HR managers. The supervisory level should also be represented. Management representation should be aimed at ensuring:

- adequate authority in order to give proper consideration to recommendations; and
- the necessary knowledge and expertise to provide accurate information to the committee on company policy, production needs and on technical matters in relation to premises, processes, plant, machinery and equipment.

Conduct of safety committees

Safety committees should meet as often as necessary. The frequency of meetings will depend upon the volume of business, which in turn is likely to depend on local conditions, the size of the workplace, numbers employed, the kind of work carried out and the degree of risk inherent.

Sufficient time should be allowed during each meeting to ensure full discussion of all business.

Meetings should feature an agenda and minutes of the meeting should be circulated as soon as possible to all members, together with the most senior executive responsible for health and safety. Arrangements should be made to ensure that the board of directors is kept informed generally of the work of the committee.

Running a health and safety committee

As with any committee, it is essential that the constitution of a safety committee be in written form. The following aspects should be considered when establishing and running a safety committee.

Objectives

These could be:

- to monitor and review the general working arrangements for health and safety; and
- to act as a focus for joint consultation between employer and employees in the prevention of accidents, incidents and occupational ill-health.

Composition

The composition of the committee should be determined by local management, but should normally include equal representation of management and employees, ensuring all functional groups are represented.

Other people may be co-opted to attend specific meetings, e.g. health and safety adviser, company engineer.

Election of committee members

The following officers should be elected for a period of 1 year:

The Chairman

The Deputy chairman

The Secretary

Nominations for these posts should be submitted by a committee member to the secretary for inclusion in the agenda of the final meeting in each yearly period. Members elected to office may be renominated or re-elected to serve for further terms. Election should be by ballot and should take place at the last meeting in each yearly period.

Frequency of meetings

Meetings should be held on a quarterly basis or according to local needs. In exceptional circumstances, extraordinary meetings may be held by agreement of the chairman.

Agenda and minutes

The agenda should be circulated to all members at least 1 week before each committee meeting. The agenda should include:

- **Apologies for absence**. Members unable to attend a meeting should notify the secretary and make arrangements for a deputy to attend on their behalf.
- **Minutes of the previous meeting**. Minutes of the meeting should be circulated as widely as possible and without delay. All members of the committee, senior managers, supervisors and trade union representatives should receive personal copies. Additional copies should be posted on notice boards.
- **Matters arising**. The minutes of each meeting should incorporate an action column in which people identified as having future action to take, as a result of the committee's decisions, are named. The named person should submit a written report to the secretary, which should be read out at the meeting and included in the minutes.
- **New items**. Items for inclusion in the agenda should be submitted to the secretary in writing, at least 7 days before the meeting. The person requesting the item for inclusion in the agenda should state in writing what action has already been taken through the normal channels of communication. The chairman should not normally accept items that have not been pursued through the normal channels of communication prior to submission to the secretary. Items requested for inclusion after the publication of the agenda should be dealt with, at the discretion of the chairman, as emergency items.
- **Safety adviser's report**. The safety adviser should submit a written report to the committee, copies of which should be issued to each member at least 2 days prior to the meeting and attached to the minutes. The safety adviser's report should include, for example:
  - a description of all reportable injuries, diseases and dangerous occurrences that have occurred since the last meeting, together with details of remedial action taken;
  - details of any new health and safety legislation directly or indirectly affecting the organization, together with details of any action that may be necessary;
  - information on the outcome of any safety monitoring activities undertaken during the month, e.g. safety inspections of specific areas; and
  - any other matters which, in the opinion of the secretary and himself, need a decision from the committee.
- **Date, time and place of the next meeting**.

Health and Safety (Consultation with Employees) Regulations 1996

These regulations brought in changes to the law with regard to the health and safety consultation process between employers and employees. Under the Safety Representatives and Safety Committees Regulations 1997, employers must consult safety representatives appointed by any trade unions they recognize. However, many employees are not members of a recognized trade union.

Under the Health and Safety (Consultation with Employees) Regulations 1996, employers must consult any employees who are not covered by the Safety Representatives and Safety Committees Regulations. This may be by direct consultation with employees or through representatives elected by the employees they are to represent (representatives of employee safety).

HSE guidance

HSE guidance accompanying the regulations details:

- which employees must be involved;
- the information they must be provided with;
- procedures for the election of representatives of employee safety;
- the training, time off and facilities they must be provided with; and
- their functions in office.

Employment Rights Act 1996

Part V of the Act deals with 'protection from suffering detriment in employment'. Section 44 covers the right not to suffer detriment in 'health and safety cases'.

Health and safety cases

An employee has the right not to be subjected to any detriment by any act or any deliberate failure to act, by his employer done on the ground that:

- having been designated by the employer to carry out activities in connection with preventing or reducing risks to health and safety at work, the employee carried out (or proposed to carry out) any such activities;
- being a representative of workers on matters relating to health and safety at work or a member of a safety committee:
  - in accordance with arrangements established under or by virtue of any enactment; or

  - by reason of being acknowledged as such by the employer,

  the employee performed (or proposed to perform) any functions as such a representative or a member of such a committee;

- being an employee at a place where:
  - there was no such representative or safety committee; or
  - there was such a representative or safety committee but it was not reasonably practicable for the employee to raise the matter by those means;

  he brought to his employer's attention, by reasonable means, circumstances connected with his work which he reasonably believed were harmful or potentially harmful to health or safety;

- in circumstances of danger which the employee reasonably believed to be serious and imminent and which he could not reasonably have been expected to avert, he left (or proposed to leave) or (while the danger persisted) refused to return to his place of work or any dangerous part of his place of work; or
- in circumstances of danger which the employee reasonably believed to be serious and imminent, he took (or proposed to take) appropriate steps to protect himself or other persons from the danger.

For the purposes of the last subsection above, whether steps which an employee took (or proposed to take) were appropriate is to be judged by reference to all the circumstances including, in particular, his knowledge and the facilities and advice available to him at the time.

An employee is not to be regarded as having been subjected to any detriment on the ground specified in the last subsection above if the employer shows that it was (or would have been) so negligent for the employee to take the steps which he took (or proposed to take) that a reasonable employer might have treated him as the employer did.

Patterns of employment

Patterns of employment vary considerably between organizations. Whilst some organizations may operate effectively from 8.30 a.m. to 5 p.m. for 5 days a week on a continuing basis, other organizations, such as those in the food industry, may rely heavily on 24-hour, 7 days a week continuing production. Inevitably, working unsocial hours, extra hours in the form of overtime and the need to operate shifts will directly affect payment systems.

Payment systems

Any form of extra payment for working unsocial hours must be seen to be fair to the employees concerned.

Shift work and atypical working

Considerable attention has been paid in recent years to the effects of work operations on certain groups of workers, namely 'atypical workers'. These are workers who are not in normal daytime employment, together with shift workers, part-time workers and night workers. Normal daytime employment implies, as a rule, working for a set number of hours between 6 a.m. and 6 p.m. Anything outside those parameters could be classed as atypical working periods.

Approximately 29 per cent of employees in the UK work some form of shift pattern and 25 per cent of employees undertake night shifts. Researchers have, over the years, studied the physical and psychological effects of atypical working on these groups of people, in particular, factory workers and transport workers, and have reported a number of findings, for instance:

- 60–80 per cent of shift workers experience long-standing sleep problems;
- shift workers are 5 to 15 times more likely to experience mood disorders as a result of poor quality sleep;
- drug and alcohol abuse are much higher among shift workers;
- 80 per cent of shift workers complain of chronic fatigue;
- approximately 75 per cent of shift workers feel isolated from family and friends;
- digestive disorders are between four and five times more likely to occur in shift workers; and
- from a safety viewpoint, more serious errors and accidents resulting from human error occur during shift work operations.

The appeal of shift work

So why are people prepared to do shift work? In some manufacturing activities, there may be no alternative to shifts due to the need, principally, to keep the production process running continuously. However, because of the disturbing effects of shift work, it has always tended to attract a higher wage than standard day work, in some cases a shift work premium. Such a premium frequently appeals to younger workers who may need the extra money at a particular point in their lives, e.g. to raise the deposit for a house, but who intend later, once this financial objective has been achieved, to revert to routine day shifts.

The problem is, of course, that once people have become used to the improved wages and accompanying lifestyle, they tend to get locked into the system of shift work. What they do not appreciate is the stress that this can create in their own lives, in the lives of their family and the changes in health state that can take place gradually over a period of years.

Shift work and stress

Fundamentally, very few people are trained to appreciate the stress that shift work can create and the strategies necessary to cope with this stress. They have probably never been advised, for instance, that adjustment to shift work is very much age-related. Younger people can adapt to the way of life better, but as people get older sleep patterns become more important and older shift workers frequently find that sleep may be lacking or frequently broken. This is a particular cause of stress.

In effect, shift workers become part of a subculture, often encountered amongst permanent night workers who, because they are asleep whilst everything within the family group is taking place, lose touch with members of the family. This can produce feelings of guilt, frustration and isolation resulting in changes in attitude, short-temperedness, personality change and loss of motivation.

Conflict can arise, particularly if the family is unaware of the stress associated with shift work, between the personal needs of the shift workers and social and family needs. Inevitably, many people try to get the best of both worlds by leading a normal everyday life during the day and working shifts at the same time. The changing of work and sleep schedules has a profound effect on their body clocks, resulting in significant physical and emotional problems over a period of time.

The principal causes of stress for shift workers

The psychological factors which affect an individual's ability to make the adjustments required by various work schedules are:

- **Age**. Young people adapt more easily to changing shift patterns due to the need for sleep being less than that of an older person.
- **Sleep needs**. Some people need less sleep than others and can adjust to shift work more easily on this basis. Sleep and sleep patterns are, however, a complex psychological phenomenon. Sleep takes place in a series of four stages and, because of daytime disturbances, many shift workers do not experience the beneficial delta or deep sleep stage, so important in terms of physical restoration.
- **Sex**. Women can experience complex problems in adapting to shift work, particularly in their reproductive cycles.
- **'Day persons' and 'night persons'**. Some people are more naturally alert in the morning and have more difficulty in adjusting to shift changes. Other people perform more efficiently at night. Much of this variation is associated with arousal levels which vary from person to person.
- **The type of work**. Generally, people experience less stress where undertaking work which requires physical activity, e.g. assembly work, than work which is inactive. This is due to the fact that sleep-deprived people are likely to

lose concentration, particularly if undertaking work of a monitoring nature, e.g. inspection work.

- **Desynchronization of body rhythms**. Long-term problems arise when people rotate schedules at a rate more quickly than that to which the body can adjust. Body rhythms get out of synchrony with the external environment and internal processes normally synchronized, such as the digestive system, begin to drift apart. This drifting happens because internal processes adjust at different rates. This gradual desynchronization leads to longstanding tiredness, a feeling of being generally 'run down', depression (a classic manifestation of stress) and lack of energy.

The health and safety implications

In many organizations, accident rates are substantially higher for shift work and night work operations. Many people would write this fact down to reduced supervision levels, a lack of training of shift workers in safe working practices or to the view that shift workers 'couldn't care less'. Very rarely do senior management endeavour to ascertain the reasons for these high accident rates, however.

Deprivation of sleep results in chronic fatigue in a substantial number of shift workers (60–80 per cent). Fatigue is frequently associated with impaired memory, judgement, reaction time and concentration. It is not uncommon for shift workers to doze off on the job or even be found asleep. 'Falling asleep at the wheel', especially between the hours of 1 a.m. and 4 a.m., is a common problem with HGV drivers. A Japanese study showed that 82 per cent of near miss train accidents occurred between midnight and 8 a.m. The *Exon Valdez* ran aground a few minutes after midnight, due to the crew having been working 'heavy overtime'. Many other major incidents, e.g. Three Mile Island, Bhopal and the *Challenger* explosion also occurred between midnight and 8 a.m.

Most of the reports on these incidents indicate 'human error due to operator fatigue' as a significant contributory cause of the accident concerned.

Reducing the stress of shift work

Strategies are available aimed at minimizing the desynchronization of body rhythms and other health problems associated with shift working. The principal objective is to stabilize body rhythms and to provide consistent time cues to the body.

Firstly, there is a need to recognize that workers must be trained to appreciate the potentially stressful effects of shift working and that there is no perfect solution to this problem. However, they do have some control over how they adjust their lives to the working arrangements and the change in lifestyle that this implies.

Secondly, they need to plan their sleeping, family and social contact schedules in such a way that the stress of this adjustment is minimized. Most health problems arise as a result of changing daily schedules at a rate quicker than that at which the body can adjust. This can result in desynchronization, with reduced efficiency generally due to sleep deprivation.

- **Sleep deprivation**. This can have long-term effects on the health of the shift worker. It is important to consider individual lifestyle, in particular diet, the actual environment in which sleep takes place, family and social relationships and, in certain cases, the use of alcohol and drugs.
- **Diet**. A sensible dietary regime, taking account of the difference between the time of eating and the timing of the digestive system, will assist the worker to minimize discomfort and digestive disorders.
- **Alcohol and drugs**. Avoidance of alcohol and drugs, e.g. caffeine and nicotine, can result in improved sleep quality. Occasional use of sleeping tablets may be beneficial, but should be used under medical supervision.
- **Family and friends**. They should appreciate the demands on the shift worker and make every effort to assist in reducing the potentially stressful effects of this type of work. In particular, better planning of family and social events is necessary to reduce the isolation frequently experienced by shift workers.

Fundamentally, shift work has been a feature of British industry for over a century. For some people, it is a way of life that they have adjusted to with ease over a period of time. For others, it can be significantly stressful, resulting in a wide range of psychological and health-related symptoms and effects, together with an increased potential for accidents.

However, a number of remedies are available to organizations. These include:

- consultation prior to the introduction of shift work;
- recognition by management that shift work can be stressful for certain groups of workers and of the need to assist in their adjustment to this type of work;
- regular health surveillance of shift workers to identify any health deterioration or change at an early stage;
- training of shift workers to recognize the potentially stressful effects and of the changes in lifestyle that may be needed to reduce these stressful effects; and
- better communication between management and shift workers aimed at reducing the feeling of isolation frequently encountered amongst such workers.

Home working

Many employees work at home with little or no direct contact with their employer. The scale of work activities undertaken by home workers is extensive, such as the packaging of items, such as screws for national store chains, assembly, finishing and packing of electrical goods, work with computers, telephone contact work, such as telesales operations and other tasks, such as ironing and repairing clothing. In the case of those involved in assembly work, components may be delivered and finished items collected on a weekly basis. Other home workers may receive information on their future work output requirements by letter or e-mail.

The hazards arising from home-working activities

The hazards to home workers are extensive. These may include electrical hazards arising from badly maintained electrical equipment, such as soldering equipment, display screen equipment and hand irons, manual handling hazards, fire hazards, chemical hazards and the potential for contracting upper limb disorders. It is common for family members to assist with the work and, in some cases, children may be illegally involved in home-working activities as a means of increasing income by the home worker.

The duties of employers

Employers have a general duty of care at common law to ensure that work activities undertaken by employees are safe and without risks to health and that employees are adequately trained to undertake the work. Under the criminal law, employers have duties under the HSWA and certain regulations towards home workers comparable with the duties owed to employees working at a specific workplace.

Whilst a home worker's house may not be under the direct control of an employer, the general and specific duties of employers towards employees apply to this type of work. Employers, therefore, must have some form of procedure for safety monitoring of home workers' work activities with particular reference to their working environment, the tasks undertaken, equipment, articles and substances used, and structural features of the room or area where the work is undertaken. Home workers should also receive appropriate information, instruction, training and supervision. In the latter case, supervision may take the form of regular visits by a home workers' co-ordinator or supervisor to ensure written safety procedures are being observed.

In particular, the employer should prepare and put into practice a statement of policy with respect to home working as part of the organization's Statement of Health and Safety Policy. Such a statement should incorporate the employer's duties towards home workers and the organization and arrangements for ensuring these duties are put into practice.

Work area inspections and risk assessment

Before employing people to undertake work at home, employers should ensure that an inspection of the home work area is undertaken, with particular reference to environmental working conditions, the safety of work equipment, including that owned by the home worker, fire safety, the storage of hazardous substances and the extent of manual handling required. Where appropriate, risk assessment should be undertaken with a view to identifying the significant hazards and the preventive and protective measures necessary. It may be necessary for the employer to designate specific areas where the work must be undertaken.

Many home workers use display screen equipment as a significant part of their work. Under the Health and Safety (Display Screen Equipment) Regulations, an employer must undertake a workstation risk analysis in the case of defined display screen equipment 'users', that is, 'employees who habitually use display screen equipment as a significant part of their normal work'. On this basis, workstations may need to be modified to meet the requirements of these regulations.

Similar provisions apply in the case of home workers involved in manual handling operations on a regular basis, where it would be necessary for a manual handling risk assessment to be undertaken by the employer with a view to preventing or controlling risks arising from this work.

Conclusions

There are many factors, features or characteristics of organizations which influence the way people behave at work, particularly with regard to the implementation of safe working procedures.

The organization has a direct effect on human behaviour at work. In particular, it endeavours to control both individual and group behaviour in many ways. This may be through the systems for supervision and control, training activities, management development systems, apprentice schemes, the reward structure and rates of pay, including bonuses, shift work and overtime payments.

Consideration of these factors is crucial to the success of health and safety management systems.

Key points

- All people operate within an organizational environment which may be affected by many factors.
- Organizations must innovate in order to survive. To do this, they must have the good will of the workforce.
- Health and safety organization should be clearly defined and laid out in an organisation's Statement of Health and Safety Policy.
- The need for organizations to integrate safety goals with organizational goals must be recognized.

- Organizations comprise integrating systems which rely heavily on people for their effective implementation.
- Feedback is an essential element of any control process in an organizational system.
- For organizations to be effective, any form of conflict must be resolved quickly and effectively.
- There must be a formal system for consultation on health and safety-related issues.
- Patterns of employment, e.g. shift working, need regular review due to the potential for error that can arise in such work.

7 Improving human reliability

Failures in human reliability are a contributory factor in many accidents, including some of the well-known major incidents resulting in multiple fatalities. One of the most important requirements for employers, therefore, is that of improving the human reliability of employees. This may entail the study of attitudes, motivation and other behavioural factors.

The significance of personal factors

Employees bring personal habits, attitudes, skills, personalities and other factors to their jobs that, in relation to the task demands, may constitute strengths or weaknesses. Individual characteristics influence behaviour in complicated and significant ways. Their effects on task performance may be negative and cannot always be mitigated by job design solutions. Some characteristics, such as personality, are fixed and largely incapable of modification. Others, such as skills and attitudes, are amenable to modification or enhancement. The person needs, therefore, to be matched to the job. Important issues in this area are:

- **Task analysis**. Through task analysis, especially in the case of critical tasks, a detailed job description should be generated. From this, a specification can be drawn up to include such factors as age, physique, skills, qualifications, experience, aptitude, knowledge, intelligence and personality. Personnel selection policies and procedures should ensure the specifications are matched by the individuals.
- **Training**. Training will produce an employee capable of working without close supervision, confident to take on responsibility and perform effectively, providing that initial selection is carried out with care. Training of all types (induction training, orientation training, on-the-job training, etc.) benefits individuals themselves, as well as their colleagues. Self-confidence and job satisfaction grow significantly when people are trained to work correctly under both routine and emergency conditions. Training should aim to give all individuals the skills to allow them to understand the working of the processes, machinery and equipment. Training is not a 'once and for all' activity, for when processes and procedures change and complicated skills, particularly when underused, deteriorate, these can directly or indirectly influence a person's performance.
- **Monitoring of performance**. The monitoring of personal performance, apart from providing feedback on that performance, includes supervision of health and safety practices and other approaches for developing an effective climate of opinion towards safety as a whole.

- **Fitness for work and health surveillance**. For certain jobs, there may be specified health standards for which pre-employment and/or periodic health surveillance is necessary. These may relate to the functional requirements for the job or the impact of specified conditions on the ability to perform the job adequately and safely. One example is the need for regular medical examinations of divers. There may also be a need for routine health surveillance to test for the effects of exposure of employees to physical hazards, such as heat stress, chemical hazards, such as those arising during the use of organophosphorus insecticides, biological hazards, such as those causing viral hepatitis or work-related hazards, such as the heat disorders. All these effects impair operational ability and it is essential that scheduled health surveillance procedures identify such conditions at a very early stage in their development.

Other occupational health procedures, such as a review of health on return to work following a period of sickness absence, the recognition of the place of counselling and advice during periods of need or personal stress and consideration of the possible need for redeployment should also be considered. In this context, alcohol or drug abuse, and the possible side effects of prescribed drugs are also relevant. Access to specialists may be required to deal with these health-related issues.

The risk assessment issues

Individual aspects of human behaviour are affected by many factors. With the increased emphasis on 'safe person' strategies in occupational health and safety, many aspects of human behaviour need to be considered when looking at the area of human capability as part of the risk assessment process.

Human reliability studies

'Human reliability' is defined as the probability that a person will correctly perform some system-required activity during a given period of time without performing any extraneous activity that can downgrade the system. Human reliability studies have, over the years, examined both correct and incorrect human actions in a range of situations using several types of operator model.

Modelling approaches

Behavioural or human factors models

These models focus on simple manifestations (error modes). Error modes are commonly described in terms of commissions, omissions and extraneous actions. These methods endeavour to derive the probability that a specific manifestation or event will occur. However, as causal models are either very basic or simple, the theoretical basis for the prediction of failures in performance is inadequate.

Information processing models

People process information in many ways. This type of model considers internal 'mechanisms', such as those for reasoning and the making of decisions. The methods endeavour to explain the flow of causes and their effects through the model. Causal models have been perceived to be complex in many cases, with little predictive power, and are better suited for retrospective analysis sooner than prediction of error or failures in human reliability.

Cognitive models

Cognitive models examine the relationship between error modes and their causes, taking into account the sociotechnical environment as a whole. Compared with information processing models, cognitive models are relatively simple and the context is explicitly represented. They are well suited for both prediction and retrospective analysis.

Cognition is the reason why performance is efficient, and why it is sometimes limited, whereby the operator is seen as not only responding to events, but also acting in anticipation of future developments.

Analytical techniques

The analysis of human reliability involves a number of techniques as indicated below.

Fault tree analysis

Fundamentally, this technique commences with consideration of a chosen 'top event', such as a major escalating fire or an explosion, and then assesses the combination of failures and conditions which could cause that event to arise. It is widely used in quantitative risk analysis, particularly where control over process controls is critical to meet safety standards.

Event tree analysis

This technique works from a selected 'initiating event', such as a pressure control failure. It is, basically, a systematic representation of all the possible states of the processing system conditional to the specific initiating event and relevant for a certain type of outcome, e.g. a pollution incident or a major fire.

It should be appreciated that any given event may be the top event in a fault tree, as well as the initiating event in an event tree.

Consequence analysis

Consequence analysis is a feature of risk analysis which considers the physical effects of a process failure and the damage caused by these physical effects. It is undertaken to form an opinion on potentially serious hazardous outcomes of accidents and their possible consequences for people and the environment.

The purpose of consequence analysis is to act as a tool in the decision-making process in a safety study which incorporates the features as described below:

- description of the process system to be investigated;
- identification of the undesirable events;
- determination of the magnitude of the resulting physical effects;
- determination of the damage;
- estimation of the probability of the occurrence of calculated damage; and
- assessment of risk against established criteria.

The outcome of consequence analysis is four-fold, thus:

- for the chemical and process industries to obtain information about all known and unknown effects that are of importance when something goes wrong in the plant and also to get information on how to deal with possible (credible) catastrophic events;
- for the designing industries to obtain information on how to minimize the consequences of accidents;
- for the workers in the processing plant and people living in the immediate vicinity to give them an understanding of their personal situation and the measures being taken to protect them; and
- for the legislative authorities to consider the adequacy of current controls and possible revisions to same.

Consequence analysis is generally undertaken by professional technologists and chemists who are experienced in the actual problems of the technical system. The logic chain of consequence analysis is shown in Figure 7.1.

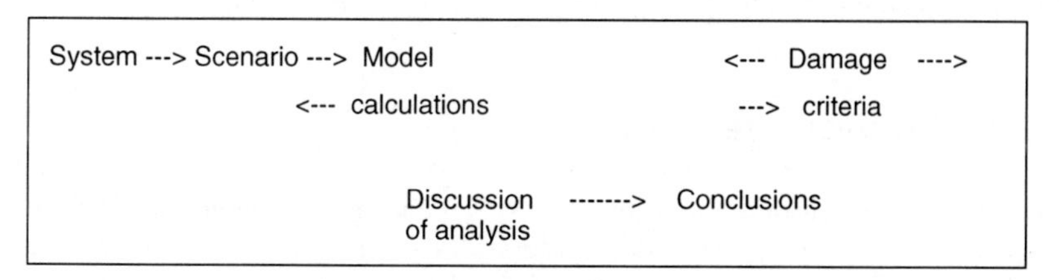

Figure 7.1 Logic chain of consequence analysis

The first step in the chain is a description of the technical system to be investigated. In order to identify the undesirable events it is necessary to construct a scenario of possible incidents. The next stage is to carry out model calculations in which damage level criteria are taken into account. Following discussions by the assessment team, conclusions can be drawn as to possible consequences.

Feedback from model calculations to the scenario is included, since the linking of the outputs from the scenario to the inputs of models may cause difficulties. There is also another feedback, that is from damage criteria to model calculations, in case these criteria should be influenced by possible threshold values of the legislative authorities.

Technique for human error rate probability (THERP)

THERP is a technique for predicting the potential for human error in an activity. It quantitatively evaluates the contribution of the human error component in the development of an untoward incident. Special emphasis is placed on the human component in product degradation.

The technique uses human behaviour as the basic unit of evaluation. It involves the concept of a basic error rate relatively consistent between tasks requiring similar human performance elements in different situations. Basic error rates are assessed in terms of contributions to specific systems failures.

The methodology of THERP entails

- selecting the system failure;
- identifying all behaviour elements;
- estimating the probability of human error; and
- computing the probabilities as to which specific human error will produce the system failure.

Following classification of probable errors, specific corrective actions are introduced to reduce the likelihood of error. The major weakness in the use of the THERP technique is the lack of sufficient error rate data.

Scientific management

F.W. Taylor established the principles of 'Scientific management' (1911) and endeavoured to dispel some of the old 'rules of thumb' about the way work should be undertaken in organizations. In 1881, as a result of his studies, he published a paper that turned the cutting of metal into a science. He subsequently examined the manual process of shovelling coal and other loose materials and, by experimenting with different designs of shovel for different materials, Taylor designed shovels that would permit the operator to shovel almost continuously for a whole working shift. As a result of his studies, the number of people involved in shovelling at the Bethlehem Steel Works was reduced from 500 to 140. This experimental work, together with his studies on the handling of pig iron, greatly contributed to the analysis of work design and introduced the concept of method study.

Taylor postulated, 'Man is a creature who does everything to maximize self-interest'. In other words, people are primarily motivated by economic gain. In establishing this viewpoint, he examined the differences between potential managers, who were involved in the organizing, planning and supervising of work operations, and the rest of the workforce, who were not particularly interested in these aspects. They preferred to have simple tasks organized for them that they could be trained to undertake without having to make decisions. Once the work had been planned, the prospect of better wages was the principal incentive for increasing production.

This viewpoint, held by many of the entrepreneurs of the nineteenth and early twentieth centuries, led to a range of variations on the theme, such as productivity bonuses, incentive schemes, piecework, payment by results and even the payment of 'danger money' for undertaking high-risk activities. Taylor's concept of Scientific management established many basic strategies and work systems, such as the division of labour, mass production, work study and great emphasis on the optimum conditions for work, training and selection of employees.

Objectives of scientific management

The four management objectives under Scientific Management were:

1. The development of a science for each element of a man's work to replace the old rule-of-thumb methods.
2. The scientific selection, training and development of workers instead of permitting them to select their own tasks and train themselves as best they could.
3. The development of a spirit of hearty co-operation between management and employees to ensure that work would be carried out in accordance with scientifically devised procedures.
4. The division of work between management and workers in almost equal shares, each group taking over work for which it was best suited instead of the previous situation in which responsibility rested largely with the workers. This resulted in organizational hierarchies, systems of abstract rules and impersonal relationships between employees.

The Hawthorne experiments (Social Man)

Studies in the 1930s by Elton Mayo at the Hawthorne Works of the Western Electric Company, Chicago endeavoured to identify the best working conditions for people with the main objective of improving productivity. Previous attempts to improve worker performance and thereby increase output were based on notions that workers should be regarded as machines whose output could be increased by attention to the physical conditions and the elimination of wasteful movements with the resulting fatigue.

Mayo's methods involved studying the output of a small group of female workers who were consulted about each of a large number of changes in working practices introduced. Nearly all these changes led to improvements in output. However, when these continuing changes eventually returned to the original physical conditions, output increased still further. Such revelations led Mayo to conclude that, as the changes observed could clearly not be attributed to the changes and the physical conditions as such, they must be due to social factors and to a change in the attitude of the workers. Because someone clearly was showing an interest in the problems of these workers, they experienced a feeling of importance and responded accordingly by giving their best. It was found, for instance, that most grievances raised were only symptoms of a general discontent, and that a chance to discuss these grievances in the presence of a sympathetic listener frequently led to resolution.

The need for their co-operation and the fact that consultation had taken place over changes, resulted in the formation of a group culture. Members valued being in the group, felt responsible to the group and the organization and, particularly, were concerned with meeting the standards of performance established by the group. The outcome of these studies was the realization that people were not directly motivated by economic gain (the financial rewards) but tended more towards a social philosophy of 'a fair day's work for a fair day's pay'.

They were particularly concerned with the social interaction that took place during work periods and took pride in belonging to an identifiable work group. These factors alone provided a high level of job satisfaction, the more significant the work group, the greater the satisfaction. They also responded to pressure from their peers and the interest shown in them individually by various levels of management. On this basis, work was seen as a social activity, not just a means of obtaining money.

A number of important points emerged from Mayo's studies into worker attitudes during the period 1927 to 1932, in particular:

- the opportunity to air grievances had a beneficial effect on morale;
- complaints were not often objective, but symptomatic of a more deep-routed disturbance;
- workers are influenced by experiences outside their workplace, as well as those inside the workplace;

- dissatisfaction was often based on what the worker saw as an underestimation of their social status in the company;
- voluntary social groups formed at the workplace had a significant effect on the behaviour of individual members, giving rise to group norms for production and group sanctions for people who went outside these norms or established informal standards.

Mayo's studies further resulted in a changed emphasis on the role of the supervisor from that of autocrat to group leader and a greater attention to the need for establishing and maintaining the morale of work groups.

Stimulus and programmed response

A stimulus is something that causes a change in behaviour. It commonly results in a response, some form of change caused by that stimulus. People are continually receiving a variety of stimuli, from the sound of a bell ringing, indicating the end of a particular time period, to the traffic lights changing from amber to red, indicating the need to stop the vehicle. This is part of the learning process which commences in infancy, a baby responding to the sound of its mother's voice, and continues throughout life.

These learned responses may, inevitably, entail some form of skill. People learning to drive acquire, over a period of time, a range of skills, such as gear changing, steering the vehicle, stopping and starting procedures. Eventually, as confidence increases, these responses become programmed to the extent that the driver no longer needs to think about the stages of changing gear, use of the rear view mirror or the giving of signals.

Programmed responses are an essential part of the learning process and of improving human reliability.

Cognitive and learning styles and problem-solving

Cognitive styles

People have particular preferences when it comes to processing information and 'cognitive styles' refer to the preferred ways that an individual processes information. As such, cognitive style is commonly seen as a dimension of personality influencing attitudes, values and social interaction. (The term 'style' describes a person's typical mode of thinking, memorizing or solving problems.)

There are a number of cognitive styles, the most well-known being 'field independence' as opposed to 'field dependence'. With the former, there is a tendency to approach the environment in an analytical, as opposed to global, fashion. Field-independent personalities can distinguish figures as discrete from their backgrounds compared with field-dependent individuals who experience events in an undifferentiated way. Field-dependent individuals, on the other hand, have a greater social orientation relative to field-independent personalities.

Other cognitive styles that have been identified include

- **Scanning**. Differences in the extent and intensity of attention resulting in variations in the vividness of experience and the span of awareness.
- **Levelling versus sharpness**. Individual variations in remembering that pertain to the distinctiveness of memories and the tendency to merge similar events.
- **Reflection versus impulsivity**. Individual consistencies in the speed and adequacy with which alternative hypotheses are formed and responses made.
- **Conceptual differentiation**. Differences in the tendency to categorize perceived similarities among stimuli in terms of separate concepts or dimensions.

Learning styles

Learning styles are concerned with characteristic styles of learning. Kolb (1984) proposed a theory of experiential learning that involves four principal stages:

1. concrete experiences;
2. reflective observation;
3. abstract conceptualization; and
4. active experimentation.

The four stages can be shown as polar opposites as shown in Figure 7.2. Kolb postulated that there are four types of learners, namely divergers, assimilators,

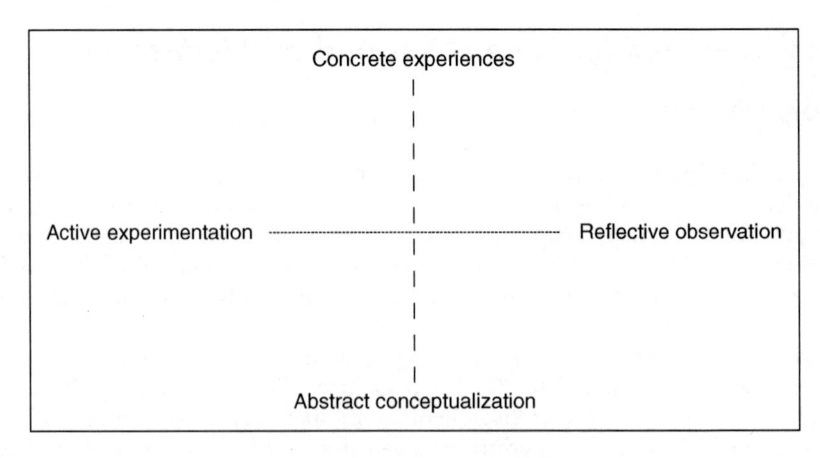

Figure 7.2 Learning styles (Kolb, 1984)

convergers and accommodators, depending upon their position on these two dimensions. For example, an accomodator prefers concrete experiences and active experimentation.

Problem-solving

A problem could be defined as 'the gap between the actual or current situation and the desired situation'. People are commonly beset by a range of problems resulting in distraction, stress and similar adverse responses. Whilst many people are good at solving problems and, indeed, many organizations train their managers in problem-solving techniques with a view to preventing hold-ups in the decision-making process, some people experience life as one continuing series of problems which seem incapable of resolution.

The problem-solving process is best carried out on a group basis and can be considered in a series of stages:

1. **Defining the problem and the desired objective**. It is essential that the problem is clearly defined and supported by facts. The desired situation should further be clearly defined as an objective which is realistic and achievable. A tendency to rush to specify solutions, before defining the actual problem, can be a contributory factor in incorrect decision-making at a later date.
2. **Identifying and defining the root causes**. Very few problems have one simple cause and the objective at this stage must be that of identifying the root causes, sooner than the symptoms of the problem. Brainstorming, a technique that allows a large number of ideas to be generated, can be beneficial at this stage.
3. **Looking to alternative solutions**. At this stage, following the identification of the root causes, a number of alternative solutions to the problem may be generated.
4. **Evaluating the alternatives**. The group needs, firstly, to establish criteria for assessing the solutions put forward. What are the general characteristics of an appropriate solution? There may be a case for developing a matrix which incorporates all the criteria against which the solutions are measured.
5. **Deciding on the best solution**. The group must now agree the best solution either by use of the evaluating criteria or, perhaps, a simple voting system. If the group is having difficulty reaching a consensus, the group leader may need to devote time to resolving controversy and disagreement.
6. **Developing an action plan**. Action planning is intended to involve people, obtain their commitment and increase the likelihood of satisfactory implementation of the solution. An action plan should indicate, on a stage-by-stage basis, what has to be done at each stage, the method, materials, equipment, etc. to be used, the time scale, individual responsibility for implementation and the expected outcome. The plan should be produced in writing identifying these various aspects of individual stages.

7. **Implementing and evaluating solution**. Steps to achieving the solution should be implemented according to the action plan. However, very few action plans proceed fully according to the written down plan. Consideration must be given, therefore, to foreseeable variations that could arise at one of the stages.

Human reliability modelling

Within the field of cognitive psychology, a number of techniques of human reliability modelling are available.

Operator action tree

People need to make decisions prior to taking action. This can be perceived as a tree with branches. The operator action tree technique represents the various stages of the decision-making process, taking into account the time available for decisions and the number of decision paths available to an individual. It is represented as a tree, commencing with an initiating event, followed on one side of the tree by the stages of observation, diagnosis of the situation, individual response and the ultimate success of following that particular series of responses. On the other side of the tree are all the other possible decisions that could have led to failure.

Operators commonly make decisions, but may cause failure by incorrect diagnosis of the situation or inadequate response to the situation.

Interpersonal relationship studies

Relationships between people are a feature of life. They commence in infancy, the baby bonding with his mother, progressing through school and work situations. They may be of a temporary or permanent nature, such as a marital relationship, and the strength of the relationship may vary or fluctuate over a period of time. They imply the sharing of common ground over a particular area of interest and vary in differing levels of intimacy.

Interpersonal relationships are social associations, connections or affiliations between two or more people. Studies into interpersonal relationships examine, for example, the stages of relationship formation, types of interpersonal relationships, the factors affecting such relationships and a number of theories, such as social exchange theory and equity theory.

In the work situation, interpersonal relationships exist at all levels and are spread through the various levels of an organizational hierarchy. Studies indicate that if managers are to motivate employees to better standards of performance, there is a need for the establishment of common ground between the parties concerned. Loss of this common ground, which may happen over a period of time, may tend to end interpersonal relationships.

Motivation and reinforcement – Workplace incentive schemes

A motivator is something that endeavours to mould behaviour or change behaviour. Motivation is important in at least three ways. It is a condition for eliciting behaviour, it is necessary for reinforcement and it controls the variability of behaviour.

Reinforcement is an essential condition of learning. It is defined as a stimulus or event that strengthens a response when it follows the response. Reinforcement implies some degree of conditioning of the individual to respond to a given stimulus and it may be positive or negative.

Positive reinforcement means that the reinforcement is something that the learner approaches as a result of the learning process. This may be a new stimulus or one which has been modified. *Negative reinforcement* occurs when the learner escapes from or avoids an unpleasant or unattractive stimulus.

Workplace incentive schemes endeavour to motivate workers to improved standards of performance. They provide rewards which are desirable and attractive to the participants in the scheme.

Reward schemes

'How do we get people to work harder?' 'If they do work harder and achieve the desired result, what sort of reward should they be provided with?' These are typical questions to which employers are continually seeking an answer. One of the solutions may be through the operation of a reward scheme.

Reward schemes follow a process of planned motivation directed at improving attitudes of workers and improving human reliability.

Planned motivation

Planned motivation is defined as a method by which the attitudes, and thereby the performance of people, can be improved.

Planned motivation schemes are, in effect, an industrial catalyst, and described as 'a tool to maximize performance'. However, one of the outcomes of planned motivation schemes is that, in many cases, they may alter behaviour, but not necessarily attitudes. In other words, people will adapt their behaviour for a limited period in order to reap the benefits offered by the scheme but, once the scheme terminates, they revert to their original standard or level of performance.

Planned motivation schemes are most effective where

- people are restricted to one area of activity, such as some aspect of production work which is measurable;
- measurement of safety performance is relatively simple;
- there is regular stimulation or rejuvenation of the scheme;
- support is provided by both management and trade unions; and
- the scheme is assisted by appropriate positive information and propaganda.

Safety incentive schemes

Safety incentive schemes follow the same pattern as planned motivation schemes. With any form of safety incentive scheme, the main objective is that of providing motivation for employees by

- identifying targets which can be rewarded if achieved; and
- making the reward meaningful and desirable to the people concerned.

Important considerations in successful safety incentive schemes

1. They should be linked with some form of safety monitoring, e.g. inspections, safety sampling exercises. This approach has been found to give the best results.
2. Correct and achievable targets should be set.
3. Under no circumstances should they be linked with accident or sickness absence rates.
4. If based on the lost time concept with respect to accidents, how quickly people return to work after an accident can be significant.
5. They tend to be short-lived and can get out of hand.
6. In some situations, they can shift management responsibility for safety to employees.

The decision-making process

Decision-making involves, in many cases, the making of a choice from a number of identified options, whereby a person reaches a conclusion as to a particular course of future action. It is the stage at which plans, policies and objectives are turned into specific actions by individuals. The decision-making process involves the direction of human behaviour towards a particular objective or goal.

Drucker's approach

Peter Drucker (1999) classified the types of decisions that people make on the basis of

1. **Tactical and strategic decisions**. Tactical decisions are those decisions that are generally routine and unimportant. They usually involve few alternative choices and mainly relate to the use of resources. The term 'strategy' implies the behaviour of the organization in endeavouring to achieve success for a set of goals within a competitive market. Strategic decisions are those management decisions involving:

   - the current situation with regard to a particular or changing situation;
   - the current state of resources or what the resources ought to be.

   They include decisions on basic goals which, in turn, can affect the organization, productivity and operation of a business.
2. **Organizational and personal decisions**. Organizational decisions are those made in the role of an official of the organization, e.g. company secretary, and reflect current policy. Personal decisions are made by a manager acting as an individual. As such, they cannot be delegated to someone else.
3. **Basic and routine decisions**. Basic decisions tend to be of a long-range nature and involve the future of an organization. Incorrect basic decisions, for example, to ignore the requirements of current health and safety legislation and 'hope for the best' on the part of employers, can be expensive. Routine decisions are those which are made on a regular basis and require limited consideration.

Simon's approach

H.A. Simon (1947) distinguished between programmed and unprogrammed decision-making.

1. **Programmed decision-making**. Programmed decisions tend to be of a routine and repetitive nature and are generally regulated by written procedures. They involve low-risk situations and, as such, can be delegated to line managers.
2. **Unprogrammed decision-making**. Unprogrammed decisions cover, in the main, 'one-off' situations and are not repetitive. They involve high-risk situations where the risks cannot easily be assessed, where there are a range of courses of action and where the decisions may involve extensive allocation of resources.

Decision-making and the operator

Workers in various parts of an organization have to make decisions on a regular basis, particularly with respect to the operation of machinery, plant and vehicles and, for instance, in the use of hazardous substances. The decision-making element of a job may well be incorporated in the job description and imparted to the operator at induction and orientation training. Clearly, any decision-making must be related to his tasks and what his employer wants him to achieve, such as

- to keep the process running smoothly with no breaks in production;
- to keep a steady supply of raw materials feeding the machinery;
- to ensure finished products are removed and stored prior to packaging;
- to make regular adjustments to machinery in line with different products being processed; and
- in the event of breakdown, to trace the fault, take remedial action and return to normal operation as quickly as possible.

No two operators may necessarily achieve the above objectives in the same way. Individual skills may vary considerably in terms of speed, machinery control and observing processing at the various stages. Control skills are particularly important. These are developed as a result of training, supervision, experience and knowledge of the process and machinery.

Control skills

- **Sensing**. This entails the ability to detect indications of sound processing activities. Sounds emitted from a machine indicate it is running correctly. Odd smells may indicate a fault which needs attention.
- **Prediction**. A trained operator should be able to predict what could happen if the controls are not monitored for a period of time.
- **Familiarity**. In-depth knowledge of the machinery, together with the most common faults that cause incorrect operation, enables the operator to know the means that can be used, their effects and the possible reaction with other processes.
- **Perceiving**. This is the ability to interpret signs and indications from the machine, together with readings, as to what stage the process is operating.
- **Decision-making**. The ability to choose the control actions which will achieve the desired result in the specific circumstances is an important element of decision-making. Decision-making entails a number of approaches.
    - The operator may follow a well-known rule-of-thumb, namely doing what he has always done in the past because it works in that given situation.
    - Based on his knowledge of the process, an operator may use a self-developed mental model of the system (the intuitive method).

- Using a logical approach and consciously working out the meaning of the elements of the system, an experienced operator can, fundamentally, assess the situation and arrive at a rational decision.

The risk of faulty decision-making

Faulty decision-making can be associated with, for example,

- failure to distinguish between the tactical or strategic decisions required, resulting in incorrect emphasis being placed on a particular decision;
- delegating organizational decisions to others, when the decision should be made by a senior manager;
- incorrect perception of the need for a particular decision;
- basic decisions being made based on incorrect information.

Decision-making skills vary considerably from person to person. The success or otherwise of an organization relies heavily on basic and routine decisions made at all levels of the organization.

Benchmarking

One of the ways of improving human reliability is through techniques such as 'benchmarking'. The term 'benchmark' originated in surveying practice and is defined as 'a mark left on a line of survey for reference at a future time'. It refers to a measure of best practice performance. Benchmarking implies the setting of a standard and comparing progress or performance against that standard. It has been variously defined as

- Improving ourselves by learning from others.
- The process of identifying, understanding and adapting outstanding practices and processes from organizations anywhere in the world to help your organization improve its performance (American Productivity and Quality Center).
- The process of measuring industrial operations against similar operations for the purpose of improving business processes by identifying best practices.
- A process used in management, and particularly strategic management, in which businesses use industry leaders as a model in developing their business practices.
- The search for the best practices that yields the benchmark performance, with emphasis on how an organization can apply the process to achieve superior results.

It entails asking questions:

How are we doing?

Are we tracking the right measures?

How do we compare with others?

Are we making progress fast enough?

Are we using the best practice?

Aims and objectives of benchmarking

- To improve performance through learning from others.
- To find out where you stand.
- To make comparisons with other organizations and then learn the lessons that those comparisons throw up.
- To develop relationships with customers and suppliers, including contractors.
- To identify gaps in performance.
- To improve the organization's reputation.
- To seek fresh approaches to bring about improvements in performance.
- To follow through with implementing improvements.
- To avoid 'reinventing the wheel'.
- To follow up by monitoring progress and reviewing the benefits.
- To ensure compliance with the law.
- To prevent or reduce exposure to adverse incidents which downgrade the system.
- To reduce legal compliance costs.
- To improve overall management performance.

Types of benchmarking

There are three main approaches to benchmarking.

1. Internal benchmarking between functions, departments or a similar organization as a means of improving performance.
2. Competitive benchmarking across industries within a given sector aimed at establishing best practice through identification of gaps in performance. This can be done on a product, functional, departmental or on a company-wide basis.
3. Comparative benchmarking across all business sectors aimed at establishing best practice in all areas of operation. This type of benchmarking is restricted to common processes or technologies.

Principal features of the benchmarking process

The benchmarking process typically includes three phases, namely planning, collection and analysis. At all these stages it is necessary to define the relevant information to be collected and the process to be followed for comparing practices.

- *Planning* involves identifying areas where benchmarking is necessary.
- *Collection of information* focusses on identifying sources of norms, that is, best practices applicable to the particular process or parameter studied. These could be regulatory authorities, sector-specific industry associations, research and development institutes or other industries.
- *Analysing the information* procured is a critical step in benchmarking. Before comparing parameters, it is necessary to check whether certain criteria are met. Once this is established, comparison may be conducted to determine deviation from the norms and identify a course of action to be taken. Comparisons are useful to measure gaps existing across organizations, to identify areas of improvements and to identify industry leaders and followers.

Benchmarking practice

Identifying best practice

Most industries, public bodies and institutions have their leaders in that particular field. Information on these 'leaders' may be available through publications, such as annual reports, newspapers and trade magazines, the local Chamber of Commerce and from the Internet.

Identifying your current position and problem areas

In order to utilize information about how other organizations are doing, it is essential that the organization knows what it is actually doing or, at least, how it would like to be doing. This requires the setting of performance indicators which are measurable.

Once these areas of performance and the relative performance indicators have been established, they can be used in the selection of benchmarking partners. In many cases, a questionnaire technique can be used to determine the best benchmarking partner.

Selecting benchmarking partners

Following preliminary enquiries with respect to leading organizations and institutions, it is necessary to select partners, ascertain their willingness to participate and develop a questionnaire aimed at providing information as to those aspects which could feature in the benchmarking process.

Preliminary discussions with potential benchmarking partners should ascertain whether they are suitable for such an exercise. It is essential that their benchmarking team comprises people who have the comparable mix of technical and people skills with the organization's own benchmarking team and that there is full agreement on the range and scale of information to be gathered.

It would be appropriate at this stage for reciprocal site visits to be made between benchmarking partners to ascertain the scale of their activities, and to compare work methods, processes and activities so that, ultimately, all members of benchmarking teams have a good understanding of their partner's operations and potential problems.

Setting performance indicators

This is, perhaps, the most important stage of the benchmarking process. A number of techniques are available for setting performance indicators. For example, key performance indicators (KPI), (also known as key success indicators), are commonly used in organizations. KPI are quantifiable measurements, agreed beforehand, which reflect the critical success factors of an organization. A company may have as one of its KPI the percentage of its income that comes from returning customers. A chain of restaurants, for example, may measure the percentage product complaints against the total number of meals served as a KPI.

Whatever KPI are selected, they must reflect the objectives of the organization, they must be crucial to its success and must be quantifiable or measurable. KPI may change in line with changes in organizational goals.

In the field of health and safety, there are a number of performance indicators available, such as a comparison with management systems outlined in Successful Health and Safety Management [HS(G)48], the HSE's Corporate Health and Safety Performance Index (CHaSPI) for larger organizations and Health and Safety Performance Indicator for small businesses.

Measuring and comparing performance

Using the agreed KPI, the benchmarking team measure performance and then compare the individual performance of partners.

Learning and acting on lessons learned

The benchmarking team commonly prepare reports indicating areas of unsatisfactory performance including recommendations for improvement on a short-, medium- and long-term basis. Progress in implementation of recommendations is monitored on a continuing basis.

Monitoring for continuous improvement

It is essential that where improvements have been made these improvements are maintained. It is far too easy for organizations to revert to former unsatisfactory practices and procedures once the monitoring process ceases. The benchmarking team should indicate where there is evidence of a reduction in standards.

Rules of benchmarking

To summarize, benchmarking methodology incorporates the following rules and steps:

- setting objectives and defining the scope of the project;
- gaining support from within the organization;
- selecting a benchmarking approach;
- identifying benchmarking partners;
- gathering information (research, surveys, questionnaires, benchmarking visits);
- distilling the learning;
- selecting ideas for performance indicators to put into practice;
- piloting the project;
- implementing the project;
- learning and acting on lessons learned.

Benchmarking costs

Inevitably costs will be incurred in the benchmarking process. These include

- **Time costs**. Members of the benchmarking team need to invest time in researching problem areas, finding exceptional organizations to study, make visits and prepare reports with recommendations for future action.
- **Visit costs**. Apart from not undertaking their normal work when visiting partners, expenses will be incurred in making visits.
- **Database costs**. It is common to create and maintain a database of best practices and the organizations associated with each best practice.

A blame-free culture

How often, in the initial stages of accident investigation, a witness to the accident will state 'It wasn't my fault!' Many people, at all levels in an organization, are more concerned that they may be found to blame for everything or something that results in an accident, sooner than on the causes of that particular downgrading incident.

In order to ensure more constructive accident investigation and the implementation of measures to prevent recurrences of accidents, organizations need to adopt a blame-free culture. Such a culture recognizes that people do make mistakes at work. In Chapter 5, Perception of risk and human error, some of the causes of human error were identified – lack of understanding, lapses of attention, mistaken actions, inadequate information, misperceptions of situations, etc. Only in clearcut cases of unsafe behaviour, such as the nullification of safety devices on machinery, horseplay situations and direct and intentional breaches by an individual of safety procedures, should blame be attached to a person.

A blame-free culture has much to offer an organization in terms of improved employer–employee relationships, better co-operation and communication on health and safety issues, a readiness on the part of employees to report hazards and shortcomings in their employer's protection arrangements and a greater readiness to participate in measures directed at reducing accidents and ill-health arising from work.

Personal error reporting

Personal error reporting by employees is an essential element of promoting and developing a blame-free culture in an organization. Whilst employees have a duty under the Management of Health and Safety at Work Regulations to report hazards and shortcomings in their employer's protection arrangements at work, such a system can be extended by the introduction of personal error reporting, including those which result in near misses, those incidents which do not result in injury damage or loss but which have the potential to result in injury or loss.

If personal error reporting is to be successful, there must be clear guidelines on the system for reporting, together with the provision of feedback to those concerned on the outcome of any investigation into the particular personal error. The need to report personal errors should be included in induction training and employees should be reminded of procedures at regular intervals.

System design and organization of work

The systems approach

Human failings can affect the safety performance of an organization, particularly where complex problems are involved. People are an important feature of the organizational system. It is necessary, therefore, to develop a systems approach to explain human behaviour.

Systems fall into two categories:

1. **Sociotechnical systems**. This system implies that an organization is a combination of technology (tasks, equipment and working arrangements) and of a social system (interpersonal relationships). The two systems are in constant interaction.
2. **Open systems**. In this system, the organization imports resources and information from the environment in which it operates and exports products and services to that environment.

According to Likert (1960), the total system is comprised of three levels:

1. society as a whole;
2. organizations of similar functions; and
3. subgroups within the larger system.

Therefore, society evolves a system incorporating feedback and its evolution affects its interaction with the organizations which develop within its environmental influence.

Types of organization

Self-regulating organizations

The human body is often considered to be a nearly perfect example of a system of self-regulating organization. The pattern of activity is self-fulfilling and balanced. Information is fed back and it becomes self-adjusting as a result of new information. The theory is that a self-regulated or self-controlled system with effective feedback mechanisms should be adaptive to change.

Therefore, an organization does not stay in the same condition for any length of time. Each organization undergoes a continual process of change, the rate of which is controlled by internal factors. The organization of a system means the development of that concept into a living, active, vital working unit. The self-regulating organization is one which is progressively changing in its evolution, recognizing that its objectives may have to be altered or the order of priority of objectives may change.

Natural organizations

Groups of people come together because they identify common needs, aims and objectives. They have an instinct to do so and they realize that in membership of a group of similarly motivated people there is an element of strength and purpose which cannot be achieved by the individual on his own.

This is where the organization's own continued 'life' depends upon the group itself continually reappraising its purpose and undergoing a process of self-regulation. Thus a group can come together for one reason but stay together for others. It is the *culture* and *mission statement* which bind the group together around shared values and goals.

This *social organization* or co-existence in groups or companies, which comes about at least partially to satisfy our gregarious instinct, is identifiable both in working groups and non-working groups.

Cybernetics

Cybernetics studies the flow of information around a system and the way in which information is used by a system as a means of controlling itself.

Management systems

To function correctly, a system must have controls, i.e. a set of rules, regulations or instructions by which it operates. Most systems known as management systems, such as production, planning, sales, budgetary control and health and safety, are readily recognized with cybernetics as subsystems, the system being the entire organization. The effectiveness of any subsystem can only be assessed when seen as part of the whole.

Successful subsystems do not necessarily add up to successful systems. The subsystem may not fit. They may not have the right elements. The measure of the extent of convergence of mutual or common objectives may not have been considered. Hence the advantage of bringing quality, efficiency and safety together into a total management system.

Feedback

Feedback is an essential part of any control process in an organizational system. Its existence marks the point of difference between *planning* and *control*. Feedback provides evidence or information for decisions which adjust the system all the time. Feedback is usually obtained by reporting variation from, for instance, objectives and also the means designed to achieve these objectives.

A thermostat provides a good example of feedback in a closed system. In open systems, people may be the instruments who measure, compare and take appropriate corrective action. Feedback can be observed as an error-correcting mechanism, a form of *negative feedback*. To give negative feedback, the controller makes a correction in the opposite direction from that in which the system is drifting. *Positive feedback* has the effect of increasing the drift and causes the system to go out of control.

Organization of work

It is essential that people at work are organized. This entails some form of organizational structure. The *formal structure* takes place as duties are assigned, procedures adopted, rules and regulations are circulated and processes are begun. The formal organization serves as a guide and tends to give an organization an identity of its own, independent of the people who join the workforce.

As it grows and more labour is added, management authority systems are devised. From a sociological viewpoint, social organization arises wherever people interact on a continuous basis in pursuit of common goals and the organization consists of the behaviour expectations which people have towards each other as group members.

Management

The evolution of management organization has accompanied the growth and expansion of industry and commerce. The expansion of technology and the profit motive led to large-scale production. Corporate business enterprises arise when individuals were unable to finance the expansion taking place and the stockholders were invited to group together and finance the developing enterprise.

Management hierarchies developed on the basis of individual responsibility and accountability to stockholders for running the business successfully and providing profit for the stockholders.

Workers

On the basis that 'Unity is strength', people in similar occupations and/or work locations established their own organizations, the objectives being clearly agreed and stipulated by the members. In the early days, groups of workers formed craftsmen's guilds which were subsequently followed by trade unions and workers' associations.

The reliability of organizational systems

Any organization is only as good as the reliability of its management systems, for example, financial control, forecasting, planning and organizational development. However, many organizations fail due to the inability of its leaders to consider human behaviour. Problems can arise in adopting a systems approach to human behaviour. In many cases there is conflict between the organizational objectives and those of individuals.

Weber and the bureaucratic model

Bureaucracy grows as enterprises grow. A bureaucracy is simply a hierarchical arrangement of unit organizations, an orderly arrangement of people and work based on divisions of function and authority. The hierarchy may be viewed as several layers of authority, which form the *pyramid of authority*.

Weber (1964) studied and analysed bureaucratic structures and functions and drew a close relationship between such structures and functions and the general theory of power and dominance. He established that the greater the degree of power and dominance exerted by management over the workers, the less reliable was the system. This view brought in the concept of *participative organizational control* or *participative management*. As long as the communication organization is appropriate and effective, it is felt that this may be the most effective system, since it is nearest to the cybernetic concept of self-regulation.

Performance monitoring and auditing

Monitoring performance

Monitoring implies the continuing assessment of the performance of the organization, of the tasks that people undertake and of operators on an individual basis against agreed standards and objectives. Performance monitoring is a common feature of most organizations in terms of financial performance, production performance and sales performance. It implies clear identification of the organization's objectives, perhaps through a written 'mission statement' and the setting of policies and objectives which are measurable and achievable by all parts of the organization, supported by adequate resources. Performance monitoring is generally accepted as a standard feature of good management practice.

The Management of Health and Safety at Work Regulations clearly emphasize the need to monitor performance. This may take place following the setting of health and safety policy, organizational development, risk assessment, establishment of the role of competent persons, or following the actual development of techniques of planning, measuring and reviewing performance.

At the organizational level, managers should be aware of their strengths and weaknesses in health and safety performance. These may be identified through various proactive forms of active safety monitoring, such as safety inspections, sampling exercises and audits, and through reactive monitoring systems, such as the analysis of accident and sickness absence returns. At task level, the implementation or otherwise of formally established safe systems of work, permit to work systems, in-company codes of practice and method statements are an important indicator of performance. Reactive monitoring through feedback from training exercises, in particular, those which are aimed at increasing people's perception of risk, improving attitudes to safe working and generally raising the level of knowledge of hazards, will indicate whether there has been an improvement in performance or not.

The concept of 'competent persons', outlined in the Management of Health and Safety at Work Regulations, raises a number of important issues. The general concept of 'competence' is based on skill, knowledge and experience, linked with the ability to discover defects and determine the consequences of such defects (Brazier v. Skipton Rock Company Ltd. [1962] 1 AER 955). In the designation of competent persons, however, not only do organizations need to consider the above factors, but also the system for monitoring and measuring their performance against agreed objectives.

Performance monitoring should not be used solely for designated competent persons, however. Most organizations practising Management by Objectives or operating performance-related pay systems have a standard form of job and career review or appraisal for staff, but how many of these systems take into account the health and safety performance of the various levels of management? Clear evidence of performance monitoring of both competent persons and managers generally will be required in future to ensure compliance with the regulations.

On-the-job performance monitoring should take into account the human decision-making components of a job, in particular the potential for human error. Is there a need for job safety analysis leading to the formulation of job safety instructions, a review of current job design and/or an examination of the environmental factors surrounding the job?

And what about the operator? The Management of Health and Safety at Work Regulations place a duty on employers to take into account the capabilities of people when entrusting employees with tasks. The whole concept of individual capability for the safe operation of a task is extremely broad. Here, there is a need to consider not only physical capability to, for instance, load and unload products in and out of vehicles, but also mental capability in terms of the degree of understanding necessary for certain tasks. How do we assess individual mental capability? The Approved Code of Practice makes no reference whatsoever to this aspect, simply referring to various aspects of training as a way of achieving competence in the operation of safe working practices. Admittedly, there are various ways of assessing the capabilities of people through techniques like human reliability assessment, but how many safety practitioners or managers are trained in this technique? There is much work to be done in this respect if human performance is to be accurately monitored.

Auditing performance

An audit is a procedure which subjects each area of an organization's activities to a critical examination with the principal objective of minimizing loss. A safety audit could be defined as 'the systematic measurement and validation of an organization's management of its health and safety programme against a series of specific and attainable standards' (RoSPA).

Safety audits, as a form of safety monitoring, should identify strengths and weaknesses in health and safety performance. Every component of the total system is included, in particular management policy and systems, group and individual attitudes, training arrangements, features of processes, emergency procedures, documentation, etc. What is important about any auditing system is that the results should be measurable against agreed standards. Many safety audits, however, simply fail to achieve this objective, being, in many cases, a series of questions which require a simple 'Yes' or 'No' response.

On this basis, many organizations tend now to use a form of health and safety review which identifies key components of performance ranked according to significance. Key components may be decided on an annual basis taking into account such factors as past accident experience, impending legislation, the identified risks and potential areas of loss. In many respects, they can be compared to a safety sampling exercise but on a much broader scale and directed at identifying strengths and weaknesses in health and safety management systems.

Health and safety reviews are undertaken by the organization's health and safety practitioners on a 6-monthly or annual basis. Fundamentally, they are directed at identifying improvement or deterioration in health and safety performance against nationally agreed standards. In many cases, they may be linked to the organization's reward structure at all levels of management, and be incorporated as a feature of Total Quality Management. As such they have much to offer as a means motivating management to ensure effective levels of health and safety management are maintained.

Monitoring and auditing performance at all levels within an organization feature strongly in the new approach to occupational health and safety brought about by the impending Management of Health and Safety at Work Regulations. It is conceivable that safety management reviews will replace the former physical inspections of workplaces carried out by the enforcement authorities. A safety management review is, fundamentally, an evaluation of the organization's management system, and what has been done to reduce the risks in its business. The systems for monitoring and auditing performance are an important feature of such a review.

Job satisfaction and appraisal schemes

Job satisfaction could be described as the process of achieving the right balance between the needs of a particular job, in terms of the demands of the job, and the personal needs of the job holder. Maslow's hierarchy of needs (1954) identified a number of stages in the process of motivation – safety and security needs, social needs and ego needs. All these needs are very important elements of job satisfaction. *Safety and security needs* are inborn, *social needs*, in terms of being accepted as one of the work group contribute greatly to job satisfaction and the *ego needs*, the need to be noticed, to receive acclaim for a good job done, are particularly significant for the more ambitious members of the organization. So how do organizations link the concept of job satisfaction with appraisal schemes?

Appraisal schemes are, fundamentally, concerned with two aspects:

1. An appraisor, such as a person's immediate manager, agreeing with the appraisee future measurable performance objectives which, in the eyes of both the appraisor and appraisee, are achievable by the appraisee.
2. Subsequent discussion with the appraisee, and assessment by the appraisor, as to whether the appraisee's performance objectives have actually been met or achieved.

Appraisal schemes can be extremely effective in increasing the motivation of all levels within an organization, thereby bringing about improvements in performance. However, it is not sufficient to consider this as a once a year exercise. Arrangements should operate whereby progress by individuals in the achievement of agreed performance objectives is discussed on a regular basis, with assistance being provided where appropriate.

Appraisal schemes, however, are only as good as the managers running them in terms of

- consistency of approach by all managers;
- clear-cut guidelines for both managers and employees on the purpose of the appraisal scheme; and, in particular,
- the need for employees to appreciate the objectives and purpose of the scheme.

Efforts and reward

One of the problems experienced by many organizations is that of linking the outcome of an individual's appraisal with the rewards being offered at the time which may not be great, particularly if the organization is not performing well financially.

Whilst many employers would agree that this is not the purpose of an appraisal scheme, employees commonly take the view that there is no point in their making extra effort to achieve set objectives if there is nothing in it for them at the end of the exercise. This can result in reduced job satisfaction or even dissatisfaction. On this basis, how an appraisal scheme is sold or promoted to employees is important. They must appreciate the purpose of the scheme and its principal objectives.

Health- and safety-related performance objectives

There is clearly a case for every manager, no matter at what level within an organization, and every employee, agreeing one or more health- and safety-related performance objectives on a year-to-year basis. As with other performance objectives, these objectives must be measurable and achievable by the parties concerned. Under no circumstances should objectives be related to accident and injury rates, however, as this can result in under-reporting of such incidents in order to meet the objective.

Typical performance objectives include

- to ensure all employees attend a health and safety awareness training programme within a specified period of time;
- to undertake risk assessments for specifically named substances hazardous to health or particular work activities, producing the completed risk assessment documentation;
- to produce a record of all portable electrical appliances on site, together with their last date of testing;
- to establish a health and safety committee, with a formally written constitution, and hold the first meeting within 6 months;
- to undertake a stress survey of all employees, including the preparation of a stress questionnaire and analysis of the outcome of questionnaires completed and returned; and
- to ensure health and safety issues are discussed fully at board and senior management meetings.

Job satisfaction and management style

Management style has a direct influence or effect on job satisfaction. Rensis Likert, an American social psychologist, researched management styles amongst a group of managers and supervisors with a view to establishing the most productive style in terms of employee performance. As a result of this research, he put forward two generalizations:

1. Managers and supervisors who achieve the highest productivity, the smallest labour turnover, lowest absence rate and lowest operating costs use a different pattern of leadership from that used by less productive managers.
2. Higher-producing managers and supervisors use individualistic styles of leadership.

High-producing groups

Likert (1960) identified and measured the difference in the pattern of operation of high- and low-producing managers (Table 7.1). He subsequently identified the general characteristics of high-producing groups thus:

Table 7.1 Classification of management styles (Likert, 1960)

The explosive authoritarian type

- Management use fear and threats
- All communication is downward
- Superior and subordinates are psychologically far apart
- The bulk of decisions are taken at the top of the organization

The benevolent authoritative type

- Management uses rewards
- Attitudes are subservient to superiors
- Information flowing upward is restricted to what the boss wants to hear
- Policy decisions are taken at the top, but decisions within a prescribed framework may be delegated to lower levels

The consultative type

- Management uses rewards
- Occasional punishments and some involvement is sought
- Communication is both up and down
- That which the boss does not want to hear is given in limited amounts and only cautiously
- Subordinates can have a moderate amount of influence on the activities of their own departments
- Broad policy decisions are taken at the top and more specific decisions at lower levels

The participative group type

- Management give economic rewards
- Full use is made of group participation
- Groups are involved in setting high performance goals
- Groups participate in improving work methods
- Communication flows downwards, upwards and with peers and is accurate
- Superiors and subordinates are very close

(continued)

Table 7.1 Continued

• Decision-making is done widely through the organization through group processes
• It is integrated into the formal structure by regarding the organization chart as a series of overlapping groups with each group linked to the rest of the organization by means of persons who are members of more than one group
• It demands greater involvement of individuals.

- High-producing groups had more favourable attitudes towards other members of the organization, towards the work they undertook and towards the organization in general.
- High-producing leaders motivated those working for them by appealing to a range of motives, that is, they did not concentrate on, for example, financial motives on their own but used factors such as *ego*, economic motives, curiosity and security motives, thereby achieving maximum motivation towards achieving the organization's goals and, at the same time, contributing to the personal goals of those involved.
- High-producing organizations tend to be tightly-knit systems which consist of interlocking groups. Each of these groups commands a high degree of loyalty amongst its members, with favourable attitudes between managers and employees. The is effective and efficient communication and each decision is based on the necessary information.
- Whilst organizational performance is measured, these measures are not used for strict control, but for self-guidance. There is mutual goal-setting and responsibility for achievement is shared between superiors and subordinates.
- High-producing managers use technical methods, such as work study, but use them to reinforce the human element of management instead of forcing management into a 'straight jacket' of techniques.
- High-producing managers recognize that people react favourably to experiences in which they feel they are supportive and are able to demonstrate a sense of personal worth. They further realize that people react unfavourably to situations which threaten their sense of dignity and worth. High-producing managers also realize that what is significant to people is not the situation as it is objectively, but rather the situation as they see it individually.
- Managers who are high producers build their organizations on the principle of supportive relationships. This means that all participants of interaction in an organization must see the interaction as supportive and must view the process as one which maintains and boosts each person's sense of personal worth.
- High-producing managers also recognize that, unless individuals are members of effective groups, these individuals will not realize their full potential. The organization must, therefore, be built up in terms of groups and, to obtain

maximum benefits, the group should be interlocking, that is, a superior in one group will be a subordinate in another group.

Job satisfaction and self-actualization

This concept, developed by Chris Argyris (1962), is based on Maslow's theory of self-actualization. Many organizations encourage an employee to seek job satisfaction on the basis that if he is not achieving job satisfaction, or at least aiming at it, he is not realizing his full potential. This represents a loss both to the organization and the individual. Many management development schemes operate on this principle.

Argyris defines an individual who aims at job satisfaction as 'a mature person who aims at self-actualization'. On this basis, a mature person is one who is aiming at mature ends and who, although striving towards growth, behaves in such a way as to allow others to achieve similar ends.

An essential feature of job satisfaction, and of self-actualization, is the interaction that an individual has with other people. Self-actualization is reached with complete maturity. However, a distinction must be drawn with happiness, as perceived, and self-actualization. Much depends on the individual's motivation and, to some extent, competitive spirit.

Job safety standards

What is a job safety standard?

This can be defined as the prescribed minimum measures necessary to ensure operator safety during a particular job. Job safety standards should be established for all high risk activities in particular.

A job safety standard should take into account:

- general and specific legal requirements relating to the job;
- specific accident prevention measures, such as the isolation of sources of energy;
- written safe systems of work, including the use of permit-to-work systems;
- environmental factors, such as lighting and ventilation;
- measures to protect operators from health risks arising from the job, such as exposure to welding fume, noise and hazardous substances;
- specific health surveillance requirements;
- the correct use of work equipment;
- fire prevention and protection measures;

- electrical safety procedures;
- safe mechanical handling operations;
- the provision and use of personal protective equipment;
- the information, instruction, training and supervision requirements for operators; and
- any emergency procedure necessary in the event of, for example, fire, explosion, premature collapse of a building during demolition and chemical spillage.

Job safety standards are commonly devised through the technique of job safety analysis.

Job safety analysis

Job safety analysis evolved from job or task analysis, and is defined as

> the identification of all the accident prevention measures appropriate to a particular job or area of work activity and the behavioural factors which most significantly influence whether or not these measures are taken.

Job/Task analysis

The identification and specification of the skill and knowledge contents of jobs.

Job/Task training analysis

The identification and specification of what has to be taught and learned for individual jobs.

Job safety analysis

The approach is both diagnostic and descriptive. It may be job-based or activity-based.

Job-based - Machinery operators; fork lift truck drivers

Activity-based - Manual handling operations; roof work; cleaning activities.

Job safety analysis evolved from:

- task analysis; and
- method study and work measurement.

The latter incorporates the SREDIM principle, namely:

Select, the work to be studied

Record, the method of doing the work

Examine, the total operating system

Develop, the optimum methods for doing the work

Install, the method into the company's operations

Maintain, the defined and measured method

Applying the SREDIM principle to job safety analysis, the procedure is as follows:-

1. Select the job to be analysed.
2. Break the job down into component parts in an orderly and chronological sequence of job steps.
3. Critically observe and examine each component part of the job to determine the risk of accident.
4. Develop control measures to eliminate or reduce the risk of accident.
5. Formulate written safe systems of work and job safety instructions.
6. Review safe systems of work and job safe practices at regular intervals to ensure utilization.

Job safety instructions

These are used as a means of communicating the safe system of work to the operator. They can be produced as:

- a general manual covering all jobs;
- specific cards attached to machines or displayed in the work area.

Criteria for selection of jobs for analysis

- Past accident and loss experience
- Maximum potential loss
- Probability of accident recurrences
- Legal requirements
- Relative newness of the job
- Number of employees at risk

A job safety analysis record is maintained for each job analysed.

Stages of job safety analysis

Job safety analysis usually takes place in two specific stages, each of which has different information requirements.

Initial job safety analysis

The following factors are considered:

Job title

Job operations

Materials used

Hazards

Work organization

Department/section

Machinery/equipment used

Protection needed

Degree of risk

Specific tasks

Total job safety analysis

The following factors are considered:

Operations

Skills required

Learning methods

Hazards

External influences on behaviour

Job safety analysis records

All jobs that have been subject to job safety analysis should be recorded. A typical record is shown in Figure 7.3.

The benefits of job safety analysis

Job safety analysis, as a technique, is very closely allied to work study in terms of eliminating unnecessary operations and by simplifying, or breaking down into elements the more difficult and complicated work activities.

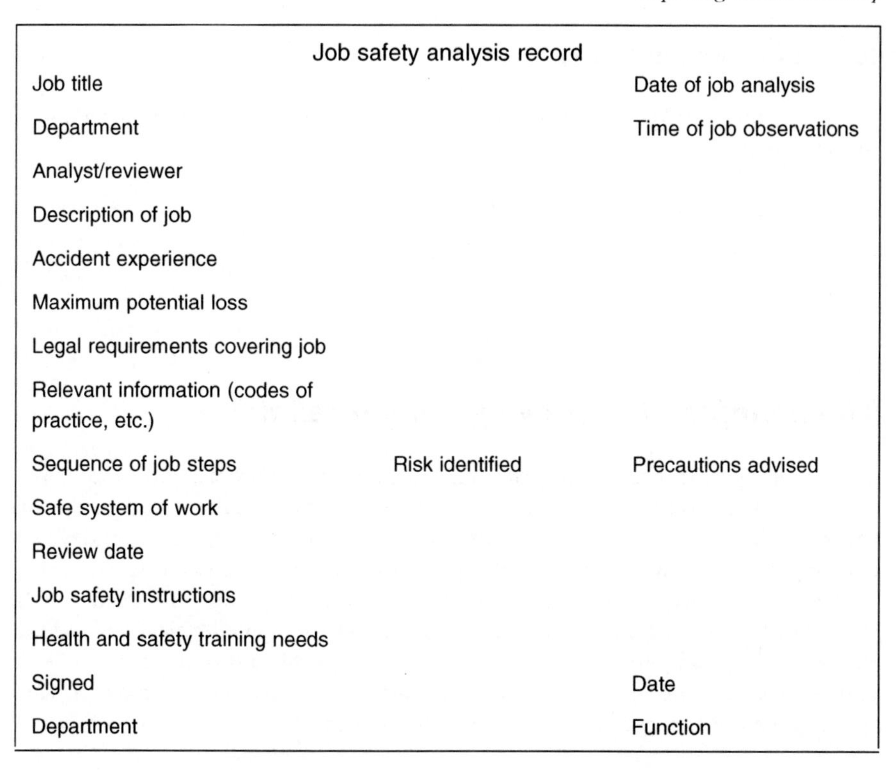

Job safety analysis record

Job title — Date of job analysis

Department — Time of job observations

Analyst/reviewer

Description of job

Accident experience

Maximum potential loss

Legal requirements covering job

Relevant information (codes of practice, etc.)

Sequence of job steps	Risk identified	Precautions advised

Safe system of work

Review date

Job safety instructions

Health and safety training needs

Signed — Date

Department — Function

Figure 7.3 Job safety analysis record

It also incorporates elements of hazard analysis and risk assessment. If, in a process such as sorting parcels, work study can reduce the number of operations in which parcels are handled, for instance by the introduction of mechanised systems, then the potential for manual handling injuries should be substantially reduced. The remaining significant risks can then be assessed and preventive measures installed.

Other benefits of a job safety analysis approach in the design of safe systems of work include:

- identification of the health and safety training needs of operators and the training procedures necessary to meet these needs;
- identification of the specific stages of a task, the existing hazards and the precautions necessary at each stage;
- identification of the correct work equipment; and
- identification of the necessary physical and mental capabilities of people undertaking the job, together with any specific skills.

Job safety instructions

These are the formal instructions arising from job safety analysis which convey to the operator:

- the step-by-step safe way of undertaking the task;
- the hazards that may arise at each step; and
- the precautions to be taken.

The benefits of improving human reliability

People, along with their individual skills, loyalty, commitment and motivation, are an organization's most valuable asset. Organizations survive and thrive on the reliability of their employees and there is a continuing need to develop and improve this aspect through the provision of information, instruction and training, consultation, the setting of standards and levels of performance in a range of tasks, the measurement of performance against agreed objectives and the rewarding of good performance. Ensuring people know and understand the best and safest way of undertaking a task adds greatly to human reliability, removing uncertainty and the potential for unsafe working practices.

The benefits of improving human reliability can be measured in terms of improvements in

- job performance through improved skills and knowledge;
- morale, commitment and loyalty to the organization;
- safety performance; and
- the safety culture within the organization.

Conclusion

This chapter has examined the aspect of human reliability and a number of methods and techniques for it. There have been a number of studies on this subject in the last century including those of F.W. Taylor and the Hawthorne experiments. More recently attention has been given to techniques such as human reliability modelling, studies of interpersonal relationships, benchmarking, the systems approach and the use of job satisfaction and appraisal schemes.

Organizations have much to gain in improving human reliability in terms of improvements in job performance and safety performance, enhanced skills and knowledge, together with better morale, commitment and loyalty to the organization on the part of employees.

Key points

- Human reliability is defined as the probability that a person will perform correctly some system required accurately during a given period of time without performing any extraneous activity that could downgrade the system.
- The analysis of human reliability involves techniques such as fault tree analysis, event tree analysis and consequence analysis.
- People have particular preferences when it comes to processing information.
- Planned motivation schemes are an industrial catalyst described as a tool to maximize performance.
- Organizations should strive towards achieving a blame-free culture.
- To function correctly, a system must have controls, that is a set of rules, regulations and instructions by which it operates.
- Feedback is an essential part of any control process in an organizational system.
- Job satisfaction can be described as the process of achieving the right balance between the needs of a particular job in terms of the demands of the job and the personal needs of the job holder.
- Managers should be required to set health- and safety-related performance objectives which are measurable and achievable on a year-to-year basis.
- Job safety standards should be established for all high risk activities in particular.
- The benefits of improving human reliability can be seen from improved job performance, safety performance and in the development of an appropriate safety culture.

8 Ergonomic principles

The scientific study of man and his work, and the concept of the 'man–machine interface' can be traced back to the work of an American engineer, Frederick W. Taylor who, in 1881, developed 'time study' as the basis for improving the efficiency of work. This was followed by the work of Frank B. Gilbreth, also an engineer in the United States who, around the turn of the twentieth century, began to use methods which led to 'motion study'. Gilbreth, assisted by his wife, Lilian, who was a trained psychologist, introduced the use of photography and particularly cinephotography, to analyse the way in which skilled work was carried out. This pioneer work of Taylor and the Gilbreths was documented by R.M. Barnes in 1958 and led to the concepts of 'time and motion study' and 'work study'. These studies have made significant contributions to the scientific study of work.

Ergonomics, as a discipline, came to prominence during the Second World War. The relationship of men to their physical, organizational and social environment became an urgent concern of applied sciences during this period when the achievement of high productivity among civilian workers and of high efficiency amongst fighting men, was at a premium. Increases in the numbers of studies of working man matched the accelerated growth of research in other fields of scientific enquiry related to the successful pursuit of war.

What is ergonomics?

Many personal factors, such as physical and mental stress, can affect human capability and limit human performance. In some cases, these factors may create hazards in the workplace. People and their work operations can impose limitations which need to be addressed. This is where the study of ergonomics and the application of ergonomic principles in the design of work interfaces are most important.

Ergonomics is variously defined as

- the scientific study of work;
- the scientific study of the relationship between man and his environment;
- the study of man in his working environment;
- fitting the man to the job or fitting the job to the worker;

- human engineering; and
- the study of the man–machine interface.

More detailed definitions include: (1) That branch of science and technology that includes what is known and theorized about human behaviour and biological characteristics that can be validly applied to the specification, design, evaluation, operation and maintenance of systems to enhance safe, effective and satisfying use by individuals, groups and organizations; and (2) The study of human abilities and characteristics which affect the design of equipment, systems and jobs . . . and its aims are to improve safety and . . . well-being.

The scope of ergonomics

Ergonomics is a multidisciplinary science which uses basic knowledge from the human, engineering, economic and social sciences. It incorporates elements of occupational medicine and health, occupational psychology, law and sociology. All these disciplines are associated with people at work and, therefore, represent an aspect of ergonomic study.

Ergonomics, as applied to the man/machine interface, has been further defined as 'the physiological, anatomical and psychological aspect of man in his working environment'. Ergonomics further embraces a number of disciplines including physiology, anatomy, psychology, engineering and environmental science. It examines, in particular, the physical and mental capacities and limitations of workers taking into account, at the same time, psychological factors, such as learning, individual skills, perception, attitudes, vigilance, information processing and memory together with physical factors, such as strength, stamina and body dimensions.

Fundamentally, ergonomics is concerned with maximizing human performance and, at the same time, eliminating, as far as possible, the potential for human error. There is an essential connection between ergonomics and human engineering.

Human engineering

Luczak *et al.* (1987) in a study in German-speaking areas defined 'human engineering' as the systematics of analysis, order and design of technical, organizational and social conditions of work processes aimed at offering human beings productive and efficient work processes with harmless, manageable and undisturbed working conditions, as well as standards regarding the content of work, work analysis, working environment and remuneration and co-operation so that they felt motivated to expand their scope of activity, acquire competence and develop and preserve their personality in co-operation with others.

Areas of ergonomic study

Broadly, ergonomics can be split into two main areas:

1. **Physical ergonomics**. This area is concerned with aspects such as working posture, materials handling, repetitive movements, work-related musculoskeletal disorders, workplace layout, safety and health. Typical examples of physical ergonomics include:
   - the design of tasks to eliminate lifting from the floor, such as the provision of a scissor lift or elevated angle conveyor, thereby enabling the work to be handled at waist height;
   - design of machinery which enables tools to be operated at a normal standing position, sooner than the bent stance required, for example, in the operation of a horizontal lathe.
2. **Cognitive ergonomics**. Cognitive ergonomics focusses on the fit between human cognitive abilities and limitations with respect to machinery and equipment used, such as with process and control panels, lift truck operation and word processors, principally directed at preventing human error. This is particularly important in the design of complex, highly technical or automated systems, such as that for the cockpit of an aeroplane or a chemical processing plant, in preventing major catastrophes. Cognitive ergonomics involves the study of topics such as mental workload, decision-making, skilled performance, human–computer interaction, human reliability, work stress and work training.

The man–machine interface

Clearly, ergonomics is a branch of science which involves the consideration of people at work in terms of human characteristics, the tasks they undertake, the design of the controls and displays to machinery they operate (the man–machine interface), the environments within which they work and how these various factors affect the total working system. A more appropriate definition is 'human factors engineering' or 'fitting the task to the individual'.

Ergonomics was developed through the need for people to operate and be in control of highly sophisticated machinery and equipment and, particularly, in the early stages of the Second World War, the need for quite technically inexperienced people to handle this equipment after a very short training period. When these groups of labour were used, both in the armed forces and in industry, the greater complexity of the machines inevitably raised the rate of operator errors and incorrect judgements, commonly resulting in personal injury and damage to plant, materials and equipment.

Even these days, it is common for management to write off damage to plant and equipment as down to the 'human factor', implying that little or no action can be taken to prevent such losses. This was particularly the case during the last war until a group of medical researchers and engineers began the systematic study of people's physical and psychological limitations. This was only the beginning, but ergonomics is now a specialized branch of study in its own right.

The team approach to ergonomic studies

Ergonomic studies have, in the past, considered a range of work situations such as the layout and design of crane cabs and London Transport bus drivers' cabs, together with display screen equipment workstations. As such, they require a team approach utilizing the skills of several specialists.

Physiologists

The task of the physiologist is to ensure that the work is within the physical capacity of the operator. A number of factors need consideration here, in particular, the strength, stamina and effort required to carry out the task together with the speed and accuracy with which bodily movements must be made.

It may be necessary to measure the amount of oxygen expended in performing the work, the amount of oxygen consumed and the rate at which blood is being circulated by the heart. These factors are affected not only by the nature of the work itself abut also by the heat created by the operator, the ambient temperature and the clothing worn by the operator.

The physiologist's contribution to a study might include recommendations with respect to temperature and ventilation requirements, the best method of doing the work with the least stress on the operator, the need for rest periods and the provision of suitable work clothing.

Anatomists

Physical and anatomical characteristics, such as the strength and physique of the operator, height, arm span, shoulder height and elbow height, may need to be studied in the case of certain tasks, particularly those of a repetitive nature, such as the operation of machinery or certain manual handling tasks.

The anatomist would wish to ascertain that the work equipment is suitable for operation with ease by the average worker, eliminating the possibility of physical stress caused by excessive bending, stretching or having to reach over other items. Similarly, in the case of work carried out while sitting down, as in the case of a supermarket checkout operator, chairs should be designed to suit the physical dimensions of the operator, with seat height and back rest adjustments, and the control for operating the moving belt should require minimum effort. A well-trained and experienced operator in this situation will very rarely lift any of the goods on the belt, using the belt to convey the goods as close as possible to the price scanner.

Psychologists

One of the objectives of ergonomic study is to design a work activity in such a way that the work is relatively easy to carry out, retains the interest of the operator and in environmentally friendly surroundings.

Psychologists consider the mental make up of operators and their ability to take in information, react to the information received and, where necessary, take avoiding action. Various signs and signals may also be associated with the work under consideration. Signs may include fixed signs which provide information on hazards and the precautions necessary, whereas signals can include visual signals, such as coloured and flashing lights, pointers on scales and dials, together with auditory signals, such as buzzers and bells. The operator should be able to respond to all this information in the quickest possible time, without fatigue, with ease and certainty.

Engineers

Engineers have the task of designing work equipment to meet the recommendations of the other members of the team. The ultimate objective is to design equipment which requires the least physical and mental effort on the part of the operator. In some cases, there may have to be some form of compromise between what is desirable and what is actually practicable.

The employee and work

People at work undertake a range of tasks which require varying degrees of mental and physical capability. In many cases, no consideration is given to 'fitting the man to the job', resulting in people having to, for instance, adopt awkward postures while undertaking tasks and, in many cases, causing fatigue. Others may be involved in manual handling activities which may be beyond their physical capacity in terms of strength, stamina and fitness.

In other cases, the layout of controls and displays to machinery may not necessarily provide the most efficient way of operating a machine in terms of, for example, the actual repetitive hand movements required, the way the machine passes information to the operator and in ensuring safe operation of that machine. Moreover, as machinery becomes more sophisticated in terms of the control elements, in particular, many people may not have the mental capability to understand, for example, these computer-driven controls.

Fitting the man to the job

'Fitting the man to the job' requires careful consideration on the part of management in terms of assessing the physical and mental capabilities of people, with particular reference to ergonomic principles. A number of factors need consideration here.

Task characteristics

All tasks, no matter how simple they appear, incorporate a number of characteristics. They may include the frequency of operation of, for example, a foot pedal, the repetitiveness of the task, the actual workload, how critical it is that the task is accurately completed according to the prescribed procedure, the duration of the task and its interaction with other tasks as part of a total manufacturing process.

Task demands

The specific characteristics of a task will make a number of physical and mental demands on the operator. Certain tasks, such as loading and unloading delivery vehicles, may require a high degree of physical strength and stamina. Other tasks, such as examination and inspection of lifting equipment and electrical equipment, may need varying degrees of vigilance and attention in ensuring that the equipment meets safety standards laid out in legislation. Most tasks of this nature require some form of memory utilization.

Instructions and procedures

The quality of both verbal and written instructions to operators has a direct relationship with the potential for human error. As such, instructions and written procedures should be clear, unambiguous, sufficient in detail, readily understood, easy to use, accurate and produced in an acceptable format.

Instructions and procedures should be subject to regular review and revision, particularly where operators are experiencing difficulties with their interpretation and use.

Potential stressors

Many tasks, due to a failure to consider the physical and mental needs of the operator in terms of his level of intelligence, attitude and motivation, can be stressful. This applies particularly in the case of repetitive and boring tasks and those which require a high degree of concentration.

With the increasing attention being given to the potential for stress at work, a number of questions need to be asked during the design of tasks and at the task analysis stage (Table 8.1).

The sociotechnical factors

Sociotechnical factors are concerned with people working in groups and cover a wide range of considerations. They include the social relationships between

Table 8.1 Task stressors

1. *Does the task isolate the operator, both visibly and audibly, from fellow operators?*
   Isolation from the working group can be a significant cause of stress.
2. *Does the task put operators under pressure due to the need to complete it within a certain time scale?*
   Such situations create stress and increase the potential for human error.
3. *Does the task impose a higher level of mental or physical workload on people than normal?*
   Current sickness absence levels may be a direct indicator of an overload situation.
4. *Is the task of a highly repetitive nature?*
   Monotony and boredom are standard features of many production processes. While some operators are quite happy to undertake repetitive tasks, others will find these tasks stressful, resulting in fatigue.
5. *Do the various tasks create conflict among operators?*
   Conflict can arise as a result of varying workloads and levels of complexity for different tasks. Similarly, rates of pay may vary according to the significance of individual tasks, resulting in conflict over the allocation of the more highly paid tasks.
6. *Do some tasks result in physical pain or discomfort, such as manual handling operations or working in low temperatures?*
   These factors should be considered as part of the risk assessment process, in job safety analysis and in the design of safe systems of work.
7. *Is there a risk of distraction during certain critical tasks?*
   Distraction is one of the principal causes of human error, particularly in various forms of assembly work, the results of which could be significant. Distraction will also increase stress on the operator due to the need to regain concentration whilst undertaking these highly critical tasks. Critical tasks should be planned with consideration being given to the potential for distraction. In certain cases, design of the working area should be such as to prevent operators from being distracted.
8. *Is there sufficient space available to undertake the work safely?*
   Congested, badly planned work area layouts create stress and increase the potential for human error and accidents.
9. *Where a shift work system is in operation, does this system take account of the physical and mental limitations of operators?*
   Working long hours on a rotating shift system can have a direct effect on individual behaviours. While many people can cope quite satisfactorily with shift work, others find it stressful for a range of reasons.
10. *Where an incentive scheme is in operation, are the incentives offered seen as fair to all concerned?*
    Incentive schemes frequently create conflict in that certain groups of workers are perceived to benefit more than others.

operators and how they work together as a group, group working practices, manning levels, working hours and the taking of meal and rest breaks, the formal and informal communications systems and the benefits and rewards available. Organizational features are also incorporated under this heading, such as the actual structure of the organization and individual work groups, the allocation of responsibilities, the identification of authority for certain actions and the interfaces between different groups.

Environmental factors

The duty on the part of an employer under the Health and Safety at Work etc. Act to provide and maintain a working environment that is, so far as is reasonably practicable, safe, without risks to health, and adequate as regards facilities and arrangements for welfare of employees at work, is well recognized. The Workplace (Health, Safety and Welfare) Regulations, together with the HSC Approved Code of Practice (ACOP), further reinforce these general duties under the Act.

The design of the working environment has a direct effect on people's performance. Poorly designed environments contribute to accidents and, in some cases, occupational ill-health manifested in diseases and conditions, such as noise-induced hearing loss, work-related upper limb disorders and thermoregulatory disorders, such as heat stroke.

Particular attention should be paid to a range of environmental stressors, such as

- extremes of temperature;
- inadequate illuminance levels;
- inadequate comfort ventilation and extract ventilation to remove airborne contaminants;
- high levels of humidity;
- excessive noise levels.

For further information, see Chapter 9, Ergonomics and human reliability.

Principal areas of ergonomic study

The study of ergonomics covers the following areas.

The human system

People are different in terms of physical and mental capacity. This is particularly apparent when considering the physical elements of body dimensions, strength and stamina, coupled with the psychological elements of learning, perception, personality, attitude, motivation and reactions to given stimuli.

Other factors which have a direct effect on performance include the level of knowledge and the degree of training received, their own personal skills and experience of the work.

Environmental factors

The consideration of the working environment in terms of layout of the working area and the amount of individual work space available, together with the need to eliminate or control environmental stressors is an important feature of ergonomic consideration. Environmental stressors, which have a direct effect on health of people at work and their subsequent performance, include noise, vibration, extremes of temperature and humidity, poor levels of lighting and ventilation.

The man–machine interface

Machines are designed to provide information to operators through various forms of display, such as a temperature gauge. Similarly, the operator must ensure the correct and safe operation of the machine through a system of controls, such as push-button 'stop' and 'start' controls, foot pedals or manual devices. The study of displays, controls and other design features of vehicles, machinery, automation and communication systems, with a view to reducing operator error and stress on the operator, is a significant feature of design ergonomics.

Factors such as the location, reliability, ease of operation and distinction of controls and the identification, ease of reading, sufficiency, meaning and compatibility of displays are all significant in ensuring correct operation of the various forms of work equipment.

The total working system

Considering the above factors, the approach to ergonomics can be summarized under the heading of the Total Working System, which is broken down into four major elements as shown in Table 8.2.

Design ergonomics

The design of equipment and work systems has a direct effect on performance. In the design of work layouts, systems and equipment the following matters should be considered.

Layout

Layout of working areas and operating positions should allow for free movement, safe access to and egress from the working area, and unhindered visual and oral

Table 8.2 The total working system

Human characteristics	*Environmental factors*
Body dimensions	Temperature
Strength	Humidity
Physical and mental limitations	Lighting
Stamina	Ventilation
Learning	Noise
Perception	Vibration
Reaction	Airborne contamination
Man–machine interface	*Total working system*
Displays	Fatigue
Controls	Work rate
Communications	Posture
Automation	Productivity
	Accidents and ill-health
	Health and safety protection

communication. Congested, badly planned layouts result in operator fatigue and increase the potential for accidents.

Vision

The operator should be able to set controls and read dials and displays with ease. This reduces fatigue and the potential for accidents arising from faulty or incorrect perception.

Posture

The more abnormal the operating posture, the greater is the potential for fatigue and long-term injury. Work processes and systems should be designed to permit a comfortable posture which reduces excessive job movements. This requirement should be considered in the siting of controls, e.g. levers, gear sticks, and in the organization of working systems, such as assembly and inspection tasks.

Comfort

The comfort of the operator, whether driving a vehicle or operating machinery, is essential for his physical and mental well-being. Environmental factors, such as temperature, lighting and humidity, directly affect comfort and should be given priority.

Principles of interface design

Design ergonomics examines the interface between the operator and the equipment used with particular reference to the design and location of controls and displays. Controls take the form of hand or foot controls, such as levers and pedals, respectively. Displays provide information to the operator as seen with the dashboards of vehicles of all types and in various forms of control panel.

In the design of work equipment, such as fixed and mobile machinery, a number of factors need consideration.

Separation

Physical controls should be separated from visual displays. The safest routine is achieved where there is no relationship between them.

Comfort

Where separation cannot be achieved, control and display elements should be mixed to produce a system which can be operated with ease.

Order of use

Controls should be located in the order in which they are used, e.g. left to right for starting up a machine or process and the reverse direction for closing down or stopping same.

Priority

Where there is no competition for space, the controls most frequently used should be sited in key positions. Important controls, such as emergency stop buttons, should be sited in a position which is most easily seen and reached.

Function

With large operating consoles, the controls can be divided according to the various functions. Such division of controls is common in power stations and highly automated manufacturing processes. This layout relies heavily on the skills of the operator and, in particular, his speed of reaction. A well-trained operator, however, benefits from such functional division and the potential for human error is greatly reduced.

Fatigue

The convenient siting of controls is paramount. In designing the layout for an operating position, the hand movements of body positions of the operator can be studied (Cyclogram Torque technique) with a view to reducing or minimizing excessive movements.

Signals

Signals on work equipment displays, such as the instrument panels to process plant and machinery, dashboards in vehicles, aircraft and earth moving equipment, have a significant influence on the making of decisions by operators. A signal, such as a red light flashing on a control panel, should direct the operator to taking appropriate action as quickly as possible. The time between the recognition of the signal and the taking of action is crucial in many cases, particularly where the control panel is isolated from the process plant in a central control booth or area with operators receiving information from localized screens.

This raises the question of individual perception and alertness, or vigilance, on the part of operators. The type of information presented from a stimulus, such as a buzzer on a control panel, depends upon a number of factors:

- the actual nature of the stimulus, e.g. flashing light, bleeping sound;
- their previous experience of this stimulus;
- current disposition to take action; and
- any prejudices, attitudes and goals with respect to the work itself.

The fact is that people respond differently to signals as seen with driving situations in particular. Much of this response is based on previous experience and, in many cases, drivers can miss or fail to respond to signals. They may misunderstand the purpose of the signal, consider it irrelevant or of no concern.

It is necessary, therefore, to increase operator vigilance to the signals. What are the best ways of presenting cautionary or warning signals to people, such as airline pilots, train drivers, machine operators and display screen equipment users? There is evidence to indicate that when multiple cautionary-warning signals are presented peripherally:

- the use of a master signal, one which draws people's attention immediately, reduces response time and the number of signals missed; (the master signal has the effect of stopping people in their tracks telling them to expect a message at any moment)
- auditory master signals are superior to visual signals;
- the use of a combined visual and auditory master signal produces the fewest missed signals;

- a two-tone auditory master signal is superior to a monotone; and
- for illuminated dark legend signals, a dark legend on an illuminated background is superior to an illuminated legend on a dark background.

Interface design audit

In the design and ergonomic assessment of interfaces, it is common to use an interface design audit. An example audit is shown in Table 8.3.

Table 8.3 Interface design audit

Work equipment	**Location**	
Date	**Assessor**	
		Yes/No
• **Operator information requirements**		
1. Is the information provided to the operator:		
(a) sufficient;		
(b) adequate;		
(c) comprehensible and unambiguous;		
(d) relevant;		
(e) provided in a satisfactory mode;		
(f) not intellectually demanding or stressful; and		
(g) such as to require extensive memory skills?		
2. Has the operator received the necessary information, instruction and training taking into account the physical and mental limitations of the operator?		
3. Has the potential for human error during operation been identified?		
• **Controls**		
4. Does use of the controls create physical stress on the operator in terms of the use of pressure exerted by the arms and hands and, in some cases, the feet?		

Table 8.3 Continued

	Yes/No
5. Are push buttons and other digital controls easy to operate?	
• **Displays**	
6. Can separate displays and instrument panels be read quickly and accurately without creating visual stress?	
7. Is a high level of accuracy in reading instrument panels required?	
8. Can variations or deviations in readings be readily observed?	
9. Is each display and instrument panel readily discernible from the others?	
10. Is information presented in a suitable format for processing and recording?	
• **Instrument panels**	
11. Is the layout of the panel clear so that each instrument can be easily read and operated?	
12. Are the processes represented symbolically by a form of spacial display?	
13. To ensure correct control and the prevention of human error, are pieces of information linked with the appropriate control?	
14. Are hand-operated knobs, buttons, etc. separated from dials?	
15. Is it possible for the operator to locate the correct knob, button, etc. quickly?	
16. Is the operation of knobs and other finger controls consistent, e.g. clockwise for INCREASE, anticlockwise for DECREASE?	

(*continued*)

Table 8.3 Continued

Recommendations

1. Immediate action
2. Short-term action (7 days)
3. Medium-term action (1 month)
4. Long-term action (6 months)

Information, instruction and training requirements
Assessor____________________

Task analysis and task design

The analysis and design of tasks should embrace the ergonomic principles outlined above. Not only does this analysis assist in the design of safe systems of work, but careful analysis and design will reduce stress on the operator and improve operator performance.

Task analysis

This is a characteristic procedure in occupational psychology, whereby a particular task is analysed with a view to devising methods for the selection and/or training of employees. The analyst seeks to investigate the actual work undertaken and analyse and assess its various elements. Whilst there are several variations of the technique, above all, the methods used should be comprehensive and systematic. The outcome of task analysis is commonly used for the determination of pay and in human resource planning.

Task analysis is, fundamentally, concerned with the identification and assessment of the skill and knowledge components of jobs. It may also be used in the appraisal and description of man–machine relationships.

Job training analysis, the identification and specification of what has to be taught and learned for individual jobs, follows on from task analysis.

Task design

A number of factors need consideration in the design of tasks including:

- identification and comprehensive analysis of the critical tasks expected of individuals and the appraisal of likely errors;

- evaluation of required operator decision-making and the optimum balance between human and automatic contributions to safety actions;
- application of ergonomic principles to the design of man–machine interfaces, including the display of plant and process information, control devices and panel layouts;
- design and presentation of procedures and operating instructions;
- organization and control of the working environment, including the extent of workspace, access for maintenance work and the effects of noise, thermal, lighting, ventilation and humidity conditions;
- provision of the correct tools and equipment;
- scheduling of work patterns, including shift organization, the control of fatigue and stress, and arrangements for emergency operations or situations; and
- efficient communications, both immediate and over periods of time.

Anthropometry

A key feature of ergonomic design is that of matching people to the equipment they use at work. In many situations, particularly with new industrial processes, where complex monitoring equipment is being designed or situations in which optimum performance of operators is vital for success, as with certain military equipment, anthropometrically designed equipment is essential. Anthropometry is the study and measurement of body dimensions, the orderly treatment of resulting data and the application of those data in the design of workspace layouts and equipment.

The position and layout of dials and controls to equipment are commonly determined by the engineering or primary output requirements for the equipment. Very little consideration, if any, may been given to man as the operator, largely due to the fact that, in the past, he has taken the trouble to adapt to controls, irrespective of the discomfort or restraints imposed by the positioning of these controls. The effects of poorly designed equipment or bad layout of controls and displays can reduce operator efficiency and increase the potential for error leading to accidents.

Anthropometric data

In the design of fixed and movable equipment, including a range of vehicles used in the workplace, designers may need to take note of a range of data on

- sitting or standing height;
- arm length;
- arm reach in a forward, sideways, upward and downwards direction;
- hand and finger size;

- knee height;
- strength of individual muscle groups;
- range of movement possible at various limb joints; and
- grip strength.

The objective of collecting anthropometric data is to obtain information on the distribution of all or some of the above physical characteristics within a specified worker population. In many cases, work equipment that has been designed to accommodate people of average reach, sitting height or arm span, may prove to be unsuitable for a substantial percentage of the worker population. On this basis it is essential that data obtained cover a range of measurements for a given dimension so that work equipment can be designed to accommodate a chosen proportion of workers within that range. That proportion may vary between 70 and 90 per cent of the worker population. However, an average value may be inadequate and knowledge of the largest and smallest values is also insufficient. The designer must know how certain characteristics, such as arm length, standing height, grip strength and degree of elbow flexion, are distributed amongst a population of workers.

Measures can be related to the body's movements in terms of range, strength or compatibility. In the case of body and limb movement, the designer may need to know, for a given population, the degree of turn possible towards the midplane of the body (medial rotation) or away from the midplane of the body (lateral rotation). Similar information may be necessary on the range of movement possible towards the midline of the body (adduction) or away from it (abduction).

There are limits, which vary considerably between people, to the angles through which parts of the body may be moved closer together (flexion) or further apart (extension), as in the case of arms and legs. In certain tasks requiring manipulative skills, consideration may need to be given to the degree through which the palm of the hand may be rotated downwards (pronation) or upwards (supination).

Other biochemical assessments may be necessary with respect to the operation of particular muscle groups that mediate hand grip or foot depression strength in the case of manual controls and foot pedals, respectively. The frequency of operation is significant in these cases in terms of a single, short-lasting operation, repeated operation at given intervals, or continuous operation over a period of time.

Anthropometric design factors

Anthropometric design is concerned with the functional space to be occupied by the human operator and the most comfortable position for the operator to be able to operate machinery and equipment, including vehicles. This leads to improved operator functioning and a reduced potential for errors. Two areas are of particular significance in the design process.

Work space

The actual functional space that an operator needs should be considered in terms of human anatomy, shape and size. A designer needs to have basic information not only about the average weight and height of the working population which will use the equipment but also about the possible range of variation. This means that provision must be made for tall thin people, as well as short fat people.

A number of questions need to be asked at this stage.

Are the controls within easy and comfortable reach of the operator, including any foot controls?

Is it possible to adjust these controls for people with short arms and/or short legs?

Can all instruments and displays be seen and understood clearly?

Will the operator be in a sitting or standing position when operating controls and reading displays?

Will the operator be comfortable in this position and enabling a correct posture to be maintained?

Will the operator have to operate two controls simultaneously which are placed at the upper and lower limits of reach?

Are control buttons and knobs correctly shaped to enable their smooth operation?

Does the force required to manipulate levers, turn control wheels, depress foot pedals cause physical stress to the operator and increase fatigue?

Is it possible for the operator to operate controls in a sitting position?

These questions are related, in part, to the 'effective work space' provided for operators. This aspect can be studied by the use of a specific suit worn by the operator. This suit incorporates small light bulbs along its extremities from head to feet. By the use of photographic techniques it is possible to study particularly the hand and arm movements of operators within a particular work space, together with body movements. This study enables a far more significant assessment of the space requirements for operator tasks than mere observation.

Other studies are based on the forces exerted or applied by operators during work activities. Here it is possible to measure the forces which need to be exerted by the foot on a pedal, with the knee joint either straight or bent at different angles. Similar studies can be made with respect to the forces exerted by the arms and shoulders in pushing or pulling activities with the body in different positions, the joints at different angles to the load, and the strength of hand grip with the arm and elbow at different angles.

Seats and sitting arrangements

Many people spend a substantial part of their time at work sitting down undertaking a range of tasks. Very little attention, however, is paid to the quality of seats provided, their relative comfort and the needs of the particular operator. Moreover, it is common practice in many organizations to relate the style, quality and comfort of chairs to the status of the individual within that organization, sooner than that person's actual needs. This results in the more lowly paid staff members undertaking, for instance, clerical or assembly work, sitting on uncomfortable moulded stacking chairs for substantial periods of time. It is not hard to understand, therefore, why such people complain of backache, parasthaesia (tingling and numbness) in the upper thighs and buttocks and postural fatigue due to the unsuitability of such chairs for that type of work.

The design of a chair with respect to shape, size, height, depth from front to back, angle and shape of back rest, the support provided by the back rest and whether supplied with arms, is of significance. An important requirement for a satisfactory seated posture is that the weight of the body should be taken by the ischial tuberosities, the bony promontories on either side of the pelvis. The soft tissue overlying these promontories is subjected to considerable pressure, which would be sufficient if applied to other regions of the body to compress blood vessels and impair the blood flow. This is one reason why prolonged sitting can be so uncomfortable. Pressure may be applied, for example, to the back of the thighs, cutting down blood flow and sometimes sufficiently on nerves to cause pain. The blood vessels in the tissue overlying the tuberosities are arranged in a special manner, so pressure exerts much less effect.

The materials used in chair construction have a direct effect on comfort. Wood is satisfactory because is has a low thermal conductivity so it does not feel cold, and water vapour can pass into and through wood. Plastics are impermeable and in a warm or hot environment sweat will collect on the thighs and buttocks, causing considerable discomfort. Where upholstery is worn or too thin, insufficient support is provided for the ischial tuberosities and pressure is distributed over all the surface of back and thighs.

The legal requirements

The Workplace (Health, Safety and Welfare) Regulations

These regulations lay down requirements with respect to workstations and seating thus:

> Every workstation shall be so arranged that it is suitable both for any person at work in the workplace who is likely to work at that workstation and for any work of the undertaking which is likely to be done there.

A suitable seat shall be provided for each person at work in the workplace whose work includes operations of a kind that the work (or a substantial part of it) can or must be done sitting.

A seat shall not be suitable for the purposes of the above paragraph unless:

(a) it is suitable for the person for whom it is provided as well as for the operations to be performed; and

(b) a suitable footrest is also provided where necessary.

The HSC Approved Code of Practice to the regulations makes the following points:

Workstations should be arranged so that each task can be carried out safely and comfortably. The worker should be at a suitable height in relation to the work surface. Work materials and frequently used equipment or controls should be within easy reach, without undue bending or stretching.

Workstations including seating and access to workstations, should be suitable for any special needs of the individual worker, including workers with disabilities.

Each workstation should allow any person who is likely to work there adequate freedom of movement and the ability to stand upright. Spells of work which unavoidably have to be carried out in cramped conditions should be kept as short as possible and there should be sufficient space nearby to relieve discomfort.

There should be sufficient clear and unobstructed space at each workstation to enable the work to be done safely. This should allow for manoeuvring and positioning of materials, for example lengths of timber.

Seating provided should, where possible, provide adequate support for the lower back, and a footrest should be provided for any worker who cannot comfortably place his or her feet on the floor.

Health and Safety (Display Screen Equipment) Regulations 1992

Both the Schedule and HSE Guidance (Annex A) to the regulations makes specific reference to the work desk or work surface used by defined display screen equipment users and operators.

The schedule

2(d) *Work desk or work surface*

The work desk or work surface shall have a sufficiently large, low-reflectance surface and allow a flexible arrangement of the screen, keyboard, documents and related equipment.

The document holder shall be stable and adjustable and shall be positioned so as to minimise the need for uncomfortable head and eye movements.

There shall be adequate space for operators or users to find a comfortable position.

2(e) *Work chair*

The work chair shall be stable and allow the operator or user easy freedom of movement and a comfortable position.

The seat shall be adjustable in height.

The seat back shall be adjustable in both height and tilt.

A footrest shall be made available to any operator or user who wishes one.

Annex A

Work desk or work surface

Work surface dimensions may need to be larger than for conventional non-screen office work, to take adequate account of:

(a) the range of tasks performed (e.g. screen viewing, keyboard input, use of other input devices, writing on paper, etc.);

(b) position and use of hands for each task; and

(c) use and storage of working materials and equipment (documents, telephones, etc.).

Document holders are useful for work with hard copy, particularly for workers who have difficulty in re-focussing. They should position working documents at height, visual plane and, where appropriate, viewing distance similar to those of the screen, be of low reflectance, be stable and not reduce the readability of source documents.

Work chair

The primary requirement here is that the work chair should allow the user to achieve a comfortable position. Seat height adjustments should accommodate the needs of users for the tasks performed. The Schedule requires the seat to be adjustable in height (i.e. relative to the ground) and the seat back to be adjustable in height (also relative to the ground) and tilt. Provided the chair design meets these requirements and allows the user to achieve a comfortable posture, it is not necessary for the height or tilt of the seat back to be adjustable independently of the seat. Automatic back rest adjustments are acceptable if they provide adequate support.

General health and safety advice and specifications for seating are given in the HSE publication *Seating at Work* [HS(G)57].

Foot rests may be necessary where individual workers are unable to rest their feet flat on the floor (e.g. where work surfaces cannot be adjusted to the right height in relation to other components of the workstation). Foot rests should not be used when they are not necessary as this can result in poor posture.

'Cranfield Man'

Few workstations are 'made to measure' owing to the wide range of human dimensions and the sheer cost of designing individual workstations and machines to conform to individual body requirements. The fact that this is not done creates many operational problems and health risks, best demonstrated by research at the Cranfield Institute of Technology, who created *Cranfield Man* (Figure 8.1).

Using a horizontal lathe, researchers examined the positions of controls and compared the locations of these controls with the physical dimensions of the average operator. Table 8.4 below shows the differences between the two.

What came out of this study was the fact that, in the original design of horizontal lathes, going back over a century or more, no consideration was given to the physical needs of the operator and the body dimensions of the average operator, particularly with respect to elbow height and arm span. Because of this fact, lathe operators continually operate lathes with a bent stance, resulting in back strain and other forms of operational fatigue.

By simply mounting the lathe on a plinth, this raises the operational level to a more suitable height, eliminating the need for bent stances and reducing postural stress on the operator.

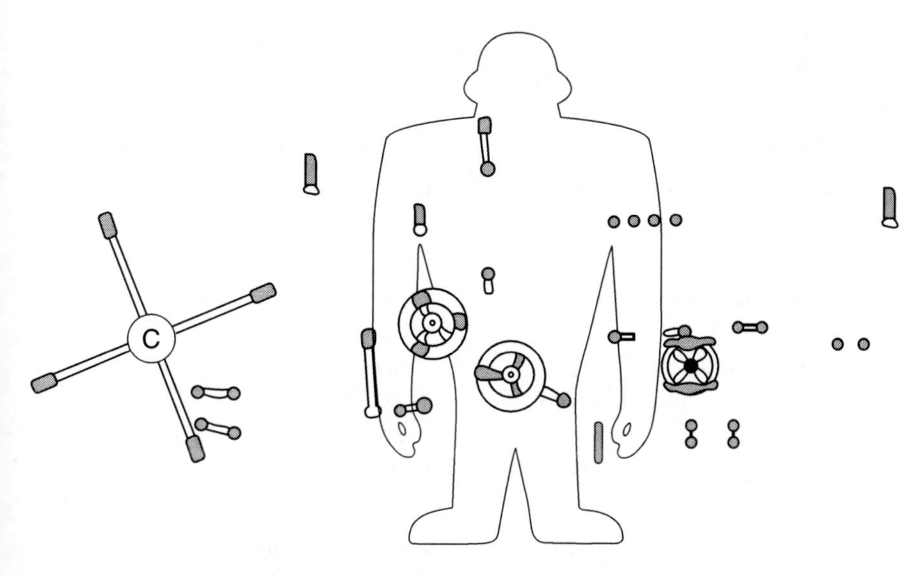

Figure 8.1 Cranfield Man – 1.35 m tall with a 2.44 m arm span

Table 8.4 Physical dimensions of the average operator compared with those of 'Cranfield Man'

Average operator (m)	Dimensions	Operator who would suit these controls (m)
1.75	Height	1.35
0.48	Shoulder width	0.61
1.83	Arm span	2.44
1.07	Elbow height	0.76

People and machines

Over the centuries, a wide range of machines has been invented largely directed at increasing production and improving and maintaining the quality of products. Prior to this, people produced goods by hand, but the trouble with people is that, in many cases, they are inconsistent in their endeavours, too slow and wasteful.

The advantages and disadvantages associated with human labour and the use of machinery are shown in Table 8.5.

Table 8.5 A comparison between people and machines

Advantages	Disadvantages
People	*Machines*
Adaptable and flexible	Relatively inflexible
Sensitive to a wide range of stimuli	Relatively insensitive
Can detect minute stimuli and assess small changes	Can detect stimuli if programmed, but cannot detect small changes
Can interpolate and use judgement	Can do neither
Can synthesize and learn from experience	Can do neither
Can react to unexpected low probability events	Cannot react in this way
Can exercise judgement where events cannot be completely defined	Cannot generally make judgements or predictions
Are good at creative tasks	Are relatively uncreative
Can be reprogrammed through training	Relatively inflexible
Can communicate with each other	Cannot communicate except through complex electronic systems

Table 8.5 Continued

Advantages	Disadvantages
Machines	***People***
Can operate in hostile environments	Lower capability
Respond quickly to emergency signals	Slow response
Can apply large forces smoothly and with precision	Can apply large forces coarsely
Information storage: large short-term memory	Limited short-term memory; easily distracted
Perform routine repetitive tasks reliably, rapidly, continuously and precisely over long periods of time	Not reliable for this work
Compute fast and accurately	Compute slowly and inaccurately
Can operate for long periods without maintenance	Suffer relatively quickly from fatigue and monotony
Insensitive to distraction	Readily distracted
Can do many different things simultaneously	Can only do one thing at a time in most cases

On this basis, there needs to be a distinction between ergonomic design aimed at

- error-free performance; and
- reducing health risks.

Prescribed and reportable diseases and conditions

Certain occupational diseases and conditions are associated with a failure to control physical agents, such as pressure and friction on the body, along with body movements, in working environments. A number of both prescribed and reportable diseases and conditions are associated with exposure to these agents.

Prescribed diseases and conditions

Typical conditions due to exposure to physical agents

Conditions	*Due to*
Cramp of the hand or forearm due to repetitive movements	Prolonged periods of hand writing, typing or other repetitive movements of the fingers, hand or arm
Subcutaneous cellulitis of the hand (beat hand)	Manual labour causing severe or prolonged friction or pressure on the hand

(Social Security (Industrial Injuries) (Prescribed Diseases) Regulations 1985)

Reportable diseases and conditions

Conditions due to physical agents and the physical demands of work

Conditions	*Due to*
Bursitis or subcutaneous cellulitis at or about the knee due to severe or prolonged external friction or pressure at or about the knee (beat knee)	Physically demanding work causing severe or prolonged friction or pressure at or about the knee
Traumatic inflammation of the tendons of the hand or forearm or or of the associated tendon sheaths	Physically demanding work, frequent or repeated movements, constrained postures or extremes of extension or flexion of the hand or wrist

(Reporting of Injuries, Diseases and Dangerous Occurrences Regulations 1995)

The principal features of these conditions are outlined below

Job movements

The type of work that people undertake is a common causative factor in the development of a range of ill-health conditions, particularly that group of disorders known as the 'beat' disorders, various work-related upper limb disorders and cramp conditions arising from work movements.

Most people will have suffered the painful condition known as 'writer's cramp' on a temporary basis, perhaps after a series of written examinations or intensive note-taking sessions. The condition soon fades, however, after completion of the writing task.

Some occupations, however, may entail prolonged periods which require repetitive movements of the hand or arm resulting in painful cramps.

Friction and pressure

Beat knee and beat elbow, commonly referred to as 'housemaid's knee' and 'tennis elbow', respectively, are two painful conditions caused by inflammation of the joint which may be followed by suppuration due to infection entering the joint.

These disorders are frequently associated with people involved in manual labour, such as paviors, gardeners, cleaning staff and horticultural workers, where the work causes severe or prolonged friction or pressure at or about the joint.

Work-related upper limb disorders include that group of disorders known as 'repetitive strain injury' (RSI), including tenosynovitis, carpal tunnel syndrome and Dupuytren's contracture. In most cases, the disorders are associated with tasks requiring the repetitive movement of the hand, elbow and/or forearm coupled with the application of force. Common symptoms include local pain, swelling, tenderness and inflammation which is aggravated by pressure or movement. Tenosynovitis, inflammation of the synovial lining of the tendon sheath, can be particularly painful and disabling (see Chapter 9, Ergonomics and human reliability).

Display screen equipment – The risks

The consideration of ergonomic principles is essential in the design of display screen equipment (DSE) workstations. The three principal risks to health associated with the use of display screen equipment are:

1. work-related upper limb disorders;
2. visual fatigue; and
3. postural fatigue.

Work-related upper limb disorders caused by repetitive strain injuries were first defined in the medical literature by Bernardo Ramazzini, the Italian father of occupational medicine, in the early eighteenth century. The International Labour Organisation recognised RSI as an occupational disease in 1960, as a condition caused by forceful, frequent, twisting and repetitive movements.

Repetitive strain injury covers some well-known conditions such as tennis elbow, flexor tenosynovitis and carpal tunnel syndrome. It is usually caused or aggravated by work and is associated with repetitive and overforceful movement, excessive workloads, inadequate rest periods and sustained or constrained postures, resulting in pain or soreness due to the inflammatory conditions of muscles and the synovial lining of the tendon sheath. Present approaches to treatment are largely effective, provided the condition is treated in its early stages. Tenosynovitis has been a prescribed industrial disease since 1975 and the HSE have proposed changing the name of the condition to 'work-related upper limb disorder' on the grounds that the disorder does not always result from repetition or strain, and is not always a visible injury.

Many people, including assembly workers, supermarket checkout assistants and keyboard operators, are affected by RSI at some point in their lives.

Clinical signs and symptoms of work-related upper limb disorders

These include local aching pain, tenderness, swelling and crepitus (a grating sensation in the joint) aggravated by pressure or movement. Tenosynovitis, affecting the hand or forearm, is the second most common prescribed industrial disease, the most common being dermatitis. True tenosynovitis, where inflammation of the synovial lining of the tendon sheath is evident, is rare and potentially serious. The more common and benign form is peritendinitis crepitans, which is associated with inflammation of the muscle–tendon joint that often extends well into the muscle.

Classification of the forms of RSI

- **Epicondylitis**. This is inflammation of the area where a muscle joins a bone.
- **Peritendinitis**. Peritendenitis results in inflammation of the area where a tendon joins a muscle.
- **Carpal tunnel syndrome**. A fibrous bridge, the *flexor retinaculum*, spans the small bones at the base of the palm of the hand. The carpal tunnel is the opening between the flexor retinaculum and these bones. The flexor tendons to the fingers pass through this tunnel, enclosed in slippery synovial sheaths which prevent friction in this narrow space. Between the tendons is the median nerve.

  This painful condition arises as a result of compression of the median nerve in the tunnel. Pressure on the nerve causes a pins and needles sensation, numbness or pain in this area and sometimes weakness of the thumb. The condition is often dealt with through splinting the wrist at night, although in some cases the condition may be cured by cutting the constricting fibres of the retinaculum.
- **Tenosynovitis**. As described above, this is inflammation of the synovial lining of the tendon sheath which most people associate with repetitive strain injury.
- **Tendinitis**. Tendinitis implies inflammation of the tendons, particularly in the fingers.
- **Dupuytren's contracture**. This is a condition affecting the palm of the hand, where it is impossible to straighten the hand and fingers due to thickening of the fibrous lining of the palm. It is common in elderly men and may be associated with other disorders of fibrous tissue or rheumatism. Many causes have been suggested, including pressure on the palms and alcoholism.
- **Writer's cramp**. Cramps in the hand, forearm and fingers.

Prevention of RSI

Injury can be prevented by:

- improved design of working areas, e.g. position of keyboard and VDU screens, heights of workbenches and chairs;
- adjustments of workloads and rest periods;
- provision of special tools;
- health surveillance aimed at detecting early stages of the disorder; and
- better training and supervision

If untreated, RSI can be seriously disabling.

Visual fatigue

Visual fatigue (eye strain) is associated with glare from the display and the continual need to focus and refocus from screen to copy material and back again. The degree of individual fatigue will vary. Vision screening of staff on a regular basis, and as part of a pre-employment health screen, is recommended.

Postural fatigue

Postural fatigue, an outcome of operational stress, may take many forms. It can include backache, neck and shoulder pains associated with poor chair and workstation design and positioning in relation to controls and displays, insufficient leg room and the need to adjust body position.

Other causes of operational stress

Operational stress can also be created by noise from the unit and ancillary equipment, excessive heat and inadequate ventilation.

The degree of operator stress may vary according to age, sex, physical build, attitude to the task, current level of visual acuity, general health and the extent of time in tasks not involving attention to a display screen. Users should be encouraged to organize their work loads to permit frequent screen breaks.

Health and Safety (Display Screen Equipment) Regulations 1992

These regulations should be read in conjunction with the Management of Health and Safety at Work Regulations 1999. The original regulations have been amended in the light of the Health and Safety (Miscellaneous Amendments) Regulations 2002 (see Figures 8.2 and 8.3).

Important definitions

Display screen equipment means an alphanumeric or graphic display screen, regardless of the display process involved.

Operator means a *self-employed* person who *habitually* uses display screen equipment as a *significant part* of his *normal* work.

User means an *employee* who *habitually* uses display screen equipment as a *significant part* of his *normal* work.

Workstation means an assembly comprising

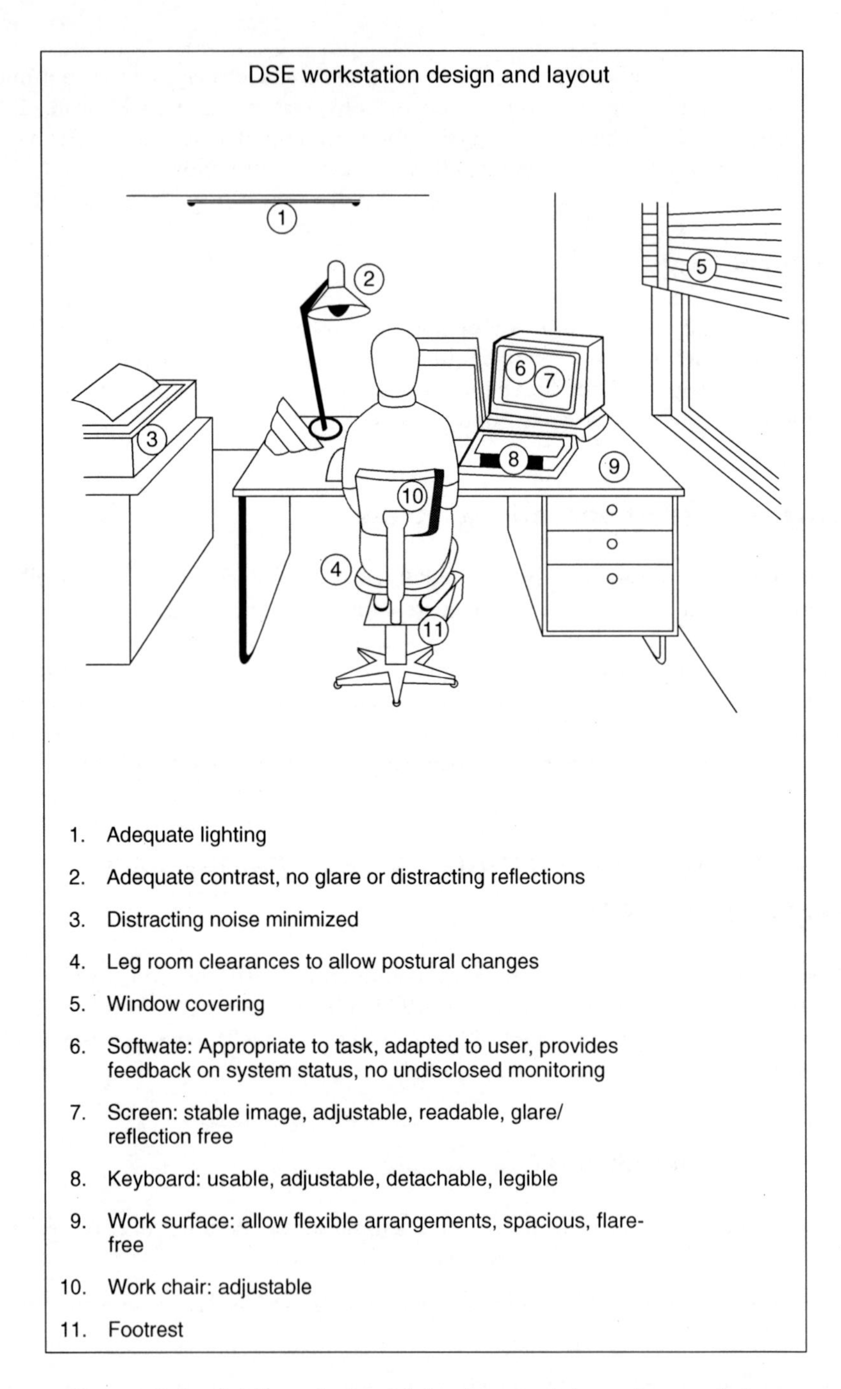

Figure 8.2 Subjects dealt with in the schedule to the regulations

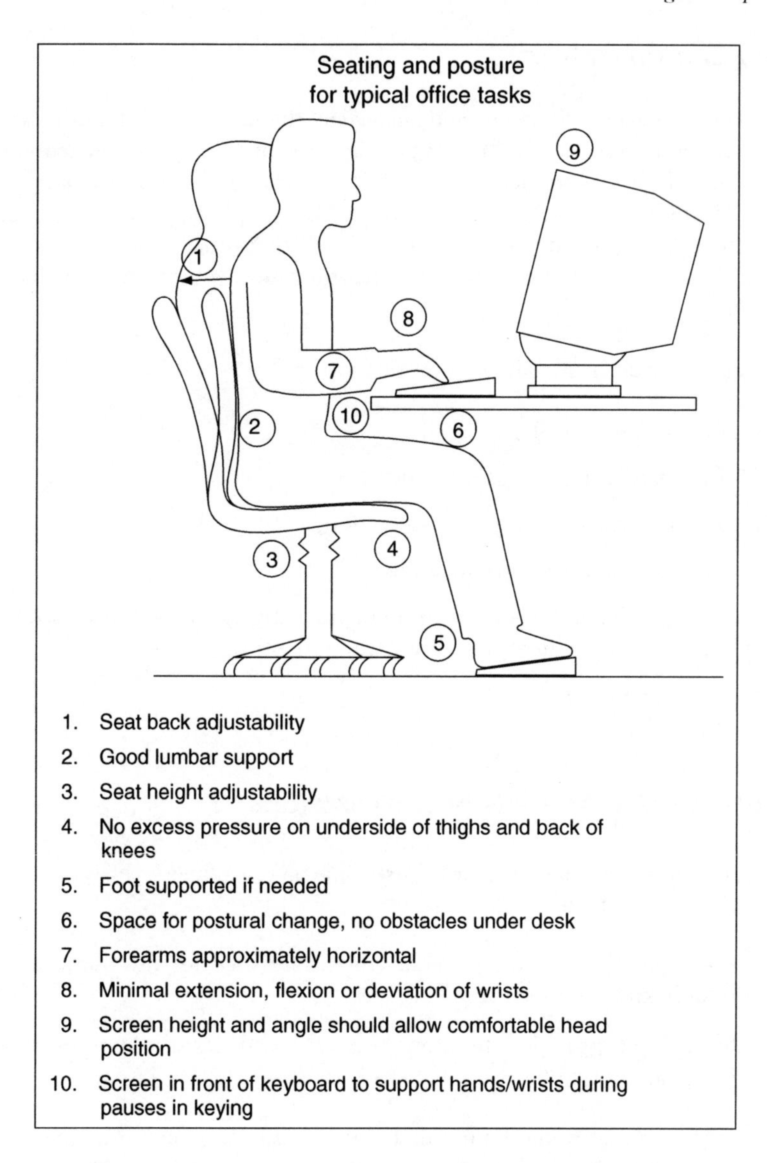

Figure 8.3 Seating and posture for typical office tasks

- display screen equipment (DSE) (whether provided with software determining the interface between the equipment and its operator or user, a keyboard or any other input device);
- any optional accessories to the DSE;
- any disk drive, modem, printer, document holder, work chair, work desk, work surface or other item peripheral to the DSE; and
- the immediate environment around the DSE.

Users and operators

The terms *habitually*, *significant* and *normal* in the definition of both *user* and *operator* are important in that many people who use display screen equipment as a feature of their work activities may not necessarily be users or operators as defined. Employers must decide which of their employees are users, together with any self-employed operators to whom they may contract work, by reference to the definitions and to the guidance which accompanies the regulations.

The regulations do not apply to or in relation to

- drivers' cabs or control cabs;
- DSE on board a means of transport;
- DSE mainly intended for public operation;
- portable systems not in prolonged use;
- calculators, cash registers or any equipment having a small data or measurement display required for direct use of the equipment; or
- window typewriters.

Regulation 2 – Analysis of workstations

1. Every employer shall perform a suitable and sufficient analysis of those workstations which

   - (regardless of who has provided them) are used for the purposes of his undertaking by users; and
   - have been provided by him and are used for the purposes of his undertaking by operators;

for the purpose of assessing the health and safety risks to which those persons are exposed in consequence of that use.

2. Any assessment made by an employer shall be reviewed by him if:

   - there is reason to suspect that it is no longer valid; or
   - there has been a significant change in the matters to which it relates

   and where as a result of any such review changes to the assessment are required, the employer concerned shall make them.

3. The employer shall reduce the risks identified in consequence of an assessment to the lowest extent reasonably practicable.

Regulation 3 – Requirements for workstations

Every employer shall ensure that any workstation which may be used for the purposes of his undertaking meets the requirements laid down in the schedule to these regulations, to the extent specified in paragraph 1 thereof.

Regulation 4 – Daily work routine of users

Every employer shall so plan the activities of users at work in his undertaking that their daily work on display screen equipment is periodically interrupted by such breaks or changes of activity as reduce their workload at that equipment.

Regulation 5 – Eyes and eyesight

1. Where a person:
   - is a user in the undertaking in which he is employed; or
   - is to become a user in the undertaking in which he is, or is to become, employed,

the employer who carries on the undertaking shall, if requested by that person, ensure that an appropriate eye and eyesight test is carried out on him by a competent person within the time specified in paragraph 2.

2. The time specified in paragraph 1 is:
   - in the case of a person mentioned in paragraph 1(a), as soon as practicable after the request; or
   - in the case of a person mentioned in paragraph 1(b), before he becomes a user.
3. At regular intervals after an employee has been provided (whether before of after becoming an employee) with an eye and eyesight test, his employer shall, subject to paragraph 6, ensure that he is provided with a further eye and eyesight test of an appropriate nature, any such test to be carried out by a competent person.
4. Where a user experiences visual difficulties which may reasonably be considered to be caused by work on display screen equipment, his employer shall ensure that he is provided at his request with an appropriate eye and eyesight test, any such test to be carried out by a competent person as soon as practicable after being requested as aforesaid.
5. Every employer shall ensure that each user employed by him is provided with special corrective appliances for the work being done by the user concerned where:
   - normal corrective appliances cannot be used; and
   - the result of any eye and eyesight test the user has been given in accordance with this regulation shows such provision to be necessary.

6. Nothing in paragraph 3 shall require an employer to provide any employee with an eye and eyesight test against that employee's will.

Regulation 6 – Provision of training

1. Where a person:
   - is a user in the undertaking in which he is employed; or
   - is to become a user in the undertaking in which he is, or is to become, employed,

the employer who carries on the undertaking shall ensure that he is provided with adequate health and safety training in the use of any workstation upon which he may be required to work.

In the case of a person mentioned in sub-paragraph (b) of paragraph 1 the training shall be provided before he becomes a user.

Regulation 7 – Provision of information

1. Every employer shall ensure that operators and users at work in his undertaking are provided with adequate information about:
   - all aspects of health and safety relating to their workstations; and
   - such measures taken by him in compliance with his duties under regulations 2 and 3 as relate to them and their work.
2. Every employer shall ensure that users at work in his undertaking are provided with adequate information about such measures taken by him in compliance with his duties under regulations 4 and 6(2) as relate to them and their work.
3. Every employer shall ensure that users employed by him are provided with adequate information about such measures taken by him in compliance with his duties under regulations 5 and 6(1) as relate to them and their work.

Display screen equipment workstation risk assessment

Risk analysis involves a consideration of the requirements laid down in the schedule to the regulations relating to the equipment, the environment and the interface between the computer and operator/user. A typical workstation risk assessment format is in Table 8.6.

Table 8.6 Display screen equipment workstation risk assessment format

In undertaking this risk assessment, the assessor should answer the questions below. Where the answer is **YES**, no further action is required.	YES/NO
The equipment	
1. The Screen	
Are the characters on the screen well-defined and clearly formed, of adequate size and with adequate spacing between the characters and lines?	
Is the image on the screen stable, with no flickering or other forms of instability?	
Are the brightness and the contrast between the characters and the background easily adjustable by the operator or user, and also easily adjustable to ambient conditions, e.g. lighting?	
Does the screen swivel and tilt easily and freely to suit the needs of the user?	
Is it possible to use a separate base for the screen or an adjustable table?	
Is the screen free of reflective glare and reflection liable to cause discomfort to the operator or user?	
2. Keyboard	
Is the keyboard tiltable and separate from the screen so as to allow the operator or user to find a comfortable working position avoiding fatigue in the arms or hands?	
Is the space in front of the keyboard sufficient to provide support for the hands and arms of the operator or user?	
Does the keyboard have a matt surface to avoid reflective glare?	
Are the arrangement of the keyboard and the characteristics of the keys such as to facilitate the use of the keyboard?	
Are the symbols on the keys adequately contrasted and legible from the design working position?	
3. Work desk or work surface	
Does the work desk or work surface have a sufficiently large, low-reflectance surface and allow a flexible arrangement of the screen, keyboard, documents and related equipment?	
Is the document holder stable and adjustable and positioned so as to minimize the need for uncomfortable head and eye movements?	

(*continued*)

Table 8.6 Continued

Is there adequate space for operators or users to find a comfortable position?	YES/NO
4. Work chair	
Is the work chair stable and does it allow the operator or user easy freedom of movement and a comfortable position?	
Is the seat adjustable in height?	
Is the seat back adjustable in both height and tilt?	
Is a footrest made available to any operator or user who requests one?	
Environment	
1. Space requirements	
Is the workstation dimensioned and designed so as to provide sufficient space for the operator or user to change position and vary movements?	
2. Lighting	
Does any room lighting or task lighting ensure satisfactory lighting conditions and an appropriate contrast between the screen and the background environment, taking into account the type of work and the vision requirements of the operator or user?	
Are possible disturbing glare and reflections on the screen or other equipment prevented by co-ordinating workplace and workstation layout with the positioning and technical characteristics of artificial light sources?	
3. Reflections and glare	
Is the workstation so designed that sources of light, such as windows and other openings, transparent or translucid walls, and brightly coloured fixtures or walls cause no direct glare and no distracting reflections on the screen?	
Are windows fitted with a suitable system of adjustable covering to attenuate the daylight that falls on the workstation?	
4. Noise	
Is the noise emitted by equipment belonging to any workstation taken into account when a workstation is being equipped, with a view, in particular, to ensuring that attention is not distracted and speech is not disturbed?	

Table 8.6 Continued

	YES/NO
5. Heat Does the equipment belonging to any workstation produce excess heat which could cause discomfort to operators or users?	
6. Radiation Is all radiation, with the exception of the visible part of the electromagnetic spectrum, reduced to negligible levels from the point of view of operators' or users' health and safety?	
7. Humidity Are adequate levels of humidity established and maintained?	
Interface between computer and operator/user In the designing, selecting, commissioning and modifying of software, and in designing tasks using display screen equipment,does the employer take into account the following principles:- • software must be suitable for the task; • software must be easy to use and, where appropriate, adaptable to the level of knowledge or experience of the user; no quantitative or qualitative checking facility may be used without the knowledge of the operators or users; • systems must provide feedback to operators or users on the performance of those systems; • systems must display information in a format and at a pace which are adapted to operators or users; • the principles of software ergonomics must be applied, in particular to human data processing?	

Comments of assessor

Date ____________________ **Signature** ____________________

Risk assessment summary

Principal risks

Specific risks

(continued)

Remedial action

1. **Immediate**
2. **Short term (28 days)**
3. **Medium term (6 months)**
4. **Long term (over 12 months)**

Information, instruction, training and supervision requirements

Date of next review Assessor

Software ergonomics

This is a specific area of ergonomic study dealing with the design and layout of display screen equipment workstations and incorporating the principles of software ergonomics.

Annex A *Guidance on minimum workstation requirements* to the Health and Safety (Display Screen Equipment) Regulations provides advice to employers on the problem of stress arising from this type of work with particular reference to task design and software.

Task design and software

Principles of task design

Inappropriate task design can be among the causes of stress at work. Stress jeopardizes employee motivation, effectiveness and efficiency and in some cases it can lead to significant health problems. The regulations are only applicable where health and safety rather than productivity is being put at risk, but employers may find it useful to consider both aspects together as task design changes put into effect for productivity reasons may also benefit health, and vice versa.

In display screen work, good design of the task can be as important as the correct choice of equipment, furniture and working environment. It is advantageous to

- design jobs in a way that offers users variety, opportunities to exercise discretion, opportunities for learning, and appropriate feedback, in preference to simple repetitive tasks whenever possible. (For example, the work of a typist can be made less repetitive and stressful if an element of clerical work is added);
- match staffing levels to volumes of work, so that individual users are not subject to stress through being either overworked or underworked;
- allow users to participate in the planning, design and implementation of work tasks whenever possible.

Principles of software ergonomics

In most display screen work, the software controls both the presentation of information on the screen and the ways in which the worker can manipulate the information. Thus software design can be an important element of task design. Software that is badly designed or inappropriate for the task will impede the efficient completion of the work and in some cases may cause sufficient stress to affect the health of a user. Involving a sample of users in the purchase or design of software can help to avoid problems.

The schedule to the regulations lists a few general principles which employers should take into account. Requirements of the organization and of display screen equipment workers should be established as the basis for designing, selecting and modifying software. In many (though not all) applications the main points are:

- **Suitability for the task**. Software should enable workers to complete the task efficiently, without presenting unnecessary problems or obstacles.
- **Ease of use and adaptability**. Workers should be able to feel that they can master the system and use it effectively following the appropriate training. The dialogue between the system and the worker should be appropriate for the worker's ability. Where appropriate, software should enable workers to adapt to the user interface to suit their ability level and preferences. The software should protect workers from the consequence of errors, for example, by providing appropriate warnings and information and by enabling 'lost' data to be recovered wherever practicable.
- **Feedback on system performance**. The system should provide appropriate feedback, which may include error messages; suitable assistance ('Help') to workers on request; and messages about changes in the system such as malfunctions or overloading. Feedback messages should be presented at the right time and in an appropriate style and format. They should not contain unnecessary information.
- **Format and pace**. Speed of response to commands and instructions should be appropriate to the task and to workers' abilities. Characters, cursor movements and position changes should where possible be shown on the screen as soon as they are input.
- **Performance monitoring facilities**. Quantitative or qualitative checking facilities built into the software can lead to stress if they have adverse results such as an overemphasis on output speed.

It is possible to design monitoring systems that avoid these drawbacks and provide information that is helpful to workers as well as managers. However, in all cases, workers should be kept informed about the introduction and operation of such systems.

Work conditions, ergonomics and health

Poor working conditions and inadequate task design can have a direct effect on the health of those involved. The various areas of ergonomic study and application, the human factors issues and the effects on health can be summarized as in Table 8.7.

Table 8.7 Anatomy, physiology and psychology: Occupational conditions, human factors and health hazards

Occupational condition	Human factors	Health hazards
• **Anatomy and physiology**		
Dimensions of seats, benches	Anthropometry • body dimensions	Bad posture Discomfort General fatigue
Motion study • workplace layout • work rate	Anthropometry • body dimensions • strengths of muscle groups Structures of joints Functions of muscles	Bad posture Discomfort General fatigue Local muscular fatigue (including tenosynovitis)
Design of hand tools	Anthropometry • body dimensions • strengths of muscle groups Structures of joints Functions of muscles	Local muscle strain Local muscular fatigue (including tenosynovitis)
Design of controls • levers • hand wheels • knobs • buttons, etc.	Anthropometry • body dimensions • strengths of muscle groups Structures of joints Functions of muscles	Local muscle strain Local muscular fatigue (including tenosynovitis)
Manual handling	Kinetic methods based on	Injuries

Table 8.7 Continued

Occupational condition	Human factors	Health hazards
	• structures of joints • strengths of muscle groups	• muscle strains • hernias • skeletal damage • slipped discs
Heavy manual work	Physical fitness Physiological cost of work • oxygen consumption • heart rate • body temperature	General fatigue
Control of • air temperature • radiant heat • humidity • air movement	Physiological cost work • heart rate • body temperature Temperature regulating mechanisms	Heat stress and disorders Cold stress and disorders
Vibration Cold		Raynaud's phenomenon
Flying	Ear anatomy	Ear damage
Diving Caisson work	Gases in blood	Bends Anoxia Oxygen poisoning
Noise	Hearing	Deafness Auditory discomfort
• **Psychology**		
Control of • air temperature • radiant heat • humidity • air movement	Subjective feelings	Thermal discomfort
Design of indicators • dials • warning lights, etc.	Sensory and perceptual abilities • especially visual 'Natural' directions	Accidents Stress

(continued)

Table 8.7 Continued

Occupational condition	Human factors	Health hazards
Design of controls • levers • buttons, etc.		
Design of • labels • notices • posters	Visual abilities • sensory • perceptual Mental abilities • learning • thinking	Accidents Stress
Lighting • quantity • distribution • glare	Visual abilities • sensory • perceptual	Accidents Stress Visual fatigue
Colour • environment	Visual abilities • sensory • perceptual	Visual discomfort Depression
• colour coding	Mental abilities • learning • thinking, etc.	Accidents Stress
Inspection Fine assembly (arrangements of: • lighting • contrasts • colours • movement, etc.)	Visual abilities • sensory • perceptual Vigilance	Boredom Stress Visual fatigue
Job design • duties	Sensory and perceptual abilities • especially visual Mental abilities • learning • thinking, etc. Motivation	Boredom Stress Accidents
Human relations	Personality	Stress Neuroses

Conclusion

Ergonomics, 'the scientific study of work', applies in many work situations. It literally means the study of the natural laws of work and is, therefore, multidisciplinary and takes an implicitly systemic approach. As such, it incorporates elements of anatomy, physiology and psychology.

The scope of ergonomics is extensive dealing with human characteristics, man–machine interface, design ergonomics, environmental factors affecting human performance, task analysis and task design, anthropometry and the relationship between working conditions, ergonomics and health.

Ergonomics has an application in a wide range of work situations and work activities, from the design of hand tools, hand and foot controls, and work stations to the specification of appropriate environmental levels with respect to lighting, temperature, ventilation and humidity.

Key points

- Ergonomics is concerned with fitting the task to the individual, maximizing human performance and eliminating, as far as possible, the potential for human error.
- Ergonomics embraces many disciplines, including physiology, anatomy, psychology, engineering and environmental science.
- The main areas for consideration are the human system, environmental factors, the man–machine interface and the total working system.
- The design of the man–machine interface is significant in terms of improved operator performance and the avoidance of accidents.
- Anthropometry, a branch of ergonomics, is the study and measurement of body dimensions, and has a significant role to play in the design of operator workstations and equipment.
- Many occupational conditions, such as work-related upper limb disorders, are associated with poor ergonomic design or the failure to take ergonomic aspects into account in the design of work stations.
- The ergonomist frequently operates as a member of a specialist team, comprising engineers, psychologists and doctors, in the solution of particular work-related problems.
- With the vast increase in the use of information technology, considerable attention has been paid in recent years to software ergonomics.

9 Ergonomics and human reliability

One of the functions of ergonomic study is that of improving human reliability. This implies, in particular, the design of the man–machine interface in such a way as to improve operator performance and reduce the potential for human error.

Ergonomically designed interfaces and control systems

An important element of the man–machine interface is the design of display and control systems.

Displays

Displays provide information to the operator. Well-designed displays improve operator performance and reduce stress. Badly designed displays, on the other hand, increase operator stress, contribute to human error and hinder performance.

The various functional characteristics of displays, based on studies by Grandjean (1980), are indicated in the Table 9.1.

A number of factors are significant in the design of displays, such as their need to attract attention in certain cases, the quality of visual representation, ease of reading, precision and acceptability to the average operator. Above all, they must be designed with a view to preventing human error. In certain cases, 'fail-safe' or overriding devices may need to be installed to prevent incorrect or dangerous operation in the event of human error.

Controls

Controls to work equipment include buttons, knobs, levers, hand wheels and foot pedals. Operators should be able to reach and manipulate controls with ease, taking into account their body dimensions and physical strength.

Badly designed and maintained controls, such as hand levers, foot pedals and large wheels which require regular rotation, can cause a range of joint and muscular disorders, together with repetitive strain injuries, including tenosynovitis.

Table 9.1 Functional characteristics of the main type of display

Display type	**Functional characteristics**			
Qualitative				
Auditory	For attracting immediate attention; annunciators are warning devices			
Visual	For representing three or more status conditions by the use of colour, shape, size or location of coding			
Representational	Provides the operator with a model of the system; good for showing spatial relationships between variables; a static example is a map, a dynamic one, process control			
Quantitative				
	Ease of reading	*Precision*	*Directing rate of change*	*Setting a reading*
Analogue (dials and meters)				
• moving pointer	Acceptable	Acceptable	Very good	Very good
• moving scale	Acceptable	Acceptable	Acceptable	Acceptable
Digital	Very good	Very good	Poor	Acceptable

Grandjean E. (1980). *Fitting the Task to the Man: An ergonomic approach*. Taylor & Francis, London.

Badly designed control systems can incorporate some or all of the following elements:

- unexpected placement of controls;
- controls that are hard to understand;
- incompatible mapping of controls to devices;
- controls that are difficult to remember;
- controls that operate in unexpected modes;
- controls that are too close together;
- controls that are some distance from devices;
- too many controls at one point;
- controls that are too easy to operate accidentally; and
- controls that operate in unexpected ways.

Classical error and systems ergonomics

One of the objectives of ergonomic study is to design tasks to fit the operator. This means taking account of differences such as size, strength and ability to handle information for a wide range of users. Subsequently, the tasks, the workplace and work equipment are designed around these differences. The benefits are improved efficiency, quality and job satisfaction. The costs of failure include increased error rates and fatigue. In some cases, failure may be catastrophic.

In Chapter 5, Perception of risk and human error, reference is made to the various forms of classical error. These may be classified as:

- unintentional error;
- mistakes;
- violations;
- skill-based errors;
- rule-based errors; or
- knowledge based errors.

Systems ergonomics endeavours to take human error into account through techniques such as systems analysis. Systems analysis allows examination of the basic structure of people within a man–machine system. The principal objective is to identify the requirements for the layout of this interaction between man and machine within the framework of the specification of the man–machine system. Systems ergonomics aims to optimize this interaction, contributing at the same time to reducing the number of errors made by operators and thereby increasing the reliability of the overall performance of the man–machine system. A basic procedure in systems ergonomics is that of defining the elements of the system and their interaction.

It should be appreciated that

- information is always transmitted on very specific channels from the exit of one element to the entry of another; and
- elements are defined by their characteristic to alter information in a specific manner which is determined by that element.

The main feature of systems ergonomics is that of disregarding the physical nature of the elements, together with their interaction, and examining the formal structure of this interaction and the means of information transmission by the element. On this basis, the results of systems ergonomics can be transferred to different man–machine systems.

Two approaches or views to systems ergonomics are taken:

- **Deterministic**. By describing the characteristics of elements within a system by means of functions, and on the basis of a given input, this leads to a prognosis of a clear starting and end function.

- **Probabilistic**. By assessing the reliability and failure probability of the elements, considering the system structure and calculating the expected total failure probability.

Physical stressors and human reliability

A broad range of physical stressors can be encountered in workplaces, such as inadequate work space, temperature extremes, noise and varying levels of ventilation.

Work space

Inadequate work space in workplaces can result in congestion and increase the potential for accidents and the transmission of minor infections, such as colds and influenza. However, despite the fact that legal requirements relating to the provision of space for employees (the old 'overcrowding' requirements under the Factories Act 1961 and the Offices, Shops and Railway Premises Act 1963) have been in operation for many years, they have attracted little or no case law.

Specific requirements relating to work space are incorporated in the Workplace (Health, Safety and Welfare) Regulations in that every room where persons work shall have sufficient floor area, height and unoccupied floor space for the purposes of health, safety and welfare. The Approved Code of Practice to the regulations stipulates that:

> Work rooms should have enough free space to allow people to get to and from workstations and to move within the room with ease. The number of people who may work in any particular room at any one time will depend not only on the size of the room, but on the space taken up by furniture, fittings, equipment, and on the layout of the room. Work rooms, except those where people only work for short periods, should be of sufficient height (from floor to ceiling) over most of the room to enable safe access to workstations. In older buildings with obstructions such as low beams the obstruction should be clearly marked.
>
> The total volume of the room, when empty, divided by the number of people normally working in it should be at least 11 cubic metres. In making this calculation a room or part of a room which is more than 3.0 m high should be counted as 3.0 m high. The figure of 11 cubic metres per person is a minimum and may be insufficient if, for example, much of the room is taken up by furniture, etc.

Temperature

Exposure to excessive workplace temperatures can result in heat stress, heat stroke, cold stress and cramps. The Workplace (Health, Safety and Welfare) Regulations place the following duties on employers:

Table 9.2 Recommended working temperatures

Type of work	Temperature
Sedentary/office work	
Comfort range	19.4–22.8°C
Light work	
Optimum temperature	18.3°C
Comfort range	15.5–20°C
Heavy work	
Comfort range	12.8–15.6°C

During working hours, the temperature in all workplaces shall be reasonable.

A method of heating or cooling shall not be used which results in the escape into the workplace of fumes, gas or vapour of such character and to such extent that they are likely to be injurious or offensive to any person.

A sufficient number of thermometers shall be provided to enable persons at work to determine the temperature in any workplace inside a building.

The Approved Code of Practice to the Workplace (Health, Safety and Welfare) Regulations advises that the temperature in workrooms should normally be at least 16°C unless much of the work involves severe physical effort in which case the temperature should be at least 13°C. These temperatures may not, however, ensure reasonable comfort, depending on other factors such as air movement and relative humidity.

Recommended working temperatures are shown in Table 9.2.

Ventilation

The Workplace (Health, Safety and Welfare) Regulations require employers to make effective and suitable provision to ensure that every enclosed workplace is ventilated by a sufficient quantity of fresh or purified air. This is the process of providing comfort ventilation, as opposed to local exhaust ventilation, which may well be required where there are hazardous airborne contaminants emitted from processes.

Enclosed workplaces should be sufficiently well ventilated so that stale air and air which is hot or humid because of the processes or equipment in the workplace is replaced at a reasonable rate (see Table 9.3).

Comfort ventilation is assessed on the basis of the number of complete air changes per hour during summer and winter months.

In many cases, windows or other openings will provide sufficient ventilation in some or all parts of the workplace. Where necessary, mechanical ventilation

Table 9.3 Recommended comfort ventilation rates

Location	Summer	Winter
Offices	6	4
Corridors	4	2
Amenity areas	6	4
Storage areas	2	2
Production areas with heat-producing plant	20	20
Production areas (assembly, finishing work)	6	4
Workshops	6	4

systems should be provided for parts of the workplace, as appropriate. Workers should not be subject to uncomfortable draughts. In the case of mechanical ventilation systems, it may be necessary to control the direction or velocity of air flow. Workstations should be resited or screened if necessary.

Where employees are exposed to airborne contaminants generated by processes, such as welding operations or those involving the interaction of hazardous chemical agents, local exhaust ventilation should be provided.

Lighting

Poor standards of lighting can cause visual fatigue and, in particular, can be a contributory factor in workplace accidents. Under the Workplace (Health, Safety and Welfare) Regulations, every workplace must have 'suitable and sufficient' lighting. In any consideration of workplace lighting, two aspects must be considered namely the quantity of light, measured in Lux, and the quality of light.

Quantitative aspects of lighting

HSE Guidance Note HS(G) 38 Lighting at work provides information on the quantitative aspects of lighting distinguishing between an 'average illuminance' level and a 'minimum measured illuminance' level according to the general work activities undertaken and the type of location (see Table 9.4).

Qualitative aspects of lighting

The following six aspects should be considered in assessing the quality of lighting provision (Table 9.5).

1. **Glare**. This is the effect of light that causes discomfort or impaired vision and is experienced when parts of the visual field are excessively bright when

Table 9.4 Average illuminances and minimum measured illuminances

General activity	Types of locations/types of work	Average illuminances in Lux	Minimum measured illuminances in Lux
Movement of machines and vehicles[a]	Lorry parks, corridors, circulation routes	20	5
Movement of people, machines and vehicles in hazardous areas; rough work not requiring any perception of detail[a]	Construction site clearances, excavation and soil work, docks, loading bays, bottling and canning plants	50	20
Work requiring limited perception of detail[b]	Kitchens, factories assembling large components, potteries	100	50
Work requiring perception of detail[b]	Offices, sheet metal work, bookbinding	200	100
Work requiring perception of fine detail[b]	Drawing offices, factories assembling electronic components, textile production	500	200

[a] Only safety has been considered, because no perception of detail is needed and visual fatigue is unlikely. However, where it is necessary to see detail, to recognize a hazard or where error in performing tasks could put someone else at risk, for safety purposes as well as to avoid visual fatigue, the figure should be increased to that for work requiring the perception of detail.

[b] The purpose is to avoid visual fatigue; the illuminances will be adequate for safety purposes.

Table 9.5 Effective reflectivity values

Given a task illuminance factor of 1, the effective reflectivity values should be:

Ceilings	0.6
Walls	0.3–0.8
Floors	0.2–0.3

compared with the general surroundings. This usually occurs when the light source is directly in line with the visual task or when light is reflected off a given surface or object. Glare is experienced in three different forms: 'Disability glare' is the visually disabling effect caused by bright bare lamps directly in the line of sight and the resulting impaired vision (dazzle) may be hazardous if experienced when working in high-risk processes, for example, at height, or when driving. 'Discomfort glare' is caused mainly by too much contrast of brightness between an object and its background, and is associated with poor lighting design. It causes visual discomfort without necessarily impairing the ability to see detail but, over a period, can cause eyestrain, headaches and fatigue. Discomfort glare can be reduced by the careful design of shades that screen the lamp, by keeping luminaires as high as practicable and maintaining luminaires parallel to the main direction of lighting. 'Reflected glare' is the reflection of bright light sources on shiny or wet work surfaces, such as glass or plated metal, which can almost entirely conceal the detail in or behind the object that is glinting. Care is necessary in the use of light sources of low brightness and in the arrangement of the geometry of the installation to avoid glint at the particular viewing position.

2. **Distribution**. The distribution of light, or the way in which light is spread, is important in lighting design. Poor distribution may result in the formation of shadowed areas which can create dangerous situations, particularly at night. For good general lighting, regularly spaced luminaires are used to give evenly distributed illumination. Evenness of illumination depends upon the height of luminaires above the working position and the spacing of fittings.
3. **Colour rendition**. Colour rendition refers to the appearance of an object under a given light source compared to its colour under a reference illuminant, such as natural light. Colour rendition enables the colour appearance to be correctly perceived. The colour-rendering properties of light fitments should not clash with those of natural light and should be equally effective at night when no daylight contributes to the illumination of the workplace.
4. **Brightness**. Brightness, or luminosity, is essentially a subjective sensation and cannot be measured. However, it is possible to consider a 'brightness ratio', which is the ratio of apparent luminosity between a task object and its surroundings. To achieve the recommended brightness ratio, the reflectivity of all surfaces in a workplace should be carefully maintained and consideration given to reflectance values in the design of interiors.
5. **Diffusion**. This is the projection of light in many directions with no direction being predominant. Diffused lighting can soften the output from a particular source and so limit the amount of glare that may be encountered from bare lamps and fittings.
6. **Stroboscopic effect**. All lamps that operate from an alternating current supply produce oscillations in light output. A stroboscopic effect occurs when the magnitude of the oscillations is great and their frequency is a multiple or submultiple of the frequency of movement of machinery, the machinery will appear to be stationary. This is uncommon with modern lighting systems but, where it does arise, it can be dangerous. Remedial measures include:

- supplying adjacent rows of light fittings from different phases of the electricity supply;
- providing a high frequency supply;
- washing out the effect with local lighting that has much less variation in light output, such as the installation of tungsten lamps; and
- using high frequency control fittings if applicable.

Humidity

Humidity in the workplace is measured in terms of relative humidity. Relative humidity is the amount of moisture present in air expressed as a percentage of that which would produce saturation. Relative humidity should be between 30 and 70 per cent. Where humidity is too low, there will be discomfort due to drying of the throat and nasal passages. Conversely, high humidity levels produce a sensation of stuffiness and the rate at which body moisture (sweat) evaporates is reduced, thereby reducing the effectiveness of the body's thermoregulatory system.

Noise

'Noise' is defined as 'unwanted sound' and measured in decibels, a logarithmic unit of sound intensity. Unwanted noise from a variety of sources can result in loss of concentration, distraction and the masking of warning signals and is a contributory factor in accidents.

However, the principal risk associated with exposure to noise is that of employees contracting noise-induced hearing loss (occupational deafness). The Control of Noise at Work Regulations require that where work is liable to expose employees to noise at or above a *lower exposure action value*, the employer shall make a suitable and sufficient assessment of the risk from the noise to the health and safety of those employees.

Assessment shall identify whether any employees are likely to be exposed to noise at or above:

A lower exposure action value (LEAV), i.e.

- a daily or weekly personal noise exposure of 80 dBA; and
- a peak sound pressure of 135 dBC;

an upper exposure action value (UEAV), i.e.

- a daily or weekly personal noise exposure of 85 dBA; and
- a peak sound pressure of 137 dBC; or

an exposure limit value (ELV), i.e.

- a daily or weekly personal noise exposure of 87 dBA; and
- a peak sound pressure of 140 dBC.

The regulations specify the factors that must be taken into account in a noise risk assessment.

Fatigue and stress

'Fatigue' is a term which is difficult to describe. It is synonymous with a feeling of tiredness, weariness or exhaustion and takes the form of both muscular (or physical) fatigue and mental fatigue. What is significant is that as fatigue increases, performance decreases.

Fatigue is a term used to cover all those determinable changes in the expression of an activity which can be traced to the continuing exercise of the activity under its normal operational conditions, and which can be shown to lead, either immediately or after delay, to deterioration in the expression of that activity, or more simply, to results within the activity that are not wanted.

Muscular fatigue

Muscle is the most abundant tissue in the body and accounts for some two-fifths of the body weight. The specialized component is the muscle fibre, a long slender cell or agglomeration of cells which become shorter and thicker in response to a stimulus. These fibres are supported and bound by ordinary connective tissue and are well supplied with blood vessels. When muscles are used, say, for manual handling operations, they are subjected to varying degrees of stress. Carrying a load generally imposes a pronounced static strain on many groups of muscles, especially those of the arms and trunk. This is a particularly unsuitable form of work for human beings because the blood vessels in the contracted muscles are compressed and the flow of blood, and with it the oxygen and sugar supply, is thereby impeded. As a result, fatigue very soon sets in, with pains in the back muscles, which perform static work only, occurring sooner than in the arm muscles, which perform essentially dynamic work. Muscular swelling is associated with the greatly increased blood flow which leads to an increased volume of tissue fluid.

Muscular fatigue is characterized by aching and even pains in the muscles concerned and increased inability and weakness of muscular movement. Although all the factors involved are not fully understood, the accumulation of lactic acid and swelling of the muscle play a significant part. Lactic acid can, and does, eventually prevent muscular contraction completely, and in lower concentration impairs it. In physical work of long duration there is invariably a build up of lactic acid.

As fatigue develops, performance of a task becomes irregular. The subjective aspects of fatigue include complaints of aches and pains, usually rather vague and generalized, malaise and sensations described as tiredness. These subjective sensations are variable and inconsistent and do not necessarily relate to other aspects of fatigue. Related to these sensations there is an increase in muscle activity, an increase in the tension exerted by muscles which should be in a state of rest or relaxation. This leads to another feature of fatigue, the increasing use of extra muscles as performance on a task continues. As the primary muscles fatigue, more and more secondary muscles are used in an attempt to relieve the load on the fatigued muscles. This would account for the fact that as fatigue develops greater effort is required to maintain performance.

To combat muscular fatigue, frequent short pauses, as opposed to occasional long pauses, from the work activity are more effective.

Mental fatigue

The causes of mental fatigue are unclear. In tasks with a considerable perceptual element, where information is derived form a number of sources, auditory, visual or tactile, as fatigue develops the field of display becomes less adequately scanned and there are lapses of attention. These are more evident in a paced task where the operator has to work at a particular rate. A lapse of attention will mean that a particular task will be missed.

Increasing mental fatigue is indicated by increased reaction times, particularly where people are being asked to make a decision. This is particularly the case with display screen equipment users where, as mental fatigue increases, attention to the screen becomes more erratic and the wrong keys may be used. Similarly, in the cockpit of an aircraft, as mental fatigue increases, the pilot's concentration may be restricted to the more important displays or too much attention paid to the peripheral elements at the expense of the central elements. The effect is that the right action may be performed at the wrong time and some actions may be omitted.

Mental fatigue is often associated, and perhaps identical, with boredom due to the nature of the work. Repetitive tasks can result in a reduction in arousal on the part of the individual, day dreaming, a degradation of performance and increased potential for errors.

As with muscular fatigue, short rest pauses should be taken to allow functioning to return to its normal level. If performance continues without the necessary break, more obvious signs of degradation can follow.

Manual handling

More than a third of all industrial injuries result from manual handling activities. Statistics from the last 60 years indicate that in almost every year the number of people injured in this way has increased. The vast majority of reported manual handling accidents result in over three-day injury, most commonly a sprain or strain, often of the back. Not only manual workers contribute to the handling injuries statistics, however. Those in sedentary occupations are similarly at risk, e.g. office workers, library staff, catering staff, hospital workers and people working in shops.

HSE Guidance accompanying the Manual Handling Operations Regulations 1992 classifies the types and sites of injury caused by handling operations as shown in Table 9.6.

Table 9.6 Manual handling operations: Types and sites of injury

	%
Types of injury	
Sprains and strains	65
Superficial injuries	9
Contusions	7
Lacerations	7
Fractures	5
Other forms of injury	5
Sites of bodily injury	
Backs	45
Fingers and thumbs	16
Arms	13
Lower limbs	9
Rest of torso	8
Hands	6
Other sites	3

What comes out of the statistical information on manual handling injuries is the fact that four out of five people will suffer some form of back condition at some time in their lives, the majority of these conditions being associated with work activities.

Typical injuries and conditions associated with manual handling can be both external and internal. External injuries include cuts, bruises, crush injuries and lacerations to fingers, forearms, ankles and feet. Generally, such injuries are not as serious as the internal forms of injury which include muscle and ligamental tears, hernias (ruptures), prolapsed intervertebral discs (slipped discs) and damage to knee, shoulder and elbow joints.

Muscle and ligamental strain

Muscular fatigue and strain were considered earlier in this chapter with respect to fatigue. Their effects can be longstanding and lead to eventual physical breakdown if the causes are not identified and preventive action taken.

Ligaments, on the other hand, are fibrous bands occurring between two bones at a joint. They are flexible but inelastic, come into play only at the extremes of movement, and cannot be stretched when they are taut. Ligaments set the limits beyond which no movement is possible in a joint. A joint can be forced beyond its normal range only by tearing a ligament, resulting in a sprain. Fibrous tissue heals reluctantly and a severe sprain can be as incapacitating as a fracture. There are many causes of torn ligament, in particular, jerky handling movements which place stress on a joint, uncoordinated team lifting, and dropping a load half way through a lift, often caused by failing to assess the load prior to lifting.

Hernia (rupture)

A hernia is a protrusion of an organ from one compartment of the body into another, e.g. a loop of intestine into the groin or through the frontal abdominal wall. Both these forms of hernia can result from incorrect handling techniques and particularly from the adoption of bent back stances, which produce compression of the abdomen and lower intestines.

The most common form of hernia or 'rupture' associated with manual handling is the inguinal hernia. The weak point is the small gap in the abdominal muscles where the testis descends to the scrotum. Its vessels pass through the gap, which therefore cannot be sealed. Excessive straining and even coughing may cause a bulge at the gap and a loop of intestine or other abdominal structure easily slips into it. An inguinal hernia sometimes causes little trouble, but it can, without warning, become strangulated, whereby the loop of intestine is pinched at the entrance to the hernia. Its contents are obstructed and fresh blood no longer reaches the area. Prompt attention is needed to preserve the patient's health and even his life may be at risk if the condition does not receive swift attention. The defect, in most cases, must be repaired surgically.

Prolapsed or 'slipped' disc

The spine consists of a number of small interlocking bones or vertebrae. There are seven neck or cervical vertebrae, twelve thoracic vertebrae, five lumbar vertebrae, five sacral vertebrae and four caudal vertebrae. The sacral vertebrae are united, as are the caudal vertebrae, the others being capable of independent but co-ordinating articulating movement. Each vertebra is separated from the next by a pad of gristle-like material (intervertebral disc). These discs act as shock absorbers to protect the spine. A prolapsed or slipped disc occurs when one of these intervertebral discs is displaced from its normal position and is no longer performing its function properly. In other cases, there may be squashing or compression of a disc. This results in a painful condition, sometimes leading to partial paralysis, which may be caused when the back is bent while lifting, as a result of falling awkwardly, getting up out of a low chair or even through over-energetic dancing.

Rheumatism

This is a painful disorder of joints or muscles not directly due to infection or injury. This rather ill-defined group includes rheumatic fever, rheumatoid arthritis, osteoarthritis, gout and 'fibrositis', itself an ill-defined group of disorders in which muscular and/or joint pain are common factors. There is much evidence to support the fact that stress on the spine, muscles, joints and ligaments during manual handling activities in early life results in rheumatic disorders as people get older.

Safe manual handling

Injuries associated with manual handling of raw materials, people, animals, goods and other items are the principal form of injury at work. Such injuries include prolapsed intervertebral (slipped) discs, hernias, ligamental strains and various forms of physical injury. It is essential, therefore, that all people required to handle items be aware of the basic principles of safe manual handling. The following principles should always be considered in handling activities and in the training of people in manual handling techniques.

- **Feet positions**. Place feet hip breadth apart to give a large base. Put one foot forward and to the side of the object to be lifted. This gives better balance.
- **Correct grip**. Ensure that the grip is by the roots of the fingers and palm of the hand. This keeps the load under control and permits the load to be better distributed over the body.
- **Arms close to body**. This reduces muscle fatigue in the arms and shoulders and the effort required by the arms. It ensures that the load, in effect, becomes part of the body and moves with the body.
- **Flat back**. This does not mean vertical, but at an angle of approximately 15°. This prevents pressure on the abdomen and ensures an even pressure on the vertebral discs. The back will take the weight but the legs do the work.
- **Chin in**. It is just as easy to damage the spine at the top as it is at the bottom. To keep the spine straight at the top, elongate the neck and pull the chin in. Do not tuck the chin on to the chest as this bends the neck.
- **Use of body weight**. Use the body weight to move the load into the lifting position and to control movement of the load.
- **Planning the lift**. When considering moving a load, make sure your route is unobstructed and that there are no obstructions or tripping hazards. Ensure there is an area cleared to receive the load. If your route requires you to wear a safety helmet, eye protection or hearing protection, put them on before you lift.

Assessing/Testing the load

The majority of injuries occur when actually lifting the load. People involved in handling operations should be instructed to assess the following:

- Are there any rotating or moving parts? If so, do not use them to lift with.
- Is the load too big to handle, e.g. a patient in a wheelchair? If so, get help.
- Is the load too heavy? Rock the load. This will give a rough idea of its weight. If too heavy, get help.

Personal protective clothing (PPE)

The provision and use of the correct PPE is an essential feature of safe manual handling. The following instructions should be incorporated in manual handling training and activities:

1. **Hand protection**. Examine the load for evidence of sharp edges, protruding wires, splinters or anything that could injure the hands. Wear the correct type of glove to prevent hand injury.
2. **Feet protection**. Wear footwear which is suitable for the job:
   - with steel toe caps to protect the feet against falling objects or if the feet could get trapped under the load;
   - steel insoles to protect against protruding nails;
   - soles that will resist heat, oil and acid.

Two-person lift

Two people should always be involved in lifting clients, for instance, from a bed to a wheelchair, or from a wheelchair to a chair. Special precautions may be necessary where clients may be liable to struggle whilst being lifted.

Use all the principles involved in a one man lift, with one variation. The leading foot should point in the direction of travel. One person should give the order to lift, ensuring that his partner understands the order. It is vital that there be unison in the movement of both people and the load.

Manual handling risk assessment

Under the Manual Handling Operations Regulations, where an employer cannot, as far as is reasonably practicable, avoid manual handling operations at work which involve a risk of employees being injured, for example, by the provision of mechanised handling aids, he must carry out a risk assessment of the operations involved.

Manual Handling Operations Regulations 1992

This risk assessment follows the format recommended in the HSE Guidance to the Manual Handling Operations Regulations 1992 (see Figure 9.1).

Manual handling of loads

EXAMPLE OF AN ASSESSMENT CHECKLIST

Note: This checklist may be copied freely. It will remind you of the main points to think about while you:
– consider the risk of injury from manual handling operations
– identify steps that can remove or reduce the risk
– decide your priorities for action.

SUMMARY OF ASSESSMENT	Overall priority for remedial action: Nil/Low/Med/High*
Operations covered by this assessment:	Remedial action to be taken: ....
....	....
....	....
Locations:	Date by which action is to be taken: ....
Personnel involved:	Date for reassessment: ....
Date of assessment:	Assessor's name: .... Signature: ....

***circle as appropriate**

Section A – Preliminary:

Q1 Do the operations involve a significant risk of injury? **Yes/No***

If '**Yes**' go to Q2. If '**No**' the assessment need go no further.

If in doubt answer '**Yes**'. You may find the guidelines in Appendix 1 helpful.

Q2 Can the operations be avoided/mechanised/automated at reasonable cost? **Yes/No***

If '**No**' go to Q3. If '**Yes**' proceed and then check that the result is satisfactory.

Q3 Are the operations clearly within the guidelines in Appendix 1? **Yes/No***

If '**No**' go to Section B. If '**Yes**' you may go straight to Section C if you wish.

Section C – Overall assessment of risk:

Q What is your overall assessment of the risk of injury? **Insignificant/Low/Med/High***

If not '**Insignificant**' go to Section D. If '**Insignificant**' the assessment need go no further.

Section D – Remedial action:

Q What remedial steps should be taken, in order of priority?

i ....

ii ....

iii ....

iv ....

v ....

And finally:

– complete the SUMMARY above
– complete it with your other manual handling assessments
– decide your priorities for action
– TAKE ACTION .... AND CHECK THAT IT HAS THE DESIRED EFFECT

Figure 9.1 Example of an assessment checklist

Section B – More detailed assessment, where necessary:

Questions to consider: (If the answer to a question is 'Yes' place a tick against it and then consider the level of risk)		**Level of risk:** (Tick as appropriate)			**Possible remedial action:** (Make rough notes in this column in preparation for completing Section D)
	Yes	Low	Med	High	
The tasks – do they involve:					
• holding loads away from trunk?					
• twisting?					
• stooping?					
• reaching upwards?					
• large vertical movement?					
• long carrying distances?					
• strenuous pushing or pulling?					
• unpredictable movement of loads?					
• repetitive handling?					
• insufficient rest or recovery?					
• a workrate imposed by a process?					
The loads – are they:					
• heavy?					
• bulky/unwieldy?					
• difficult to grasp?					
• unstable/unpredictable?					
• intrinsically harmful (e.g. sharp/hot?)					
The working environment – are there:					
• constraints on posture?					
• poor floors?					
• variations in levels?					
• hot/cold/humid conditions?					
• strong air movements?					
• poor lighting conditions?					
Individual capability – does the job:					
• require unusual capability?					
• hazard those with a health problem?					
• hazard those who are pregnant?					
• call for special information/training?					
Other factors –					
Is movement or posture hindered by clothing or personal protective equipment?					

Deciding the level of risk will inevitably call for judgement. The guidelines in Appendix 1 may provide a useful yardstick.

When you have completed Section B go to Section C.

Figure 9.1 (Continued)

Lifting and lowering loads

The HSE have issued guidelines for lifting and lowering based on the actual weight of a load and the distance from the body in which the load is lifted (see Figure 9.2).

The guidelines that the load is readily grasped with both hands and that the operation takes place in reasonable working conditions with the handler in a stable body position. The guidelines take into consideration the vertical and horizontal position of the hands as they move the load during the handling operation, as well as the height and reach of the individual handler. It will be apparent that the capability to lift or lower is reduced significantly if, for example, the load is held at arm's length or the hands pass above shoulder height.

If the hands enter one of the box zones during the operation, the smallest weight figure should be used. The transition from one box zone to another is not abrupt;

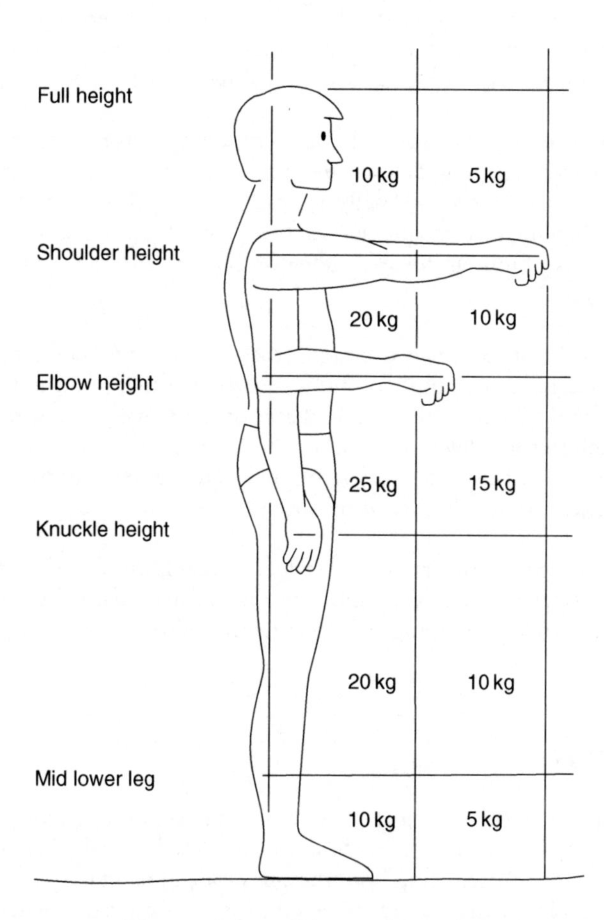

Figure 9.2 Lifting and lowering

an intermediate figure may be chosen where the hands are close to the boundary. Where lifting or lowering with the hands beyond the box zones is unavoidable a more detailed assessment should be made.

Performance shaping factors

Traditionally, the main objective of ergonomic study is that of improving the performance of the total working system together with reducing stress on operators through analysing tasks, the working environment and the man–machine interface.

The stress–strain concept

People talk about the 'stress and strain' of work. This stress–strain concept is well accepted in ergonomic study. The basic concept is that each workplace is characterized by external factors which are identical for all people working in that workplace (stress). The individual reacts differently to these stressors depending upon his individual characteristics, skills and abilities (strain).

In terms of human performance and reliability, many factors affect performance. These performance-shaping factors are associated with both external (stress-related) and internal (strain-related) factors, as shown in Table 9.7. These performance-shaping factors interact in the case of individuals producing an individual who is unique in his own level of performance, style of performance and motivation.

By reference to Table 9.7, it can be seen that *external factors* are associated with the organizational and technical conditions existing at that point in time, namely the organizational structure, the organization of the work process, task difficulty and situational factors. *Internal factors* cover aspects such as work capacity in terms of physical and psychical capacity, and achievement motivation, namely the physiological and psychological readiness to perform.

External factors, in particular work process organization, task difficulty and situational factors, have a direct influence on occupational safety. In the same way, these performance shaping factors directly influence internal performance shaping factors.

Conclusions

Ergonomic factors in the workplace have a direct effect on operator performance. Ergonomically designed interfaces should be directed at reducing the potential for human error and in reducing both the physical and mental stress on operators. Current legislation places considerable emphasis on environmental control of the workplace and in the design of work stations. Moreover, ergonomic design

Table 9.7 Performance shaping factors

External factors

- Organizational conditions

  Organizational structure

  - Management style
  - Method of payment
  - Vocational training

  Work process organization

  - Working time
  - Work scheduling
  - Job instructions and procedures in terms of
    - accuracy
    - ease of use
    - sufficiency
    - applicability
    - clarity
    - format
    - meaning
    - level of detail
    - readability
    - selection and location
    - revision
  - Manning
  - Resource availability
  - Social pressures
  - Team structure
  - Communication systems
  - Authority and responsibility
  - Group practices
  - Rewards and benefits

- Technical conditions

  Task difficulty

  - Equipment design
  - Work content
  - Task design
  - Task characteristics
    - frequency
    - repetitiveness
    - workload
    - criticality
    - continuity

(*continued*)

Table 9.7 (Continued)

duration
interaction with other tasks

- Task demands
  perceptual
  physical
  memory
  attention
  vigilance

Situational Factors

- Workplace design
- Anthropometric design
- Man–machine interface characteristics (displays and controls)
  compatibility
  ease of use/operation
  sufficiency
  location
  reliability
  readability
  meaning
  feedback
  distinguishability
  feedback
  identification
- Environmental design
  temperature
  humidity
  noise
  lighting
  work space

Internal factors

- Work capacity

Physical capacity

- Constitution
- Sex
- Age
- Physical condition
- Physical capacity

Psychological capacity

- Mental talents
- Level of education

Table 9.7 (Continued)

- Exercise
- Training
- Attitudes
- Experience
- Personality
- Skills
- Knowledge
- Motivation
- Achievement motivation

Physiological readiness to perform

- Disposition
- Circadian Rhythm
- Illness
- Emotional state

Psychological readiness to perform

- Internal motivation
  - interests
  - likings
  - mood
- External motivation

should take into account the risk of fatigue associated with activities such as manual handling.

Employers would do well to pay attention to the 'stress–strain' concept and the various factors which shape individual performance.

Key points

- Ergonomically designed interfaces should be directed at reducing the potential for human error and in reducing both the mental and physical stress on operators.
- The design of display and control systems for work equipment is an essential element of the man–machine interface.
- One of the objectives of ergonomic study is to design tasks to fit the mental and physical capabilities of the operator.
- Systems ergonomics endeavours to take human error into account through techniques such as systems analysis.
- A broad range of physical stressors affect human reliability.

- The design, specification and control of environmental factors in the workplace, such as temperature, lighting, ventilation and humidity, is an essential element of ergonomic design.
- Manual handling-related injuries, such as prolapsed intervertebral discs and hernias, are one of the principal causes of lost time and can have longstanding effects on those suffering such injury.
- The Manual Handling Operations Regulations require employers to undertake manual handling risk assessments wherever there is foreseeable risk of injury to employees arising from such operations.
- In terms of human performance and reliability, many factors affect performance. Consideration needs to be given to these performance-shaping factors in the design of work systems, working environments and the workplace.

10 Principles of communication

Communication is the principal element of all human contact and can take many forms. It provides information to people and tells them about our feelings, emotions, sentiments, thoughts and ideas. Employers need to regularly examine their communication processes particularly where there is evidence of a communications gap.

What is communication?

Communication is variously defined as:

- imparting or transmitting information (Oxford Dictionary);
- the transfer of information, ideas, feelings, knowledge and emotions between one individual or group of individuals and another.

The basic function is to convey meanings. Good standards of communication on health and safety issues are essential in the prevention of accidents and ill-health at work. Whilst much communication takes place verbally, nonverbal communication has an important part to play in the overall process.

The purpose of communication

There are a number of reasons why people communicate, for example,

- to improve knowledge and understanding;
- to influence people, change attitudes and raise awareness; and
- to bring about change, instigate some form of action and influence behaviour.

The communication process

Communication involves both a communicator (the originator of the message) and a receiver (the recipient of the message). Communication commonly takes place in four phases or stages, thus

1. transmission or sending out of data, both cognitive data and emotional data;
2. receiving or perceiving the data;

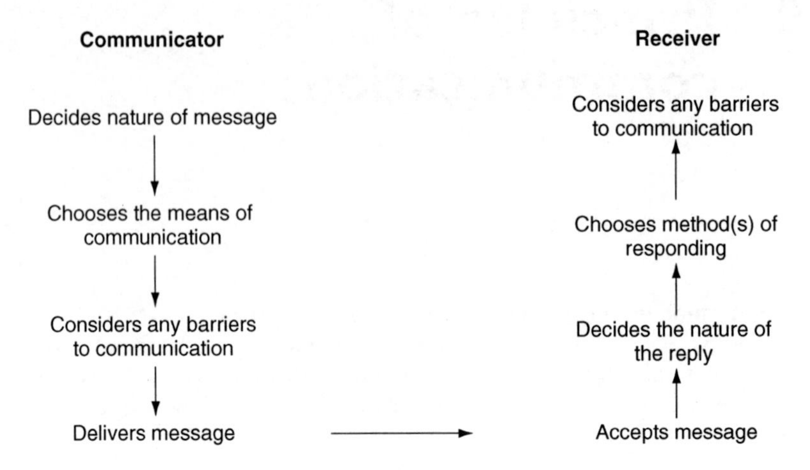

Figure 10.1 The communication process

3. understanding the data; and
4. acceptance of the data.

The various stages of the communication process are shown in Figure 10.1. Three important aspects must be considered in the communication process:

The communicator

The specific objectives of the communicator must be clear. For example, a health and safety specialist may wish to acquaint a particular group of maintenance staff in the need for precautions in working at height. In deciding on the appropriate approach, he may need to consider the risks involved in this sort of work, the need to increase awareness of safe working practices and current legal requirements.

The recipients

The recipients of a message, in certain cases, may not want to hear what the communicator has to say. They may feel that they are being 'picked on', they already know about safe working practices at height and do not perceive any benefit from this communication. Fundamentally, the recipients have established their particular barriers to communication on this issue. The health and safety specialist has the task of considering these barriers and offering an approach directed at removing these barriers.

The purpose of the communication

The communication must have meaning and significance to recipients. Where this is not the case, people will feel it is not relevant to their work activities and disregard the communication. On this basis, the purpose of the communication should be clearly stated and, where appropriate, specifying individual responsibilities of people with respect to the issues raised in the communication.

Forms of communication

Many forms of communication operate within organizations. That which is officially inspired is often referred to as formal communication, while communication which is unofficial, unplanned and spontaneous can be classified as informal communication. Communication can also be by verbal or nonverbal means.

Interpersonal communication consists of a complex fabric of interacting cues or signals of moving kinds including, for example, the sequence of words coloured by voice tone, pitch, stress or rhythm, and movement of hands, eyes and body. Communication almost always requires the use of codes of some kind, e.g. language, nonverbal cues, evidence of an emotional state, specific relationships and understandings.

Good standards of communication are an essential feature of the health and safety management process. This may take place through training, the provision of information and instructions, joint consultation through, perhaps, a safety committee, and various techniques designed to obtain feedback from employees as to the success or otherwise of health and safety measures installed.

Communication within organizations

Many forms of communication exist within organizations.

- **Formal and informal communication**. That communication which is officially inspired by directors and managers is often referred to as formal communication. Communication which is, on the other hand, unofficial, unplanned and spontaneous is classified as informal communication.
- **One-way and two-way communication**. Communication may be one way, e.g. an instruction to staff from a senior manager, or two-way, where, following transmission of a message, the views of certain persons are sought prior to the decision-making process.
- **Verbal, nonverbal and written communication**. The form of communication is significant. Communication may be by word of mouth (verbal communication), by the use of gestures, eye contact and terminal glances (nonverbal communication) and by the use of memoranda, letters and reports (written communication).
- **Intentional or unintentional**. Communication may be intentional, i.e. consciously transmitted, or unintentional, where the information is involuntarily transmitted.

Significant features of the communication process are, firstly, the actual lines or channels of communication and, secondly, the relative effectiveness of the different forms of communication.

Functions of communication

There are many functions of communication, which may be summarized as follows:

- **Instrumental**. To achieve or obtain something, e.g. improved safety performance or greater commitment to ensuring safe working practices.
- **Control**. To get someone to behave in a particular way, e.g. a person who may not be following a safe system of work.
- **Information**. To establish facts or, alternatively, to explain something to a person.
- **Expression**. To express feelings, e.g. fear, anxiety, happiness, guilt or to put oneself over in a particular way.
- **Social contact**. To enjoy another person's company.
- **Alleviation of anxiety**. To sort out a problem to ease a worry.
- **Stimulation**. To increase or raise the interest of an individual or group.
- **Role-related**. To give instructions, advice or a warning to a subordinate because the situation required it.

The direction of communication

One-way communication

One-way communication may be appropriate in certain circumstances and is certainly faster than two-way communication. As such it does not permit any form of feedback from the receiver of the communication. It is used principally in the giving of directions, instructions and orders by senior management.

Two-way communication

This has been found to be far more effective. As such it gives people the chance to use their intelligence, to contribute knowledge, to participate in the decision-making process, to fulfil their creative needs and to express agreement or disagreement. It helps both the sender and the receiver to measure their standard of achievement and when they both see that they are making progress, their joint commitment to a task will be greater. In some circumstances, the sender may feel that he is under attack as the receiver will identify his errors and inform him of them. However, this tends to be more helpful in that a frank discussion will lead to a higher level of understanding and acceptance.

Employees frequently complain about the lack of two-way communication, in particular upward communication. They feel that they have a great deal to contribute in terms of skills, knowledge, experience and goodwill which, if correctly taken on board by management, can have a significant effect on performance.

In the field of health and safety, the communication process can be improved by, for instance, the setting up of working parties with a view to achieving particular targets and implementing programmes of improvement. Working parties must be seen in a different light to health and safety committees, however.

The communications gap

Employees are interested in matters such as wages, working hours, benefits, employee relations issues, job security, job changes, working conditions, the future plans of the organization and the reasons for certain management decisions. What they are not interested in are statistical data on output, sales information, quality, costs and other aspects of the organization's performance. In many cases, they may not be interested in health and safety requirements unless they can see the clear purpose and intention.

Fundamentally, the organization's information is concerned with the well-being of the organization generally compared with the information that employees need, which is people orientated. Thus arises a communications gap in that the organization's information is largely concerned with what has happened in the past, whereas employees want to know what is happening in the future with respect to earnings, security and the continued operation of the organization.

The provision of information

Providing information is an important feature of the communication process. The sources of information on occupational health and safety are many and varied. Most of the examples quoted are, however, of a textual or written nature.

The process of giving and receiving information, however, takes a particular pathway commencing with the information source through the information provider to the actual information format to the information user. The actual format can vary substantially, thus,

- textual – books, letters, diagrams, charts, etc.;
- verbal – conversation face-to-face, lectures, etc.;

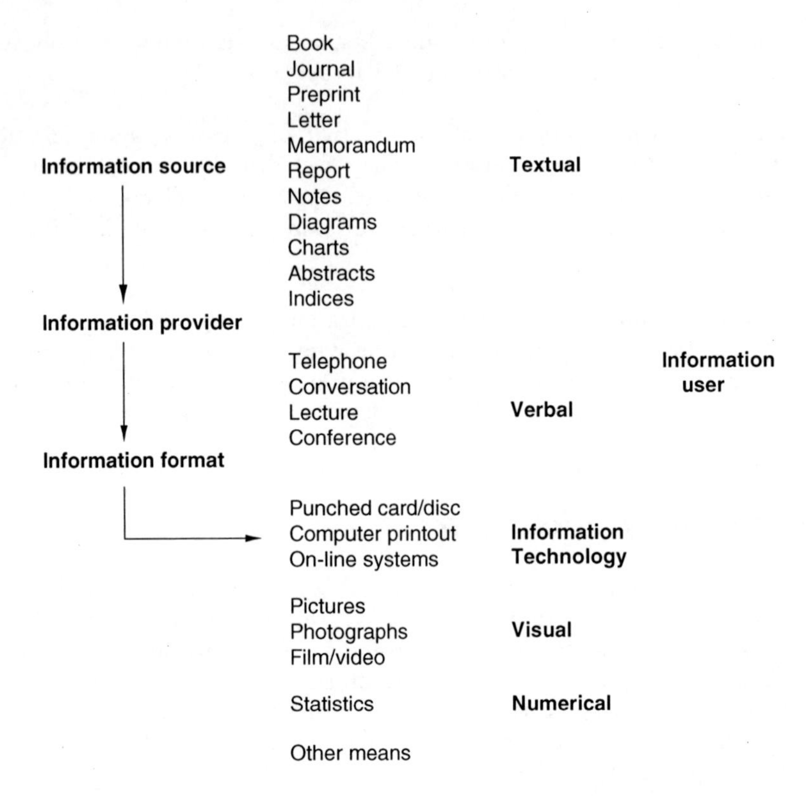

Figure 10.2 The provision of information

- information technology – computer printouts, on-line systems;
- visual aids – photographs, films, videos; and
- numerical – statistical information, bar charts, pie charts.

Other means of information transfer can be of an auditory and tactile nature.

Information retrieval systems

These systems provide information under a combination of formats, e.g. information technology, textual and visual. They all, however, incorporate the basic pathway mentioned above and are shown in Figure 10.2.

Communication on health and safety issues

Communication of the right kind has a vital part to play in health and safety as a participative process. The following aspects of communication are significant.

Safety propaganda

The use of posters, films, exhibitions and other forms of repeating a specific message are important features of the safety communication process. Safety posters should be used to reinforce current health and safety themes, e.g. the use of eye protection, correct manual handling techniques, and should be changed on a regular basis. To have the most impact, videos and films should be used as part of scheduled training activities and not shown in isolation.

Safety incentive schemes

Various forms of planned motivation directed at rewarding good safety behaviour on the basis of formally agreed objectives and criteria have proved successful. Safety incentive schemes should not be based on a reduction in accident rates, however, as this can reduce or restrict accident reporting by employees. All safety incentive schemes must be accompanied by efficient communication of results and information surrounding the scheme.

Effective health and safety training

Health and safety communication featured in training exercises should incorporate sincerity, authority, confidence, accuracy and humour. There may be a need for training of trainers in the various aspects of presentation.

Management example

This is perhaps the strongest form of nonverbal communication and has a direct effect on attitudes to health and safety held by operators. Management should openly display and promote, where appropriate:

- a desire to attain a common goal, e.g. clearly identified and agreed safety objectives;
- insight into ever-changing situations, particularly in the case of potentially hazardous operations;
- alertness to the needs and motives of others;
- an ability to bear responsibility with regard to health and safety procedures;
- competence in initiating and planning action to bring about health and safety improvements;
- social interaction aimed at promoting the health and safety objectives of the organization;
- communication, both upwards and downwards, through a wide variety of communication activities;

- clear identification by senior management with the health and safety promotional activities undertaken by the organization; and
- by setting an example to staff, visitors and others at all times on health and safety-related procedures.

Safety propaganda as a form of communication

Safety propaganda is a form of communication fundamentally concerned with

- increasing the awareness of people to hazards and the precautions necessary;
- changing attitudes; and
- imparting information on health- and safety-related issues.

Safety propaganda may include:

- publications;
- posters and signs;
- slide/tape programmes; and
- films and videos.

In order to decide which aspect of safety propaganda is best suited to the needs of the organization and specific groups within that organization, it is useful to note that a person tends to remember:

10 per cent of what he reads;

20 per cent of what he hears;

30 per cent of what he sees;

50 per cent of what he sees and hears;

70 per cent of what he says in conversation; and

90 per cent of what he says as he does a thing.

It follows, therefore, that the relative usefulness of safety propaganda will depend to a large extent on the method of communication used.

There are a variety of media through which safety propaganda can be communicated:

- publications, programmed learning (reads) – 10 per cent;
- taped commentaries, lectures (hears) – 20 per cent;
- slides, posters, signs, viewfoils (sees) – 30 per cent;
- films, videos, slide/tape programmes (sees and hears) – 50 per cent;

- discussion groups (says in conversation) – 70 per cent;
- on the job training, simulation exercises, role playing, (says as he does a thing) – 90 per cent.

Methods of communication

Pirani and Reynolds (1976) examined the use of five different methods of communication, using different facets of safety propaganda designed to persuade employees to utilize protective equipment. Their results in rank order from best (highest percentage improvement in utilization) to worst (little or no improvement) were

- role playing;
- films;
- posters and signs;
- discussion;
- discipline.

These results show that for safety propaganda to be effective, active participation is required in order to create the necessary changes in behaviour.

The importance of communication

Communication is the crucial link in successful integrative management. An essential requirement is that the channels of communication remain open and are used.

Effective communication is also an important feature of health and safety practice. Lack of communication is commonly a contributory feature to accidents and other forms of adverse incidents. Barriers to effective communication must be overcome by attention to the points raised in this chapter.

The legal duty of managers to communicate with staff and other people, through the provision of information, instruction and training, is clearly identified in the HSWA and in most of the regulations made under the HSWA.

Communications failure

Failures in the communication process are commonly associated with:-

- time defects – there may be a time lag between the communication going out and the recipient receiving it;
- spacial segregation – where the communicators may be some distance apart physically;

- work groups who, for a variety of reasons, fail to co-ordinate their various activities; and
- various degrees of conflict between the staff organization, specialists and line management.

Lack of understanding commonly arises as a result of failure to communicate accurately and fully the circumstances surrounding a situation.

Conclusions

Communication implies the imparting or transmitting of information and the principal function is to convey meaning. There must be a communicator and a recipient and, to be effective, communication should take place in a series of stages. Communication takes place in organizations in many ways. It may be formal or informal, one-way or two-way, verbal or nonverbal.

There are a variety of ways of communicating on health and safety issues. Much will depend upon the material to be communicated and the target audience. Communications failures, however, are common in many organizations and may be associated with time defects, spacial segregation, work groups who fail to co-ordinate their activities and varying degrees of conflict at work.

Key points

- One of the problems that many organizations need to solve is that of reducing the communications gap which exists between employers and employees.
- The provision of information is an important feature of the communication process and information can take many forms.
- The use of safety propaganda is one of the most important ways of communicating health and safety messages and increasing awareness in the workplace.
- Safety propaganda can have a direct effect on attitudes provided it is targeted at the right people at the right time and for the right reasons.
- Two-way communication is far more effective than one-way communication.
- Information retrieval systems tend to operate within a basic pathway.
- Safety incentive schemes and other forms of planned motivation directed at rewarding appropriate safe behaviour can be effective means of communication.
- Lack of communication is commonly a contributory factor in accidents and other forms of adverse incident.

11 Verbal and nonverbal communication

Both verbal and nonverbal communication feature significantly in the lives of people. Together, they are the principal form of communication. They are a means of indicating to people about the way they feel, for giving and receiving information and for telling people what to do.

Verbal communication

Verbal communication is the process of communication by mouth. It should not be treated solely in terms of the verbal content of what is being said, however. It should incorporate other things being communicated, for instance,

- the variation in social status of the participants;
- their emotional attitudes;
- the various nonverbal signs, such as eye gaze, body signals, etc.;
- the social context of the communication; and
- each individual's perception of the other.

Features of verbal communication

Interpersonal verbal communication is characterized by

- the way the words are spoken;
- the accompanying nonverbal information;
- facial expression, gestures and posture;
- the expectation of participants;
- the context in which the transaction occurs.

Verbal communication incorporates a number of features which assist in the actual communication process. These include

- reflexes, e.g. coughing;
- nonverbal noises, e.g. grunts, which need nonverbal accompaniment to clarify their meanings;

- voice quality, e.g. accents, which are significant in categorizing the individual and the social group to which he belongs; and
- linguistic aspects, in terms of the choice of words; and
- paralinguistic aspects, namely timing, speech, rhythm, tone of voice and pitch of voice.

Interpersonal communication

Interpersonal communication, thus, consists of a complex fabric of interacting cues or signals of moving kinds, in particular the sequence of words coloured by voice, tone, pitch, stress or rhythm, and movement of eyes, hands and body.

Communications almost always requires the use of a code of some kind, e.g. language, nonverbal cues, emotional states, relationships and understandings.

Language

This is concerned with the three kinds of meaning attached to words:

1. Denotive – the key features that distinguish it;
2. Connotive – varies according to experience, association and context; and
3. Indexical – provide an indication of the nature of the speaker.

Understanding

This is a critical factor in the communication process. It is essential that people understand the messages that are being put across. Vital information with respect to hazards, for instance, should be repeated at intervals. The FIDO principle is worth noting. It says that learning by communication is enhanced by:

Frequency

Intensity

Duration

Over again

- **Frequency**. The frequency with which a particular message is repeated during conversation is an indication of its significance and the need for the recipient of the message to understand the message and retain it.
- **Intensity**. The intensity with which words are spoken is an indicator of the importance of the message being communicated to the recipient.

- **Duration**. There must be time for the message to be received and understood by the recipient.
- **Over again**. Repeating the message is a means of ensuring the message is understood, reinforcing the key points and that ensuring the various elements of the message are clear in the mind of the recipient.

Fundamentally,

- A manager must make sure he tells his employees what they need to know. He should not leave it to them to 'read his mind' or to 'pick up' the necessary facts. He should ensure they are promptly and accurately informed of matters relevant to their work.
- Communication should be dispensed in small 'doses' as most people can absorb only a limited amount of information at one time. Long and involved communications are seldom read or listened to and, if they are, they are rarely digested. Only a few important points should be communicated at a time.
- The manager should learn to phrase his communications in simple direct style. Consideration must be given to the level of education and experience of employees and others. Even well-educated employees, however, are more likely to perceive the intended message correctly if it is phrased in the most straightforward manner.

The art of plain talk

In a particular study into human understanding, Flesch (1946) analysed the elements of language expression that, generally, make for ease of comprehension.

These are

- *the number of words in a sentence* – the shorter the sentence, the more easily it will be understood;
- *the number of syllables* – the shorter the length of the words used, the easier they are understood; and
- *the number of personal words and sentences* – the greater the number or percentage of personal words and/or sentences, the more comprehensive the language.

People involved in the provision of information, instruction and training to employees and others would do well to remember the above principles.

Barriers to verbal communication

A number of pitfalls hinder effective verbal communication. For instance,

- the communicator is unable to think clearly;
- there may be problems or difficulties in encoding the message by the receivers;

- transmission of the message can be interrupted by noise or distractions;
- the receiver may exercise selectivity in reception, interpretation and retention of the information in the message;
- the receiver may simply not be listening to the message;
- an unsuitable environment, e.g. workshop or construction site, will act as an impediment to good communication;
- a misunderstanding of feedback from the receiver of the message arises;
- there is no perceptible reaction from the receiver; and
- rumour fills the gap in the formal communication system, which is normally associated with the 'grapevine' operating within an organization.

(see Chapter 15, Presentation skills).

Listening skills

Listening is an essential element of the verbal communication process. It is an active process that has three basic steps:

1. **Hearing**. This entails listening to such an extent to grasp what the speaker is saying.
2. **Understanding**. This second stage of the listening process takes place when the listener takes in what has been heard and understands it in his own particular way.
3. **Judging**. After understanding what a person has said, judging is the process of deciding whether what that person has said makes sense.

A good listener

A good listener:

- gives full attention to the speaker, ignoring distractions and concentrating on what is being said;
- focusses his mind on the matter in question and does not let his mind wander;
- allows the speaker to finish before making a reply;
- allows time to consider what has been said before deciding on the content of his reply;
- listens for the main ideas, the principal points which the speaker wants to get across, separating opinions expressed by the speaker from actual facts;

- asks questions with a view to reinforcing his understanding of what has been said; and
- gives feedback to the speaker through various forms of nonverbal communication, (such as nodding) to indicate understanding, and facial expression (such as frowning) to indicate a lack of understanding.

Listening skills are an essential element of, for example, chairing a meeting. Many people will be expressing opinions, making points, agreeing or disagreeing with what has been said. The chairman needs to listen carefully and seek clarification from members where necessary.

Questioning techniques

People ask questions in order to gain information from other people. One of the most common situations involving the asking of questions is in the process of interviewing witnesses to an accident. In order to obtain the facts surrounding the period prior to the accident, it is important that witnesses do most of the talking to allow the investigator to elicit as much information as possible.

Forms of question

Questions can take a number of forms:

- **Open questions**. An open question introduces a general area of questioning which requires an explanation. Open questions commence with 'why', 'how', 'explain', 'describe', etc. Such questions encourage the recipient to provide a detailed response.
- **Probing questions**. This sort of question can be asked to obtain more specific information and enable the person asking the questions to explore important and uncertain points.
- **Analytical questions**. These questions elicit evidence of a person's ability to analyse the situation in question.
- **Closed questions**. This sort of question is more direct, seeking a simple answer. A typical closed question might be, 'Where were you standing at the time of the accident?' This sort of question can help focus on evasive or confused responses to previous questions.

The following forms of question should be avoided.

- **Leading questions**. This sort of question encourages the recipient to answer in a particular way, usually by a 'Yes' or 'No' response to the question. It is commonly preceded by some form of statement, such as 'The accident was caused by carelessness on the part of the operator. Do you agree?' In this case, the answer to the question may be a foregone conclusion.

- **Multiple questions**. This question actually asks several questions at once, such as 'How did the accident happen? Did anybody see anything? Who was driving the fork lift truck at the time?' These questions can be confusing to the recipient and he is unlikely to be able to remember everything that was asked.
- **Hypothetical questions**. This sort of question asks the recipient how he would have handled the situation. For example, 'How would you have prevented this accident taking place?' The recipient is unlikely to be able to imagine the situation in its real context and his answer will be based on his perception of the situation, not what should happen in reality.
- **Discriminatory questions**. Questions about personal circumstances, such as 'Do women make good fork lift truck drivers?' could be perceived as discriminatory, the assumption being that women may have a lesser aptitude for this sort of work.

The use of open and closed questions

Open questions

An open question is one that is likely to receive a long answer. As such, they deliberately seek long answers and have the following characteristics:

- they require the respondent to think and consider certain matters;
- they elicit feelings and opinions on the respondent; and
- they transfer control of the conversation to the respondent.

Open questions can be used

- as a follow-on from closed questions in order to develop a discussion and get someone, who may be somewhat reticent, to impart information;
- to find out more about an individual, his needs, problems and opinions;
- to get someone to realize the extent of a particular problem to which the questioner may have the solution; and
- to get them to feel happy towards the questioner who may have asked them about their well-being or demonstrated concern about their situation.

Open questions commence with words like 'who', 'when', 'what' and 'how'. In most cases they hand control of the discussion to the other person. Here the quality of questions is important. Well-placed questions leave the questioner in control, however. He must steer the discussion, engaging the respondent when appropriate.

Closed questions

A closed question can be answered with either a single word, such as 'Yes' or 'No', or a short phrase. Closed questions have the following characteristics:

- they provide facts;
- they are simple to answer;
- they are quick to answer; and
- they enable the questioner to maintain control of the conversation.

They can be used for:

- opening questions in a conversation, e.g. 'What is your name?';
- testing understanding, e.g. 'So you know the risks involved in this situation';
- for establishing the right frame of mind in people, e.g. 'Are you happy with this safety procedure?'
- for bringing about closure of an agreed objective, e.g. 'So you will ensure the risk assessment is completed by next Tuesday'.

It should be noted that the first words of a closed question, such as 'do', 'would' and 'will' set up the dynamic of that question thereby signalling to the recipient the easy answer ahead.

Verbal communication on health and safety issues

The legal duties of employers, including managers and supervisors, to inform, instruct and train staff and others, in other words, to communicate, are clearly identified in current health and safety legislation, such as the Health and Safety at Work, etc. Act and the Control of Substances Hazardous to Health Regulations.

Increasingly, health and safety specialists, supervisors and line managers are required to prepare and undertake short training sessions on health and safety issues for the purpose of briefing staff. The following matters need careful consideration if such training activities are to be successful and impart the appropriate messages to their staff.

- A list of topics to be covered should be developed, followed by the formulation of a specific training programme.
- Training sessions should be frequent but should not last longer than 30 minutes.
- Extensive use should be made of visual aids – videos, films, slides, flip charts, etc.
- Topics should, as far as possible, be of direct relevance to the group.

- Participation should be encouraged with a view to identifying possible misunderstandings or concerns that people may have. This is particularly important when introducing a new safe system of work or operating procedure.
- Consideration must be given to eliminating any boredom, loss of interest or adverse response from participants. For this reason, sessions should operate on as friendly and informal basis as possible and in a relatively informal atmosphere. Many people respond adversely to formal classroom situations commonly encountered in staff training exercises.

For more information, see Chapter 15, Presentation skills.

Meetings

People come together in meetings usually for a specific purpose. This may be to make decisions, to provide information, to seek individual views on a topic, to plan ahead or review progress in a particular project. A meeting is generally taken to imply the coming together of people on a formal basis at a predetermined date and time. In ensuring satisfactory meetings, the following points must be followed.

Setting up the meeting

To ensure satisfactory arrangements, a number of factors need consideration prior to the commencement of a meeting.

- *Is the room large enough to accommodate everybody?* Holding a meeting in a room which is too small can result in cramped conditions, general discomfort and loss of attention on the part of participants.
- *Is the layout right? Can people see the chairman and other people addressing the meeting?* Theatre-style layouts with rows of chairs one behind the other are not recommended in that people are generally uncomfortable with this style of layout. There may be insufficient space between rows of seats and the actual seats themselves, creating cramped conditions. Moreover, people need something on which to rest their arms, such as a table, if the meeting is of long duration. Cramped conditions result in loss of concentration and fatigue. A U-shaped arrangement of desks with chairs removes the fatigue and enables better participation as people can make eye contact with the chairman and other people running the meeting.
- *Are environmental factors, such as temperature, lighting, ventilation and humidity, adequately controlled?* These factors need attention prior to the commencement of the meeting. The temperature should be around 20–24°C depending upon the season. Ventilation should be provided by open windows with the provision of fan ventilation in hot summer conditions. Overall levels of illuminance should be around 200–300 Lux, particularly where people need to refer to written documents during the meeting. Relative humidity should be around 70 per cent.

Running a meeting

Chairmanship skills

Any meeting is only as good as the skills of the chairman in running the meeting. Even if all the physical arrangements are ideal and the room is filled with delegates, the meeting will not be successful if the chairmanship is poor. How often do people attend meetings when, after 10 minutes from commencement, it becomes blatantly obvious that the chairman knows little or nothing about the subject of the meeting, leaving fellow directors and senior managers to run the meeting on his behalf. This can result in significant loss of credibility for the chairman, together with discontent on the part of people attending.

Anyone invited to chair a meeting, whether on a one-off or regular basis, needs to consider the following points:

- the chairman should be well informed about the business of the meeting; this may entail reading reports, studying documents and ensuring a good understanding of the purpose of the meeting;
- meetings should start and finish on time;
- the agenda should be followed closely, allowing no deviations from the agenda;
- the chairman must be impartial, alert, firm and not allow himself to be side-tracked;
- meetings within a meeting should not be permitted, all discussion and questions passing through the chair;
- those more reticent and shy delegates should be encouraged to express their views;
- humour should be permitted in limited amounts;
- a decision should be taken on every item on the agenda for future recording in the minutes of the meeting;
- immediately prior to the end of the meeting people should be reminded of the decisions made, tasks which may have been delegated to individuals as a result of the meeting, together with the time scale for completion of those tasks; and
- thanked for their attendance.

Agendas

An agenda is fundamentally a list of items to be discussed at a meeting. It is the chairman's job, assisted by a secretary in some cases, to prepare the agenda for a forthcoming meeting. Agendas should take a prescribed format, with people being invited to submit items for the agenda prior to a meeting. Ideally, an agenda should be circulated to all participants at least a week prior to a meeting to enable delegates to prepare material and responses that may be required at a forthcoming meeting.

Preparation for a meeting

Successful meetings depend upon the degree of preparation by those attending the meeting prior to the meeting taking place. The chairman needs to know what the meeting is about, the objectives of the meeting and what input may be required from him in terms of making decisions.

It may be necessary for everybody concerned to study proposals for future operations, plans of proposed developments, reports from specialists, such as human resources managers, accountants, health and safety specialists, engineering managers and others, and make preliminary decisions on these issues or consider the need for further clarification on issues to be provided at the meeting.

It may well be that a senior manager or specialist is asked to make a presentation to the meeting. In this case, that person needs to decide on the nature of the presentation, the visual aids to be provided, whether an overview or detailed report on recommendations should be circulated prior to the meeting, and the particular interests of those attending the meeting with a view to steering his recommendations to gain their support (see Chapter 15, Presentation skills).

Nonverbal communication

Nonverbal communication is an important feature of the total communication process. It has several functions.

- Nonverbal communication can give support to verbal communication in that
  - gestures can add to or emphasize words;
  - terminal glances help with speech synchronization;
  - tone of voice and facial expression indicate the mood in which remarks are intended to be taken; and
  - feedback on how others are responding to what is being said is obtained by nonverbal devices e.g. facial expressions.
- It can replace speech where speech is not possible.
- It can perform ritualistic functions in everyday life and can communicate complex messages in greeting and farewell ceremonies.
- It can express feelings we have about others, e.g. like or dislike.
- It can express what condition we are in or feelings we have about others – happiness, anger, anxiety – although we may attempt to control them.
- It can be used to convey how we would like other people to see us, for example, by the way in which we present ourselves for public scrutiny.

Aspects of nonverbal communication

Visual aspects

These aspects cover what is actually seen during the communication process, such as:

- involuntary (noncontrollable) features – blushing, pallor, perspiration;
- physical appearance – gestures, facial expression, gaze;
- posture – static, incorporating movement and/or change of posture;
- orientation; and
- proximity or closeness between the communicators.

Tactile aspects

The tactile aspects of communication are associated with touch and physical contact, for example,

- aggression – hitting and striking;
- caressing and stroking;
- guiding;
- greeting (often formalized).

Olfactory aspects

These factors are associated with how people smell and can influence the quality of communication.

Auditory aspects

These features of communication are significant. They are concerned with how people hear things and their nonverbal responses. People commonly mishear auditory communications as a result of failing hearing ability, which may be associated with varying stages of deafness, or through the communicator not enunciating the message effectively.

Conclusions

Both verbal and nonverbal communication are essential elements of the communication process. The old maxim, 'It ain't what you say, it's the way that you say it!' is significant in the verbal communication process. Similarly, the type of nonverbal information being given out by an individual says much about his sincerity, mood, confidence and the emotional state of that individual. Much is to be learned from the nonverbal messages, such as eye contact, hand gestures, body movements and posture, which accompany speech.

Both verbal and nonverbal communication skills are essential for the successful running of training sessions and meetings. These forms of communication are still the most common form of communication despite advances in information technology and the use of the Internet.

Key points

- Verbal communication is the most important source of communication between individuals.
- Important features of verbal communication include the way the words are spoken, the accompanying nonverbal information, such as gestures, mannerisms, posture and the expectations of the participants.
- Interpersonal communication consists of a complex fabric of interacting cues or signals of moving kinds, in particular the sequence of words coloured by voice tone, pitch, stress or rhythm and the movements of hands, eyes and body.
- Managers should consider 'the art of plain talk' when it comes to discussions with employees and other managers.
- There are many barriers to verbal communication associated with, for instance, people not listening, transmission being interrupted by noise or distractions, misunderstanding of feedback or transmission of the message in an unsuitable environment.
- In order to improve communication managers need to develop their listening skills in particular.
- A wide variety of communication skills are required when undertaking training activities.
- For meetings to be successful, many factors need taking into account, such as the environment of the training room, layout and size of the room.
- Meetings are only as good as the chairmanship skills of the chair.
- Nonverbal communication is an important feature of the total communication process in terms of reinforcing messages, indicating the sincerity of the communicator and the importance of the message being given out.

12 Written communication

All organizations use written communication as a form of providing and exchanging information. Effective business writing is a vital communication skill, not only as a means of delivering clear and concise information, but also, and more importantly, as a means of influencing people and making things happen.

The nature and form of the written communication is important. It will depend on a range of circumstances, such as the relative formality of the communication, the information to be presented and the response, in some cases, that is expected.

Written communication may be of a formal or informal nature and takes the form of business letters, memoranda and reports. However, the way written communication is presented by individuals varies significantly. Moreover, in the last decade, a considerable amount of written communication has been replaced by electronic mail.

Irrespective of the form of written communication, a number of factors need consideration in the process of written communication, in particular, style, sentence construction, the use of grammar, the choice of words and the importance of active words. In particular, one of the basic tenets must always be remembered, i.e. 'keep it short and simple'. The first task of the writer is to make the reader's job as easy as possible. Planning and organizing the piece before starting writing is recommended.

Writing style

The manner or style of writing varies considerably from person to person. In many cases, it reflects the way people speak, their upbringing, education and their attitudes to a range of matters. Several points are important with respect to style.

The use of plain English

This is, perhaps, the most important feature of written communication. Writers should stick to words and terms that people understand and avoid complicated technical and scientific terms (in some cases, jargon) that the recipients may not understand.

Jargon

People frequently complain about the use of 'scientific jargon', 'medical jargon' and 'legal jargon' in written communications between an organization and themselves. But what is 'jargon'? Jargon is defined as 'distorted language', 'gibberish', 'speech that is unintelligible and as good as inarticulate'. People pick up jargon at all stages in life, in childhood, at school and during subsequent training for a profession or trade, and use it in all sorts of ways.

Every specialist within an organization, such as lawyers, sales and marketing, engineering, human resources, etc., has his own way of expressing things related to that specialism. Problems arise when specialists write to nonspecialists using specialized terms or jargon that the others do not understand. This can result in confusion, misunderstanding and a request for further simplified information in some cases.

Health and safety specialists are frequently accused of using jargon in written reports, including the use of abbreviated terms, such as 'COSHH', 'RIDDOR', 'PPE', etc., which other people reading a report do not understand.

The receiver

The receiver(s) of the written communication must be kept in mind at all times. The writer wants the receiver to read the communication in question from start to finish. He does not want the receiver to put it down half way through for some particular reason and turn to some other task. The needs, views and interests of the receiver, therefore, must be borne in mind in the preparation of the written communication.

Sentence construction

A sentence is a combination of words containing complete sense and should contain a subject and a verb. Sentences should be short and to the point, and on average no more than 20 words. Most long sentences can be broken up in some way, for example by the use of bullet points or lists. Lists are an ideal way of separating information and presenting it in an understandable format.

Grammar

Grammar is the science of the structure and usages of language, a system of general principles for speaking or writing a language. Written communication should concentrate on correct grammar.

Choice, economy and simplicity of words

Far too many written communications are verbose in style, riddled with technical jargon and, in parts, difficult to grasp in terms of what the writer is endeavouring to put across. It should be appreciated that many people, particularly line managers, receive large quantities of written information and instructions on a day-to-day basis. To achieve maximum attention, communications should be kept as short as possible and written in simple terms that everyone understands.

Importance of active words

Written communication should, in the majority of cases, provide the motivation for someone to do something. The use of active words and terms is considered crisp and interesting, while passive words tend to be dull and bureaucratic. An example of active writing is 'John made this decision'. In passive style, it would be stated, 'this decision was made by John'. The passive style may be appropriate in some cases where, for instance, the writer may wish to soften an approach.

Structure and layout of written documents

The majority of written documents should comprise three main elements:

- a starting or introductory sentence or sentences;
- one or more paragraphs covering the information presented; and
- a concluding sentence or paragraph, summarizing the points made.

Business letters

Letters are generally sent to external organizations either giving or requesting information. As a formal means of communication, they may also be sent internally to people conveying information or to elicit information.

Business letters should be laid out in a logical and concise manner commencing with an opening paragraph followed by the main body of the letter and a concluding paragraph. The opening paragraph should identify the main theme and purpose of the letter. The body of the letter should provide detailed information in an orderly sequence, possibly broken down into paragraphs, in order to provide the reader with information provided. The concluding paragraph should indicate further action required by the recipient.

Memoranda

A memorandum is defined as 'notes to help the memory; records of events, etc., for future use; an informal form of letter without signature, usually on specifically headed paper'.

Generally, memoranda comprise a few paragraphs and are an informal means of communication between members of an organization. They are a quick and convenient means of communicating policy, decisions and instructions. They should be simple and to the point.

In many organizations, memoranda have been replaced by internal e-mails and, with the extensive use of mobile telephones, texted messages between individuals or groups of individuals.

Electronic mail

With advances in internet technology in the last decade, the use of electronic mail (e-mails) has increased significantly. E-mails are used for a variety of purposes, for example, for sending messages, reports and generally exchanging information on a wide range of matters.

Texted messages on mobile telephones are a well-established means of electronic communication. They are quick and enable people to communicate effectively.

Reports

A report is defined as 'a written record of activities based on authoritative sources, written by a qualified person and directed towards a predetermined group'. Reports tend to be of an impersonal nature. They state the facts or findings of the author of the report, for example, following an accident investigation, make recommendations and, in some cases, seek approval for, say, expenditure, certain actions to be taken or not taken. Reports provide information, formulate opinions and are directed at assisting people to make decisions. Reports should follow a logical sequence. They should be written with clarity and may be accompanied by diagrams, tables and photographs. The principal objective of a report is to enable the reader to reach a conclusion as to future action in certain situations. A report should generally terminate with a recommendation or series of recommendations.

The preparation of reports

Reports are produced for a range of purposes. In the preparation of a report, the following aspects should be considered.

The purpose or objective of the report

Organizations commonly perceive a report as a means of bringing together a range of information and facts on which they may be required to make some form of decision. Health and safety practitioners frequently report on a range of matters such as accidents arising at work, the procedures and systems necessary to implement new legislation, the need to raise health and safety awareness throughout the organization and the need to install a system for consultation with employees.

In all circumstances, the objective of a report should be clearly established in terms of

- identifying the subject of the report;
- a statement of the important findings;
- the initial purpose of the report;
- the method of investigation used to establish the findings;
- the actual outcome or results;
- the situation which emerges from the findings; and
- the action necessary to prevent or control exposure to the situation which has emerged.

Report layout and content

The actual layout of a report may vary to some extent, but they should incorporate a number of specific elements:

- *Title* – the nature of the report;
- *Table of contents* – the specific sections of the report;
- *Introduction* – a statement of the purpose of the report, the method of approach and the reason for the study undertaken;
- *Methodology* – the techniques used to acquire the information;
- *Outcome* – a statement of the results and facts arising from the study or investigation;
- *Conclusions* – an overview of the facts established;
- *Recommendations* – short-, medium- and long-term recommendations for future action with, if appropriate, broad indications of cost;
- *Bibliography and references* – the sources of information used in the preparation of the report;
- *Appendices* – documents attached to the report which are relevant to the findings and on which findings and recommendations, to some extent, may be based.

Where reports are extensive, it may be appropriate to incorporate an index.

Stages of report preparation

Stages in the preparation of a report include the acquisition and assembly of relevant information, arranging the information in a logical order of presentation and the actual writing of the report.

Areas for investigation

Here it is necessary to consider the actual purpose and objectives of the report, the extent of the investigation or study and the current circumstances requiring the preparation of the report.

Information searching

This is the research aspect of a report whereby the writer acquires information which may be necessary to support recommendations.

- **Internal sources of information**. Internal documents, such as risk assessments, accident reports, witness statements, maintenance records and reports of inspections are all typical sources of internal information.
- **External sources of information**. External sources include current legal requirements, approved codes of practice, HSE guidance notes, industry-based recommendations and government publications.

Report preparation

Preparation of the report should preferably take place where there is little chance of interruption to enable clear thinking on the part of the report writer. Reference should be made to Report layout and content above.

Appendices and abstracts

A report may incorporate a number of appendices dealing with specific matters. Appendices may incorporate, for example, statistical data, abstracts of impending legislation, specific data on technical standards and other information which is not incorporated in the main body of the report. Cross references to the appendices should be incorporated in the report.

Presentation of the report

On completion of the report, arrangements should be made to formally present the report to senior management at a pre-arranged date and time. Alternatively, the report may be circulated to specific people some time before a pre-arranged meeting to allow them time to read the report.

One of the problems that health and safety practitioners experience is, that whilst there may be tacit agreement by senior management with the recommendations arising from a report, there may be difficulty in obtaining the necessary commitment from them to implement these recommendations. In the preparation prior to a meeting, these difficulties should be anticipated. It may be necessary to provide further information and suggestions available to support recommendations made. The benefits arising from the recommendations, such as improved working practices, in many cases, need to be spelled out to managers who may be disinterested, unconvinced or openly hostile to spending money on health- and safety-related improvements.

Telling people that they could be prosecuted for breaches of health and safety law identified in a report is not a good way of motivating them to do something about it!

Other forms of written communication

Other forms of communication are currently used in organizations, as follows.

Agenda and minutes of meetings

Agenda and minutes of, for example, health and safety committee meetings, are produced on a regular basis. They should be brought to the attention of employees at exercises such as team briefing, departmental meetings and board meetings.

Messages

A message is a written or oral communication sent by one individual to another. It is a means of using appropriate language to transmit an idea, thought or fact. In the preparation of messages, a number of points must be considered in terms of

- the thought or idea to be transmitted;
- the purpose for transmitting that idea or thought;
- the person or group of persons to whom the message is to be transmitted;
- the location for the message;
- the means of communicating the message; and
- the appropriate feedback envisaged by the sender in terms of having the desired effect on the recipient.

A system for relaying short health- and safety-related messages is commonly used in some organizations. Health and safety messages are largely directed at telling people about hazards and the precautions they need to take. They may take the form of small posters placed on desks and work benches.

In-house journals, newsletters and information sheets

One of the functions of the health and safety specialist is the preparation of in-house information on health and safety related topics for employees. In-house broadsheet or tabloid-style journals, such as *Health and Safety News*, circulated on a monthly basis, are a useful way of transmitting health and safety information to employees. They should be produced in a form which is attractive to the reader.

Newsletters are effective means of targeting specific groups of employees on health and safety issues relevant to that group. They can be used to convey information, seek the views of employees on specific issues and report on recent decisions of the health and safety committee.

Information sheets covering particular safety topics, such as safe manual handling or the safe use of hazardous substances, can be particularly successful in focussing the attention of employees on these issues. They should be circulated at health and safety meetings and used as a standard guide to safe working practices.

Electronic communications

E-mails are used to send messages, communicate on particular issues and convey information, including safety information. In these cases, employees are required to confirm receipt of the e-mail indicating they have received the information and, presumably, read it.

Use of information technology

Information technology has enabled written documents to be produced in more attractive and interest-raising ways, by the use of

- **Bullet points**. Bullet points can be used in the preparation of lists.
- **Graphics**. A range of graphics is available to enable the presentation of diagrams, charts and tables.
- **Typefaces**. Different typefaces enable points of specific importance to be highlighted or emphasized.

With the increased use of computers throughout organizations, much health- and safety-related information can be transmitted to employees by this means. These systems can be used for accessing health and safety information, operator training using specific modular packages and conducting attitude surveys on a range of issues.

Safety signs

Safety signs are an important feature of written and pictorial communication in the workplace. The Health and Safety (Safety Signs and Signals) Regulations 1996 apply to all workplaces, including offshore installations. The regulations cover various means of communicating health and safety information, including the use of illuminated signs, hand and acoustic signals (e.g. fire alarms), spoken communication and the marking of pipework containing dangerous substances. These requirements are in addition to traditional signboards such as prohibition and warning signs. Fire safety signs, e.g. signs for fire exit and fire-fighting equipment, are also covered. The signboards specified in the regulations are covered in BS 5378: Parts 1 and 3: 1980 Safety signs and colours.

Under the regulations, employers must use a safety sign where a risk cannot be adequately avoided or controlled by other means. Where a safety sign would not help to reduce that risk, or where the risk is not significant, a sign is not required.

The objective of the regulations is to promote a general move towards symbol-based signs. The term 'safety sign' was extended to include hand signals, pipeline marking, acoustic signals and illuminated signs. The regulations require, where necessary, the use of road traffic signs within workplaces to regulate road traffic.

Employers are required to:

- maintain the safety signs which are provided by them; and
- explain unfamiliar signs to their employees and tell them what they need to do when they see a safety sign.

The regulations apply to all places and activities where people are employed, but exclude signs and labels used in connection with the supply of substances, products and equipment or the transport of dangerous goods.

Employers are required to mark pipework containing dangerous substances, for example, by identifying and marking the pipework at sampling and discharge points. The same symbols or pictograms need to be shown as those commonly seen on containers of dangerous substances, but using the triangular-shaped warning signs.

Although the regulations specify a code of hand signals for mechanical handling and directing vehicles, they permit other equivalent codes to be used such as BS 6736: 1986 Code of practice for hand signalling use in agricultural operations and BS 7121: Part 1: 1989 Code of practice for safe use of cranes.

Dangerous locations, for example, where people may slip, fall from heights or where there is low headroom, and traffic routes may need to be marked to meet requirements under the Workplace (Health, Safety and Welfare) Regulations 1992.

Although these regulations require stores and areas containing significant quantities of dangerous substances to be identified by the appropriate warning sign, i.e. the same signs as are used for marking pipework, they will mainly impact on smaller stores. This is because the majority of sites on which 25 tonnes or more of dangerous substances are stored can be expected to be marked in accordance with the Dangerous Substances (Notification and Marking of Sites) Regulations 1990. These have similar marking requirements for storage of most dangerous substances.

Stores need not be marked if

- they hold very small quantities;
- the labels on the containers can be seen clearly from outside the store.

Conclusions

Written communication is an essential element of the communication process and all organizations have formal and informal means of written communication. This can take the form of letters, memoranda, reports and electronic mail. Writing style should incorporate plain English, be free from jargon and unexplained technical terms.

For health and safety specialists, the report is, perhaps, the most common form of written communication. Reports need to have impact, and the content and, in many cases, the way a report is presented, can determine whether or not recommendations in that report are likely to be implemented by management.

Written communication is also used as means of getting messages across to employees, together with the display of safety signs.

Key points

- Effective business writing is a vital communications skill.
- The recipient of written communication should be in a position whereby he can understand the purpose of the communication and what is expected of him.
- The construction of sentences, the use of correct grammar, the choice, economy and simplicity of words used are essential elements of written communication.
- Written documents should incorporate a specific structure and layout.
- Reports should be directed at providing people with suitable and sufficient information as to enable them to make decisions.
- The increasing use of texted messages between individuals and the use of information technology, through e-mails, is an increasing area of written communication.

13 Interpersonal skills

Interpersonal skills are those skills which enable people to co-exist with others. They entail, for example,

- an appreciation of how other people see that individual;
- the ability to handle, with grace, but with authority and self-confidence, surprising or awkward people or situations arising from the interactions between people;
- effective communication skills, even in the most difficult or complicated situations;
- the ability to make advance preparation for potentially difficult situations which could arise when working with a wide range of people and personalities; and
- an understanding of how to relate more effectively to colleagues, superiors and subordinates

The interrelationship between individuals and the influence of individual differences on relationships

Many forms of relationship exist between people. These relationships may be associated with age, religion, love, education, upbringing, culture, sex, marriage, attractiveness, the jobs people do and shared interests. However, everyone is different in terms of genetic factors, intelligence, physical build, personality, appearance, thoughts, ideas, emotions and feelings. So how do these individual differences interact on relationships? This is particularly appropriate when considering the need for people to work in groups and for the successful and harmonious operation of a group.

Generally, most relationships are affected by factors such as:

- the warmth of greeting when meeting;
- listening skills possessed by each party;
- the clarity of expression, including voice tone and manner of speaking;
- confidence;
- respect;
- the ability to express concerns or fears;

- the amount of time people are prepared to share with each other; and
- the connection between thoughts, feelings and behaviour.

These are all important elements in an individual's ability to communicate, one of the most important interpersonal skills.

Important interpersonal skills

People working in groups contribute a number of interpersonal skills, including

- **Leadership**. This is the process of successfully influencing the activities of a group towards the achievement of a goal or objective. A good leader has the ability to stimulate and influence others by his expertise, presence, the ability to engender respect and his manner and style of speaking. Secondary, or background, skills include delegating tasks where appropriate, motivating others, the use of clear communication techniques, mentoring performance and decision-making.
- **Teamwork**. Teamwork involves co-operation with the others, problem-solving, each individual taking the responsibility for the development and attainment of group objectives and responding to group members' ideas and suggestions.
- **Decision-making**. This entails examination of situations and selecting options for improvement, taking responsibility for decisions and their outcomes and assisting others to make decisions.
- **Networking and network building**. Networking is the ability to seek, identify and create effective contacts with others and to maintain those contacts for mutual benefit. Over a period of time an individual can build a network of reliable contacts. Secondary skills include self-confidence, effective communication and the ability to expand the network.
- **Collaboration**. This implies the ability to work in a co-operative way with other members of the team to meet defined objectives. Collaboration may entail working with peers in, for example, the sharing of information, assisting each other in the completion of tasks and establishing group interdependence. Any collaborative process may entail dealing with difficult people. A good leader anticipates trouble, listens to the feelings of people, empathizing where appropriate. This may entail the giving of constructive criticism, deflecting troublesome behaviour and anger, and dealing with insulting, aggressive and other forms of adverse behaviour. It may also be necessary to reject ideas and suggestions from subordinates without a feeling of guilt or the risk of offending subordinates.
- **Delegation**. Delegation is the act of taking responsibility for deciding when it is appropriate to ask someone to do something, make a decision or give an opinion where the group leader would normally take that responsibility himself. It is the process by which authority and responsibility is distributed from the group leader to a specific member of the group.

- **Motivating**. This is one of the most important skills for team members and entails
  - maintaining a positive attitude, even when things are going wrong;
  - convincing people and winning their commitment to the achievement of objectives;
  - setting people new challenges and encouraging them to meet those challenges;
  - generating enthusiasm amongst a group by positive encouragement;
  - having an 'open door' attitude to enable people to suggest new ideas and solutions to problems;
  - involving the whole team in new ideas, projects, problems and their solution;
  - involving everyone, including listening and taking their views on board;
  - being prepared to support team members in taking calculated and agreed risks; and
  - not blaming other team members when things go wrong, thereby avoiding a 'blame culture' which can be a significant cause of demotivation of team members.

  Teamwork depends on the development of interpersonal skills amongst members. It is, therefore, important for individual members of a group to have a good understanding of the fundamental aspects of interpersonal relationships.
- **Understanding others**. The ability to understand types and patterns of behaviour, in particular, the behaviour that individuals display towards each other, is an important interpersonal skill. People need to understand the difference between 'influence' and 'control', how people display their feelings, what drives them forward and factors such as anger and fear.

Moreover, as part of the process of understanding others, an individual should be able to predict and avoid controversy arising from his actions and decisions.

Interpersonal relationships

Interpersonal relationships are connections, social associations or affiliations between two or more people. They vary to a great extent in the levels of intimacy and sharing, implying the discovery or establishment of common ground and may be centred around some common interest. From birth, people go through the process of developing relationships, the child bonding with his mother and father, in the playground at school and later, in the work context.

The formation and development of relationships go through a number of stages, commencing with initial contact and going through varying stages of involvement.

- **Contact**. People make contact with each other in a variety of ways, but largely through a series of cues or signals, such as initial eye contact, noticing how a person looks at the other person, their body language and interactional cues, such as nodding, to indicate agreement, maintaining eye contact. This may proceed to the invitational stage, encouraging the relationship, where further contact is requested by one individual with the other. A desire not to make contact can be expressed by avoiding eye contact, looking the other way and discouraging verbal communication.
- **Becoming involved**. People become involved with other people in many situations and circumstances – at work, whilst travelling, attending a club or by accident. This 'involvement' stage entails the putting out of 'feelers', hints or questions directed at getting to know and understand the other person better. Typical questions include asking about a person's family, origins and current interests. As the involvement increases, this may be indicated by occasional displays of affection, laughing at each other's jokes, showing an interest in the other person's job and home activities and generally intensifying contact.
- **Intensification**. As a personal relationship continues to develop, there may be varying displays of intimacy, such as holding hands in public, exchanging gifts, regular reference to the new partner whilst with friends and frequent communication between partners. The degree of intensification will depend upon the willingness of each partner to allow various forms of behaviour to develop, such as hugging, kissing and sexual relationships.
- **Deteriorating relationships**. Some relationships do not develop to the intensification stage. Conversely, the intensification stage may not be accepted by one partner. Deteriorating relationships are indicated by things going downhill and falling apart. After the initial 'honeymoon' stage, one partner may come to notice flaws in the other partner. How each individual deals with this sort of situation will determine whether the relationship improves or starts to deteriorate, resulting in eventual loss of contact.

Types of interpersonal relationship

These relationships can be broadly classified as:

- **Family relationships**. A person may be related to someone else by blood (consanguinity), e.g. father, mother, through marriage (affinity), e.g. mother-in-law, brother-in-law.
- **Formalized and/or long-term relationships**. This may arise through the process of law and as a result of some form of public ceremony, e.g. marriage, adoption.
- **Non-formalized and/or long-term relationships**. People, during the course of a lifetime, may have lovers, mistresses, girl friends and boy friends. In some cases, they may live together without going through the formal marriage process.

- **Friendships**. These relationships consist of mutual love, trust, respect and unconditional acceptance between two people or a group of people. The bond is maintained through common ground in terms of interests, shared past experiences and aspirations.
- **Fraternities**. These are individuals united in a common cause or having a common interest. This may entail membership of a club, society, army brigade, fraternity or sorority.
- **Partnerships**. Partnerships include people united as partners and joint operators of a business.
- **Casual relationships**. These include those relationships which might develop at a particular point in time, such as with people attending a course or travelling on a train. They tend to be limited in duration and intensity. Other casual relationships are of a sexual nature extending beyond a 'one night stand'.
- **Soul mates**. Soul mates are people intimately drawn together through a favourable meeting of the minds and who find mutual understanding, respect and acceptance with each other.
- **Associates**. Associates are people who interact as a result of introduction by someone.

Factors affecting interpersonal relationships

Many factors affect the relationships between people. One of the most important factors is that of establishing common ground which takes place with people who, for instance, work together, learn together, play a sport together, such as a football team, and enter some form of loving relationship, such as marriage. Loss of common ground, which can take place over a period of time, will tend to end most forms of interpersonal relationship.

Interpersonal relationships through consanguinity and affinity generally persist at varying levels of intensity despite the absence of affection or common ground. Much will depend upon the bonds established in early childhood and in formative years.

With the various types of interpersonal relationship described above, a range of essential or basic skills are needed and, without these skills, more closely developed relationships are not possible. Partnerships, for instance, require both friendship and teamwork skills. It is possible to consider a hierarchy of relationships from friendships at one end of the hierarchy to soul mates at the other end. Sociology identifies a hierarchy based on forms of activity and interpersonal relations, which divides them into general behaviour, action, social behaviour, social contact, social interaction and social relations.

Marriage is, of course, the most common interpersonal relationship. This relationship is reinforced and regulated by the state. In this form of intimate relationship, a monogamy, there is generally some form of implicit agreement that both partners will retain their marriage vows at all times and not participate in sex or amorous activities with other people.

Decriminalization of homosexual and lesbian activities, and the potential for the legalization of same-sex marriages, has meant that this form of interpersonal relationship is accepted by society.

With friendships, there is a fair degree of flexibility and intensity in the relationship. Friendships within a group of people will vary significantly between the individuals involved in terms of regularity of contact, respect and common ground that may exist between two or more people in the group. In some cases, competition for the friendship of a particular individual may arise due to that individual's, attractiveness, manner of dress, skills and intelligence. Sexual relationships between two friends can alter the former relationship, taking it to a higher level or the reverse.

Not all relationships are healthy relationships, for example, abusive and violent relationships, and co-dependent relationships.

Theories of interpersonal relationships

A number of theories can be considered with respect to interpersonal relationships.

Social exchange theory

Social exchange theory considers relationships by way of the benefits exchanged between partners. The way people react to a relationship is influenced by the rewards derived from the relationship, such as love, affection, trust, money, influence, etc. Relationships can, however, become confused by entanglements, substitutions and transferences. Partners may also consider the benefits to be derived from an alternative relationship in some cases.

Equity theory

This theory is, fundamentally, based on criticism of the above theory. It proposes that people care more than merely maximizing the rewards that are available in a relationship. They expect equity and fairness in their relationships with people.

Relationship dialectics

This theory is based on the notion that a relationship is not a static item, that it is a continuing process which changes at a fairly rapid rate. In this relationship, there is some degree of stress associated with three main issues in the relationship, autonomy versus connection, novelty versus predictability and openness versus closedness. Clearly, different people have different expectations of a relationship.

Attachment styles

This technique is a totally different approach to relationships. It is based on the theory that attachment styles, which are developed during childhood, continue to exert influence on the individual right into, and during, adulthood. As such, they influence the roles people take on in relationships.

Socionics

Socionics is based on the idea that a person's character acts like a set of blocks known as 'psychological functions'. Different ways of combining these functions result in different ways of accepting and producing information which, in turn, results in different behaviour patterns and, thus, different character types.

Socionics and a number of other theories of psychological compatibility, take the view that interpersonal relationships are, to some extent, dependent on the psychological types of the partners involved. The main advantage of socionics is that it can anticipate the development in human relationships with considerable accuracy, and this makes it a very powerful tool when dealing with problems of interpersonal relationships.

Communication and interpersonal skills

Performance management systems pay considerable attention to communication and interpersonal skills. All employees need to be able to understand and use appropriate methods and style of communication in order to meet the organization's objectives, provide a service to, relate to and influence people, such as customers, clients, members of the public and others with whom they may come into contact. Such skills on the part of employees say a great deal about the organization and affect its reputation. Moreover, employees should be able to interact with others and develop appropriate relationships with them.

Many organizations categorize the extent and quality of interpersonal skill required by different levels within the organizational hierarchy. These skills are taken into account in the appointment of new people at all levels and in the promotion system.

A typical hierarchy of interpersonal skills takes into account the following skills and attributes of individuals.

- **All employees**

  Pass on information or advice, orally or in writing, clearly and accurately using plain English.

  Maintain good working relationships with all contacts and keep other people informed as necessary.

- **Supervisors**

  Build and maintain constructive working relationships with members of their section.

  Brief other people openly and consistently in a manner that enables their understanding of instructions, avoiding the possibility of misinterpretation.

  Structure and present all forms of information in a logical manner.

  Take into account the impact of information and communication on others and consider their possible reaction, adapting the method and style of communication accordingly.

- **Departmental managers, professional staff, members of the staff organization**

  Vary the style and content of communication methods according to situations, facilitating open debate to ensure satisfactory understanding and the need for action in certain cases.

  Pay considerable attention to ensuring line managers understand complex information and instructions.

  Pay considerable attention to the views of others and provide an open, supportive and learning environment for line managers.

  Represent the functions, objectives, interests and professionalism of their own group in a constructive manner.

- **Senior managers and directors**

  Take responsibility to ensure key information, messages, directions and instructions passed to departmental managers and others are fully understood, encouraging feedback, discussion and assistance in the decision-making process.

  Use a range of communication techniques with a view to ensuring understanding by managers of identified priorities and strategic issues.

  Encourage and support managers to build relationships and effective working environments.

  Interpret the behaviour of managers, predict potential reactions and deal with a variety of situations that may arise from decisions made.

  Represent the interests of the organization with external authorities, agencies, trade associations, trade unions, etc.

Styles of management and leadership and their influences on the individual

In any organization, individual management styles vary tremendously. Management styles have been categorized in the past by a grid of five different approaches ranging through the 'autocratic' to the 'oligarchical' and the 'democratic' (see Chapter 3, Organizations and groups).

One of the problems of, particularly, newly appointed managers is that of adjusting individual management style to, firstly, their perceived role in the organization, details of which might be incorporated in a job description, and secondly, to the demands made upon them by senior management. Adjusting to a different management culture, compared with that of a previous employer, may also create difficulties This need to adjust can result in both role ambiguity and role conflict (see Chapter 18, Stress and stress management).

Management styles

A number of management styles can be considered.

- The manager perceives himself as the one who solves the problem or makes the decision, using the information available, and without recourse to any other persons.
- The manager perceives himself as being in the position to obtain information from subordinates and subsequently make the decision independently. On this basis, subordinates are seen purely as the providers of information, having no role in the decision-making process.
- The manager perceives himself as sharing the information with relevant subordinates on a one-to-one basis, seeking their ideas, suggestions and solutions to the problem, without bringing everyone together as a group. He subsequently makes a decision which may or may not reflect the views expressed individually by subordinates.
- The manager perceives himself as being a sharer of information, sharing the problem with all members of his group of subordinates, collectively getting their ideas and recommendations. He then makes the decision himself, which may or may not reflect the views of subordinates.
- The manager perceives himself as sharing the problem with subordinates as a group. The group then generates and evaluates alternatives and endeavours to reach agreement on a final solution. In effect, the manager perceives himself as 'chairman' of the group who does not try to influence the group and is ready to accept any solution which has the support of the group.

Leadership

This is the process of successfully influencing the activities and aims of a group towards achievement of goals. A true leader has the ability to influence others by his sheer presence, style of language and expression, expertise and charisma. He engenders respect amongst subordinates with a view to achieving group objectives in a fair and straightforward manner.

A leader will have a number of basic or background skills including decision-making, mentoring progress, delegation and the ability to motivate and inspire. He uses a clear and persuasive approach to others, commands respect on the basis of his knowledge of the field and is an excellent communicator at all levels of the organization.

One of the deficiencies noted with many senior managers is their inability to communicate with lesser mortals in the organization. This can be seen in both formal and informal situations and has a direct effect on their credibility, confidence and competence as a leader. Good communication skills are one of the most important attributes for those in leadership positions.

Assessing and controlling conflict

Conflict can arise in a range of situations – at home, at work, on the train, etc. Managers need a range of interpersonal skills in the resolution of the conflicts that can arise at work. In certain cases, conflict may need to be resolved through a trained facilitator.

What is 'conflict'?

This term is variously defined as 'a trial of strength', 'to be at odds with' and 'to clash'. Conflict costs employers time, money, employee commitment and reputation. The hidden cost of unresolved conflict in organizations is enormous. Identifying effective ways to manage and resolve organizational conflict can have a significant effect on productivity and profitability.

Forms of conflict

Conflict can arise at a number of levels and in a range of circumstances, for example,

- **Intrapersonal conflict**. People can be at conflict within themselves over a range of matters – money, decisions and individual personality factors, such as lack of confidence.
- **Interpersonal conflict**. This form of conflict is primarily between two or more individuals, that is, at a team level within an organization or involving, for example, friends, neighbours and contemporaries.
- **Systemic conflict**. Systemic conflict is a symptom of a wider organizational issue that needs to be addressed, such as pay rates, working hours, staff turnover and changes in working practices.
- **Industrial conflict**. An industry may be at conflict with, for example, the government or the trade unions over a range of issues.

- **Global conflict**. Individual countries can be in conflict with other countries over, for example, commodity prices, armaments, poverty, social norms and economic factors. Whilst very few organizations are unlikely to be involved in global conflict situations, all the other forms of conflict can affect organizations.

Conflict resolution

Conflicts generally involve a group or party at odds with another group or party. For instance, a householder may be at odds with his next door neighbour over the height of a hedge dividing two gardens, smoke nuisance created by one party burning garden refuse or playing loud music late at night. Many conflicts of this type end up in both the civil and criminal courts.

Conflicts at work tend to be between management and employees on a range of issues, such as working hours and conditions, rates of pay, the introduction of new working practices, the division of labour and welfare-related issues. In many cases, they can be associated with:

- personal or group rivalries, disputes or jealousies;
- inadequate communications;
- impending change and resistance to change; and
- frequent appeals to senior management to resolve disputes between employees and junior managers.

The conflict cycle

Conflict situations can be perceived as a cycle in many cases. The initial stage involves emotions being disturbed or triggered by some incident, management decision or event. This creates tension which escalates over a short period of time and with people taking sides. Both sides then assess and judge the other side's power, goals and objectives with respect to the dispute issues. Where the outcome is not resolved by some form of agreement or compromise, or a situation where both sides benefit, then the group that has not been successful will wait for the opportunity to restart the conflict at a later date, thereby completing the cycle.

Many conflicts go on for years and can be associated with poor management style, the aims of the organization and fluctuations in the success of the organization. In most cases, a more participative approach to resolving conflict is required, sooner than the 'us and them' approach adopted by management in some organizations.

Whatever the conflict may be, organizations need to have a well-established procedure, which is agreed with work force representatives, on the resolution of conflicts. Failure to resolve conflicts with disgruntled employees can be expensive for the organization in terms of reduced efficiency of working, loss of goodwill and time spent in argument.

Conflict resolution commences with a diagnosis of the causes of the conflict. This must be followed by a structured and procedural solution to these causes. The most effective methods of resolution aim at a 'win–win' outcome in which both sides perceive they have gained something which they did not have before the conflict.

The way for conflict to be resolved entails the parties coming together, working together on the issues involved and under the guidance of a trained facilitator, the process of collaborative conflict resolution. The facilitator may need to be from outside the organization where, for instance, employee representatives are not prepared to accept a senior manager, such as a human relations manager, undertaking this role.

The following points need consideration in a step-by-step approach to resolving conflict.

A voluntary process

Conflict resolution should be a voluntary process that represents the organization's values if applied throughout the organization. It should be modelled and followed by managers, employees and their representatives.

Coming together

Most facilitators recognize the need for the parties concerned to be brought together, preferably away from the workplace location. The facilitator should, by this time, have gathered the appropriate information with respect to the key issues involved.

It is standard practice for the facilitator to run such a meeting with the intention of being impartial. At the commencement of the meeting, key issues should be stated without making accusations, finding fault or naming names. A standardized procedure, agreed between parties and, preferably, incorporated in the organization's Human Relations Manual, should then follow with:

- each party stating their position and its effects on them without interruption from the other party;
- each party, in turn, repeating or describing as best they can the other's position to the listeners' satisfaction;
- parties endeavouring to consider the issues from other points of view besides the two conflicting parties;
- parties brainstorming with a view to finding middle ground, creative solutions and recommendations for future action;
- parties subsequently volunteering what they can do to resolve the conflict or solve the problem;

- a formal agreement being drawn up with agreed-upon actions for both parties;
- identification and specification of a procedure to be followed should future disagreement arise;
- careful monitoring of progress in respect of actions agreed by both parties; and
- recognition of progress by some form of reward or celebration.

Freedom of speech

Each party in collaborative conflict resolution should feel empowered to speak their mind, feel listened to and feel they are a critical part of the solution. Similarly, each party is obliged to respect and listen to others, trying to understand their point of view, and actively working towards a mutual decision.

Mediation or arbitration

This is the last resort where conflict cannot be resolved by the process described above. This entails mediation by a third party, who adopts a neutral stance or, arbitration, the enforced resolution of the dispute by a neutral authority.

Conflict resolution training

Any management development programme should incorporate sessions on problem-solving and conflict resolution strategies. Irrespective of the conflict situation, learning ways to resolve issues and differences and to collaborate through responses and solutions will improve interpersonal skills that can be applied in other settings.

This form of training has many benefits. It enables people to

- accept differences;
- recognize mutual interests;
- improve persuasion skills;
- improve listening skills;
- break the reactive cycle or routine;
- learn to disagree without animosity;
- build confidence in recognizing winning solutions;
- recognize, admit to or process anger and other emotions; and
- solve problems.

The relationship between skills and accidents

Basic skills have a direct relationship with the pre-accident situation. Factors affecting individual skills in relation to accidents include

- **Reaction time**. Simple speed of reaction bears little relationship to accidents. Choice reaction, on the other hand, where there are several stimuli and several available responses, can have a direct effect.
- **Co-ordination**. Eye co-ordination tests can be used to predict accidents in certain cases, e.g. driving activities. Similarly, manipulation and dexterity tests, e.g. running a loop around a distorted electric wire, which require hand and eye co-ordination, can identify the potential for errors leading to accidents in some cases, such as with driving activities.
- **Attention**. Where attention is divided between several stimuli and the operator must respond to all at the same time, a considerable degree of skill is needed. This is an important feature of many accidents.

Many of the tests used in the selection of personnel are based on the above factors. However, the correlation between skills and accidents is only partial. Generally, the closer the test parallels the task to be undertaken, the more accurate will be any prediction as to future accident potential.

Conclusions

Interpersonal skills are an important feature of human behaviour. Everyone has their own particular set of interpersonal skills, characterized by features such as confidence in reaching decisions, the ability to motivate people and resolve conflicts amongst members of their group. People rely on these skills in order to be accepted in society and co-exist with others.

Interpersonal skills depend upon relationships between people which may take a variety of forms. They are affected by a number of factors, such as individual listening skills, confidence, respect for others, the amount of time people are prepared to share with each other and the ability to communicate well.

Key points

- Important interpersonal skills include leadership, team work, the ability to make decisions and social skills where, for example, a group of people need to collaborate in project work.
- In particular, managers need to be capable of delegating and motivating their subordinates in all sorts of ways through the display of leadership and communication skills.

- Interpersonal skills are based on the relationships between people which can take many forms from family relationships, friendships and partnerships.
- A number of theories of interpersonal relationships have been proposed in recent years including the social exchange theory, the equity theory and the concept of 'socionomics'.
- Interpersonal skills are very much connected with communication and understanding.
- There is a need for organizations to identify the extent of interpersonal skills required by people at different levels of the organization.
- Management styles are particularly important, together with the quality of leadership.
- Conflict is a situation which can arise as a result of both poor interpersonal skills and inadequate communication on the part of the parties concerned, and there is a need for some form of conflict resolution in these cases.
- There is a relationship between skills and accidents, particularly with respect to factors such as reaction time and the inability of some people to co-ordinate aspects of the work they undertake.

14 Systematic training

The duty on the part of employers to provide information, instruction and training is well established under health and safety law. Indeed, the majority of regulations made under the Health and Safety at Work Act, such as the Control of Substances Hazardous to Health Regulations, the Management of Health and Safety at Work Regulations and the Manual Handling Operations Regulations, impose some form of requirement for the training of employees in safe working procedures.

Education and training

People receive education and training at all stages of their lives. It is important to distinguish, however, between these terms.

Education

'Education' implies the 'drawing out' of a person to identify and develop his abilities and skills. It is a term which is given to activities which aim at developing the knowledge, moral values and understanding required in all walks of life, rather than knowledge and skills relating to only a limited field of activity.

The purpose of education is to provide the conditions essential for young people and adults to develop an understanding of the traditions and ideas influencing the society in which they live, of their own and other cultures and of the laws of nature, and to acquire linguistic and other skills which are basic to communication and learning.

Education, if authentic, is more than the giving and receiving of information. It is, as Maritain said, a 'human awakening'.

Training

'Training' has been defined by the Department of Employment as:

> the systematic development of attitudes, knowledge and skill patterns required by the individual to perform adequately a given task or job. It is often integrated with further education.

When one considers the number and levels of jobs in the various occupations, it is easy to see how complex and many faceted is the term 'training' and how, inevitably, it overlaps with education.

The acid test

The acid test of both training and education is the degree and quality of skill that it finally engenders – conceptual skills, numerical skills, verbal skills, social skills, administrative and managerial skills.

Training differs from education in that training

- is aimed at achieving work-related objectives;
- is based upon an identified training need; and
- involves a greater emphasis on skills development.

Systematic training

The term 'systematic' used in the definition immediately distinguishes this form of development from the traditional apprenticeship consisting more often of 'sitting by Nellie' and picking up, haphazardly, what one could through listening and observation.

What is 'systematic training'?

Systematic training implies:

- the presence of a trained and competent instructor and suitable trainees;
- a defined training objective;
- a content of knowledge broken down into learnable sequential units;
- a content of skills analysed into elements;
- a clear and orderly programme;
- an appropriate place in which to learn;
- suitable equipment and visual aids; and
- sufficient time to attain a desired standard of knowledge and competence.

Systematic training, in effect, makes full utilization of skills available in training all grades of personnel. It

- attracts recruits;
- achieves the target of an experienced operator's skill in one half or one third of the traditional time;
- creates confidence in the learners that they can acquire diverse skills through application and training;
- guarantees better safety performance and morale;

- results in greater earnings and productivity, in ease, basic mental security and contentment at work;
- excludes misfits and diminishes unrest; and
- facilitates the understanding and acceptance of change.

A well-designed training programme can be truly educational. Firstly, it is possible to situate the specific job knowledge or skill in a wider context, relating it to affiliated topics, to what is already known by the learner, and to possible future trends and developments. Secondly, any programme of technical instruction in industry must be supplemented by other courses which enable an operator to relate to his entire sociopolitical environment, which can answer his questions about the structure and meaning of society, the destinies of Man, the meaning of law and the lessons of history.

Training objectives

A training objective is a statement of trainee behaviour that is largely observable and results from a learning situation. On this basis, learning situations should provide the trainee with the appropriate behaviour to undertake a task correctly. In some cases, evidence of satisfactory completion of the task will be under observation by the trainer. Training objectives can be classified as:

The cognitive domain

This domain involves intellectual skills. Bloom *et al.* established a hierarchy of major categories in this domain. They looked at 'knowledge' as the lowest level of activity and 'evaluation' as the highest, thus:

- **Knowledge** – the ability to remember information that has already been learnt;
- **Comprehension** – the ability to take in the meaning of information;
- **Application** – the ability to use information that has been learnt;
- **Analysis** – the ability to break up information and thereby understand its structure and organization;
- **Synthesis** – the ability to link or put together information to form a whole out of individual parts; and
- **Evaluation** – the ability to judge the value of information for a specific purpose.

The affective domain

The affective (or attitude) domain is concerned with attitude skills. It includes objectives that deal with emotions, feelings, appreciation and attitudes to work activities, professional and technical activities, together with inputs received during the training process.

In this domain, educational objectives can be classified as

- **Receiving**. This demonstrates a trainee's willingness to deal with inputs from lectures, demonstrations, books, etc., together with tolerating different training approaches and methods during the training process.
- **Responding**. Responding implies trainee participation in terms of interest and reaction to inputs during the training process. At this stage, a trainee is developing his own knowledge and skills.
- **Valuing**. Trainees place value of a particular behaviour or object. They may wish to improve, for instance, group skills or actually wish to take some form of responsibility for certain group activities.
- **Organization**. The trainee syntheses different values and considers the potential for conflict in these values. The trainee will make judgements with respect to his future development of skills as they affect his trade or profession.
- **Characterization**. By this stage, the trainee has developed a system of values that has controlled his behaviour for a period of time, allowing for a characteristic and predictable way of life.

The psychomotor domain

This domain includes objectives with respect to motor skills, such as keyboard work, and the use of hand and foot controls in machinery and vehicles. The training of people in keyboard skills, for instance, is associated with the concept of 'touch typing' whereby a typist should be able to work at a keyboard without necessarily looking at the keys. Training will have set the pattern and layout of the keyboard in the trainee's mind.

In driving situations, the use of hand and foot controls are motor skills which develop from the continued use of these controls during the learning process.

Health and safety training objectives

In the consideration of health and safety training objectives, reference should be made to the features of training objectives outlined above.

Cognitive skills

These skills can be broadly divided into four groups:

1. **Information-related objectives**. These are the skills concerned with the recognition or recall of previously learned material.

2. **Comprehension objectives**. The skills cover the ability to understand or grasp the meaning of the information presented.
3. **Application objectives**. At this stage, the trainee should be able to bring a number of concepts or ideas together, such as the application of legal requirements to practical situations. The trainee should also be in the position to predict the outcome of, for example, unsafe working practices or the failure to follow safe systems of work.
4. **Invention objectives**. These objectives may include all or some of the above objectives. A trainee should be able to recognize and evaluate the significance of data presented, formulate new ideas and principles, judge the significance of a report and appraise the performance of people in work situations.

Attitude (affective) skills

Attitude skills cover an individual's attitude to safety together with the ability to participate in discussions on the subject. Health and safety training programmes should incorporate group discussion sessions to enable people to develop these skills.

Motor skills

These skills relate to the physical aspects of a person's capability in performing a given task. These include, for instance, drawing and sketching skills which enable a trainee to produce a simple line diagram of an accident situation or the design of a guard to a particular machine danger point.

The training process – Systematic training

The training process commonly takes place in four stages.

Identification of training needs

A training need is said to exist when the optimum solution to an organization's problem is through some form of training. For training to be effective, it must be integrated to some extent with the selection and placement policies of the organization. Selection, for instance, must ensure that the trainees are capable of learning what is to be taught.

Training needs should be identified with regard to retraining or reinforcement of training of existing personnel, or the induction training of new recruits. This identification should show

- what kind of training is needed in terms of the requirements of individual tasks;
- when such training will be needed – the immediate, medium- and long-term training requirements;
- for how many people this training is required; and
- the standard of performance to be attained by trainees.

Development of the training plan and programme

Training programmes must be co-ordinated with the current human resources needs of the organization, and the first step in the development of the training programme is the definition of training objectives. Such objectives or aims may best be designed by job specification in the case of new training or by detailed task analysis in respect of existing jobs. By such methods the skills required and the behaviour expected of the people concerned is determined.

Factors which need to be considered in the development of the training programme include

- what has to be taught with respect to theoretical and practical areas; and
- how it can best be taught in terms of training techniques.

These are two important factors and, as such, should receive careful consideration. What has to be taught will be determined by a number of factors – legal requirements, production and sales requirements, health and safety requirements, employment relations requirements. There is no doubt that the availability of a comprehensive job description or specification for the people concerned is of great value here.

How is the new knowledge and/or skill to be imparted? Is it to be a process of 'sitting by Nellie' or by means of a planned training programme? The disadvantages of 'sitting by Nellie' greatly outweigh the advantages due to:

- the trainee acquiring Nellie's bad habits, particularly in relation to safety performance;
- Nellie not being able to explain what she is doing or may not know the best or safest way to do the job;
- the problems of the skilled operator are dissimilar to those of the trainee;
- the size of the learning stages may be too large for the trainee; and
- personality clashes can develop if the trainer and trainee are unsuited.

This system, however, does have some advantages, in particular:

- such a system can fit into a planned training programme; and

- it involves a one-to-one relationship, which is a good training relationship, and can fit into a company promotion scheme.

Generally, much depends upon Nellie's ability as a trainer, the complexity of the task, Nellie's attitude to her use and role as a trainer and her degree of perceptual ability.

The Experienced Worker Standard (EWS) is achieved far more quickly off the job in most cases where the job objectives are clearly defined. In this case, the reliability of the trainer's own expertise is important. In this type of training, the needs of training have to be tailored to a particular situation.

Implementation of the training plan and programme

It is important to distinguish at this stage between 'learning' and 'training'.

Learning

This goes on all the time in an organization. People are continually learning from each other and from their own experience. It is the process by which manpower becomes and remains effective. If learning does not occur, an organization will stagnate and soon cease to exist. Learning can take the form of active or passive learning.

- **Active learning**. Examples of active learning systems are group discussion, programmed learning, practical field work, such as safety inspections, audits and surveys, field exposure, grand tours and in-house projects. In general, active learning methods are particularly useful in reinforcing what has already been learnt on a passive basis. Active learning systems are most effective once the framework is established and where there is plenty of time available in the training programme. It is suitable for a subject where there are no 100 per cent correct answers. There is more chance of producing a change in attitude on the part of trainees and the interest and level of arousal of trainees is maintained.
- **Passive learning**. Lectures and the use of videos tend to be more passive, although they may incorporate active components, such as syndicate exercises or role play activities. With passive learning, the basic objective must be that of imparting knowledge. The principal advantage is that this system provides frameworks and can be used where large numbers of trainees are involved. It should be incorporated as an initial introduction to a subject and as a means of imparting knowledge, including rules and procedures, providing it is relatively simple. A lecture can produce a diversity of opinions at the end, as opposed to one single idea, which is the main objective, whereas an active presentation should produce a convergence of ideas. Furthermore, there can be problems if no feedback is arranged after a passive presentation. Passive methods tend to work best with younger people, but are less acceptable to

older people. The problem of maintaining interest increases with age, the level of arousal generally reducing as people get older. A passive method is also more appropriate for intelligent people and, in the majority of cases, the time taken to present a subject is less. Generally, the amount learnt as a function of time is not as much as with an active form of presentation.

Training

Fundamentally, training is simply a way of helping people to learn. It may impart skills and/or alter a person's basic characteristics. To be effective, it must be planned and organized, i.e. systematic training.

In the implementation of the training programme, there are essentially three basic components:

1. **Giving the training**. Here it is necessary to use trained trainers, to decide on either an active or passive learning system, or combination of both, and to decide on the system for monitoring the effectiveness of trainers.
2. **Recording the results**. The results to be recorded may include improvements in operator performance and productivity, safety awareness or simply an increased use of personal protective equipment. This is basically the system for measuring the effectiveness of the training undertaken, that is, the system for obtaining feedback, perhaps by direct observation or by the use of questionnaires, together with the examination of results.
3. **Training staff and training equipment**. This area includes the training, certification and appointment of trained trainers, such as fork lift truck trainers. It may be undertaken on an internal basis using the organization's own staff or by the use of an external training organization. The means for undertaking the training also need consideration, such as the provision of a suitable training room or area together with projectors, flip charts, etc. A good standard of lighting, ventilation and environmental comfort should be maintained.

Evaluation of the results

There are two questions that need to be asked at this stage: (1) Have the training objectives been met? and (2) If they have been met, could they have been met more effectively?

Operator training in most organizations will need an appraisal of the skills necessary to perform a given task satisfactorily, that is, efficiently and safely. It is normal, therefore, to incorporate the results of such appraisal in the basic training objectives.

A further objective of any training is to bring about long-term changes in attitude on the part of trainees, which must be linked with job performance. Any decision, therefore, as to whether training objectives have been met, cannot be taken immediately the trainee returns to work or after only a short period of time. It may be several months, or even years, before a valid evaluation can be made and such observation can only be made after continuous assessment of the trainee.

A significant factor in the effectiveness of training is the extent of the *transfer of training* achieved by the trainee. This means carrying over the knowledge acquired in one task to another task. There are basically two types of transfer:

1. *positive transfer of training* which occurs where more rapid learning in one situation is achieved because of previous learning in another situation; and
2. *negative transfer of training* where learning is slower because of the previous learning in another situation interfering with the learning process.

It is essential that people be capable of transferring knowledge and skills acquired in one situation to another situation. Any training programme, whether directed to health and safety improvement or for other purposes, should endeavour to bring about this positive transfer of training.

The answer to the second question above can only be achieved through feedback from personnel monitoring the performance of trainees, and from the trainees themselves. Such feedback can usefully be employed in setting objectives for further training, in the revision of training content and in the analysis of training needs for all groups within the organization.

The learning method

The identification of the appropriate learning method is important in the design of training. Once the predominant component of learning necessary for each 'learning unit' has been identified, it is then possible to select the most suitable learning method.

Learning can be divided into five specific areas, each of which has its own objective, as follows.

Comprehending

Objective – *To develop a general understanding of information and principles and their application so that the trainee is capable of applying what has been learned in new ways and to different circumstances, because the underlying principles have been grasped.*

A classic case here is in accident investigation which requires a knowledge of certain basic principles, such as the relative toxicity of certain hazardous substances or the particular mode of operation of a machine.

Reflex development

Objective – *To develop fast reliable patterns of activity in response to correctly received signals.*

The actions required often involve the development of perceptual skills, manual dexterity and the co-ordination of movements, such as handling a fork lift truck in a congested area.

Attitude formation and the development of skills

Objective – *To change current attitudes or to develop new attitudes.*

This aspect is particularly important in accident prevention and most safety training incorporates elements of attitude formation or correction.

Memorizing

Objective – *To develop the facility to remember specific facts and figures so that they can be recalled as required.*

In the safety field people may have to remember certain matters such as safety rules, colour codes or safety signs.

Procedural learning

Objective – *To develop the ability to follow correctly a wide range of procedures that are easy to follow but nonetheless important.*

Many tasks follow a certain procedure, such as stripping down an engine with the subsequent reassembly. It is important that such an operation is carried out in the correct order. The same principles apply to an isolation procedure as part of a permit to work system.

Health and safety training and education

In the case of employees, training and education in this field should be directed at ensuring:

- employees recognize the importance of safe working procedures;
- their awareness to sources of danger and their responsibilities for safe working, including their responsibilities to other employees;

- an understanding of safety procedures, systems and rules; and
- they understand the dangers of unsafe behaviour, horseplay, etc.

The process commonly takes place through a formal training programme directed at all employees on a staged basis commencing with induction training and followed by orientation training.

Managers, including directors and senior managers, also need education and training. Directors need to be aware of their duties and responsibilities under health and safety law, the concept of 'corporate liability' and that, as the 'ruling mind' (*mens rea*), they are ultimately responsible for the health and safety of the organization. Senior managers and line managers, such as supervisors, further need to be aware of their duties and responsibilities, particularly with respect to promoting the organization's Statement of Health and Safety Policy and ensuring the operation of safe systems of work. Both groups need specific health and safety training on appointment and at regular intervals, particularly following the introduction of new health and safety legislation or changes to existing legislation.

Identification of training needs

A number of techniques are available for identifying the training needs of individuals.

Job safety analysis

This technique, used in the design of safe working procedures and in the preparation of job safety instructions, identifies specific hazards at work, influences on behaviour and the training requirements for ensuring the safe system of work is properly understood by employees.

The influences on behaviour are particularly significant in this approach. People can be influenced by many factors, such as rates of pay, production targets, continuing harassment from line management to keep things moving and the relative ease of undertaking the work. These influences shape attitudes both to working practices and the safety elements of those practices.

Risk assessment

One of the outcomes of the risk assessment process under, for example, the Management of Health and Safety at Work Regulations or the Control of Substances Hazardous to Health (COSHH) Regulations, is the recommendation for the provision of information, instruction and training of employees with respect to the significant hazards identified in the assessment and the precautions necessary on the part of all persons.

This is an important aspect of the identification of training needs and the successful outcome of risk assessment should be the establishment of a training programme to meet these needs.

Statements of Health and Safety Policy

One of the concerns expressed by directors and managers, in particular, is that they have not been trained to discharge their duties with respect to health and safety outlined in the organization's Statement of Health and Safety Policy. The training needs of these individuals should be clearly identified and attendance at formal training sessions should be obligatory. Much of this relates to the safety culture within the organization and the significance attached to the establishment and promotion of safety objectives.

In the past, some directors have frequently perceived their stated duties as just a means of ensuring legal compliance and have not taken these duties seriously. In some cases, they have totally disregarded their duties. This failure to undertake their duties has resulted in individual prosecutions of directors, particularly in the case of major incidents involving not only employees but members of the public.

It is common practice for a Statement of Health and Safety Policy to incorporate a specific 'Statement of Policy on Health and Safety Training' which is produced as an appendix to the main Statement. Such a policy could be outlined as indicated in Table 14.1.

Specific areas of training

In the establishment of policies and procedures with respect to training of employees, and particularly in the case of young persons, it is important to distinguish between induction training and orientation training.

Induction training

'Induction' implies the process of installing someone in a particular job for the first time and imparting the appropriate information and knowledge to enable that person to perform the task efficiently and safely.

Health and safety induction training should introduce new employees to the organisation's Statement of Health and Safety Policy and should cover the following principles and issues:

- the safe way is the right way;
- observance of safety rules and procedures;
- the employee's responsibility towards others;

Table 14.1 Statement of policy on health and safety training

This organization recognizes its duties under the Health and Safety at Work etc. Act 1974 and Regulations made under the Act to provide health and safety training for all employees.

Appropriate health and safety training will be provided for all employees in the following circumstances:

- on recruitment (induction training);
- on transfer of job;
- on change of responsibilities, e.g. on promotion or change of job;
- on the introduction of new work equipment or a change respecting work equipment already in use;
- on the introduction of new technology;
- on the introduction of a new system of work or a change respecting an existing system of work;
- in the correct and safe use of hazardous substances;
- in correct manual handling techniques;
- in the correct use of personal protective equipment; and
- with respect to any other health and safety-related issue considered necessary by the organisation.

Duties on all employees

All employees are required to attend designated health and safety training sessions.

- the need to report hazards and shortcomings in the employer's protection arrangements;
- the procedure for reporting injuries and ill-health arising from work;
- good housekeeping;
- the operation of safe systems of work;
- fire hazards and the precautions necessary to avoid the risk of fire;
- specific hazards that could arise in the job and the precautions necessary;
- personal behavioural aspects with respect to:
  - the need to wear suitable clothing and footwear at work, together with any personal protective equipment provided by the organization;
  - the need for good levels of personal hygiene;

  - avoidance of horseplay and the organization's approach towards unsafe behaviour, with particular respect to disciplinary action;
- the location and use of welfare amenity provisions, i.e. sanitation, hand washing, clothing storage, facilities to rest and take meals; and
- current occupational health procedures, including those relating to sickness absence.

Induction training should be supported by a tour of the workplace in which aspects such as:

- fire and emergency procedures;
- designated controlled areas;
- machinery safety;
- manual handling activities;
- personal protective equipment; and
- safe working practices,

are pointed out to new employees, supported by discussion on the issues involved.

Most organizations operate a formal health and safety induction check list procedure. New employees are advised of this procedure prior to the commencement of induction training and to the fact that they may be required to sign to the effect that they have been instructed in the safety items listed.

Orientation training

Orientation training is that area of training concerned with enabling an individual to obtain the necessary skills and aptitudes for undertaking his work safely and effectively. This may include one-to-one training in the safe use of work equipment, safe manual handling procedures, the correct use of hazardous substances and dealing with emergencies that may arise in the use of equipment and substances.

Follow-up training

Follow-up training, for both new and existing employees, is a vital element of health and safety training. This can take a number of forms:

- attendance at accredited training courses, such as those run under the auspices of the Institute of Occupational Safety and Health (IOSH) and the Chartered Institute of Environmental Health (CIEH);
- 'tool-box talks' given by supervisors or health and safety specialists to specific groups and covering the hazards specific to that work activity and the precautions necessary on the part of members of the group;

- discussion on the outcome of safety inspections, risk assessment and accident investigation; and
- briefing on safe working procedures following the installation of new machinery and plant, the restructuring of workplaces and the introduction of new hazardous substances.

It is essential that those providing follow-up training are trained in training and presentation techniques.

Health and safety specialists

Despite the duty on an employer under the Management of Health and Safety at Work Regulations to appoint one or more *competent persons* 'to assist him in undertaking the measures he needs to take to comply with the requirements and prohibitions imposed upon him by or under the relevant statutory provisions', such persons do not exist in many organizations, the health and safety advisory function being allocated to human resources managers, engineers and company secretaries, many of whom have received little or no training in the subject and, in many cases, can allocate limited time to health- and safety-related duties.

A 'competent person' is defined as a person appointed by an employer with the appropriate level of skill, knowledge and experience to be able to advise on general and specific health- and safety-related issues. This definition, which is based on case law (Brazier v Skipton Rock Company Ltd [1962] 1 AER 955) implies that any health and safety specialist, such as a health and safety adviser or safety officer should be appropriately trained to meet the health and safety needs of an organization.

It is recommended that health and safety specialists receive formal training and hold membership of a professional body such as the Institute of Occupational Safety and Health (IOSH).

Safety representatives and representatives of employee safety

Under the Safety Representatives and Safety Committees Regulations 1977, an employer must permit a trade union-appointed safety representative to take such time off with pay during the employee's working hours as shall be necessary for undergoing such training in aspects of the safety representative's functions as may be reasonable in all the circumstances having regard to any relevant provisions of a code of practice relating to time off for training approved for the time being by the Health and Safety Commission.

According to the Approved Code of Practice accompanying these regulations, training arrangements and facilities should be approved by the TUC or by an independent union or unions which appointed the safety representative. Further training, similarly approved, should be undertaken where the safety representative has special responsibilities or where such training is necessary to meet changes in circumstances or relevant legislation.

With regard to the length of training required, this cannot be rigidly prescribed, but basic training should take into account the functions of safety representatives placed on them by the regulations. In particular, basic training should provide an understanding of the role of the safety representative, of safety committees and of trade unions' policies and practices in relation to

- the legal requirements relating to the health and safety of persons at work, particularly the group or class of persons they directly represent;
- the nature and extent of workplace hazards, and the measures necessary to eliminate or minimize them;
- the health and safety policy of employers, and the organization and arrangements for fulfilling those policies.

Additionally, safety representatives will need to acquire new skills in order to carry out their functions, including safety inspections, and in using basic sources of legal and official information provided by or through the employer on health and safety matters.

The Health and Safety (Consultation with Employees) Regulations 1996 added an additional group, that is, 'representatives of employees safety', to the Safety Representatives and Safety Committees Regulations. On this basis, such non-trade union representatives must be treated in the same way with respect to training in their functions.

Contractors and their employees

Many employees work away from their main workplace, such as those involved in the servicing of plant and equipment, construction work, contract catering and cleaning activities. Apart from the general duty on employers under the Health and Safety at Work Act for such employees to be provided with health and safety training, many regulations, such as the Control of Substances Hazardous to Health (COSHH) Regulations and the Construction (Design and Management) Regulations, lay down specific requirements for the training of employees.

For an employer to implement his general duties to persons not in his employment under section 3 of the Health and Safety at Work Act, it is also necessary for him to provide such persons with information about such aspects of the way in which he conducts his undertaking as might affect their health and safety.

In the majority of cases, this implies the provision of varying levels of health and safety training for contractors and their employees. In the broadest sense, the employees of contractors, should be treated in the same way as the employers' employees in terms of the provision of information, instruction and training. Alternatively, a client, for whom project work is being undertaken by a principal contractor, should ensure that the principal contractor has a system for ensuring the provision of training for both his own employees and those of subcontractors.

So how does a client, in selecting a competent contractor for construction work at his premises, or prior to taking on a contract catering service, ensure that the employees of that contractor are adequately trained in health and safety procedures and the precautions necessary to ensure safe working? One of the ways is through the operation of a Safety Passport Scheme.

Passport schemes

Passport schemes ensure that workers have received health and safety awareness training, and are particularly useful for workers and contractors who work in more than one industry or organization. Passport schemes operate in a number of ways. In the majority of cases, such schemes are driven by a particular industry based on the need to ensure that the employees in that industry, suppliers of services, contractors, self-employed people and agency staff meet a particular training standard. On this basis, the industry may decide:

- what training is required, particularly core syllabus requirements;
- the qualifications and resources needed by trainers;
- how training will be delivered and assessed, perhaps through passing an accredited training course;
- for how long a passport will be valid;
- the need for refresher training before renewal; and
- the system for keeping records.

Passport schemes ensure that workers have basic health, safety and environmental awareness training. They are welcomed by both the HSC and the HSE and the environment agencies, as they are a way of improving health and safety performance. They also help promote good practice and can help reduce accidents and ill-health caused by work. They are especially useful for workers and contractors who work in more than one industry or firm. Workers can hold more than one passport if they have been trained for work in more than one industry.

Passports are a way of controlling access to sites, in that only workers with valid passports are allowed to work. On completion of training, passport holders should know:

- the hazards and risks they may face;
- the hazards and risks they can cause for other people;

- how to identify relevant hazards and potential risks;
- how to assess what to do to eliminate the hazards and control the risk;
- how to take steps to control the risks to themselves and others;
- their safety and environmental responsibilities, and those of the people they work with;
- where to find extra information they need to do their job safely; and
- how to follow a safe system of work.

For further information, see *Passport schemes for health, safety and the environment: A good practice guide INDG381* HSE Books.

Training methods and techniques

Training methods are directly connected to the changing of attitudes and behaviour. Methods and techniques of training are related to the way we would like people to learn things. A wide range of techniques and methods are available, depending very much on the learning system adopted, that is, active or passive.

Guided reading

With this method the trainee is given standard literature or company material, such as a procedure for ensuring safe working, to read and comment upon in a structured situation. For self-motivated trainees it can be an effective means of knowledge transfer, but there is always the danger of self-deception on the part of trainees. Guided reading forms an integral part of most training courses.

Lectures

The Department of Employment's 'Glossary of Training Terms' defines the lecture as 'a straight talk or exposition, possibly using visual or other aids, but without group participation other than through questions at the conclusion'. Group participation is thus of an auditory nature.

A lecture can appropriately function to

- indicate rules, regulations, policies and course resources to trainees;
- introduce and provide a general survey of a subject, its scope and its value;
- provide a brief on procedures to be adopted in subsequent learning activities;
- set the scene for demonstration, discussion or presentation;
- illustrate the application of rules, principles or concepts; and
- recapitulate, add emphasis or summarize.

It must be appreciated that a lecture, as a passive form of training, is only as good as the lecturer's ability to present and maintain the interest of trainees. If well done and appropriately illustrated, it can be an effective means of communication, especially when groups are large. A good lecturer can convey very swiftly the essentials of a topic, lead the group members to understanding key concepts, fascinate, amuse, provoke, challenge and involve them. The following points, however, should be considered.

- Communication is largely one-way, with little or no interchange between trainees and lecturer. The lecturer prepares and presents his material in his own way and the trainees sit, listen and possibly make notes.
- Lectures are inappropriate for teaching specific skills.
- A lecture has limited sense appeal which is chiefly aural. To hold a trainee's attention, the content of the lecture must be interesting and challenging.
- Lecturing encourages passivity. It is difficult to retain a trainee's attention and he is highly susceptible to distraction.
- It is not easy to gauge reaction in the lecture situation. If a lecturer is to do more than merely present information, he must be aware of the trainees' reactions, misconceptions, inattention, etc. and must do something about it immediately.
- Effective lecturing is a highly-skilled task. Because trainee interest and attention has to be generated by the lecturer, the latter's vocabulary, enthusiasm, planning, speech techniques, class sensitivity, etc. are essential.

Demonstration

In a demonstration, the instructor, by actual performance, shows the trainees what to do and how to do it and, with his associated explanations, indicates why, when and where it is done. The demonstration rarely stands alone, but is invariably combined with some other form of training. In essence, a demonstration shows someone how something is done or how something works, for example, the use of a fire appliance. Some of a demonstration's more important applications illustrate

- manipulative operations and procedures;
- construction and principles of operation and use;
- functioning and operation of equipment;
- standards, such as workmanship, operating efficiency; and
- operational safety procedures.

The well-constructed demonstration has several advantages in that it

- has dramatic appeal which can arouse interest and sustain attention;
- provide perspective by showing the interrelationship between steps in a procedure;
- reduces waste and damage by illustrating the correct handling of equipment, etc; and
- saves time by reducing explanation and preventing misunderstandings.

The only limit to class size is the ability of everyone to see the demonstration. A demonstration must work and work well. It should set a standard of performance for trainees and all procedures must be correct.

Guided practice

Guided practice can be defined as 'a method in which the trainee has to perform the operation or procedure being taught under controlled conditions'. There are four basic categories of guided practice:

1. independent practice, in which trainees set their own pace and work individually;
2. controlled practice, in which trainees work together at a pace set by the trainer;
3. team performance, which involves a group of trainees performing together as a team; and
4. coach and pupil, which is a method requiring paired trainees who perform alternately as trainer and trainee.

The main applications of guided practice are, in general, those of the demonstration. It is normally used as follow-up instruction in teaching manipulative operations and procedures, the functioning and operation of equipment, team skills and safety procedures. Its disadvantages are the constraints imposed by time, cost, organization, etc. It has many advantages, however.

- The trainee is given an opportunity to apply his knowledge in a realistic situation, which is possibly the best way of developing in the trainee confidence in his own ability and a positive attitude.
- Maximum active participation is possible which should increase both the quantity of learning and retention.
- The correct method of operation, etc., can be emphasized, thereby contributing to accident prevention.
- Wastage and damage is reduced by limiting the likelihood of trainee error.
- A method of validation is provided in that the trainer can observe whether the desired objective has been achieved and can identify difficulties and weaknesses in the instruction.

Group discussion

Discussion methods are often classified in three categories: directed discussion, developmental discussion and problem-solving discussion. No sharp boundaries exist between them, but they differ in their objectives.

- **Directed discussion**. The object here is to assist trainees to acquire a better understanding and the ability to apply known facts, principles, concepts, policies and procedures, and to provide trainees with an opportunity of applying this knowledge. The trainer attempts to guide the discussion so that the facts, principles, concepts, procedures, etc. are clearly interlinked and applied.
- **Developmental discussion**. The object is to pool the knowledge and past experience of the trainees to develop improved or better stated principles, concepts, policies and procedures. Topics for developmental discussion are less likely to have clear-cut answers than those in directed discussion. The trainer's task is to elicit contributions from members of the trainee group, based on past experience and bearing on the topic in hand. He should aim for balanced participation.
- **Problem-solving discussion**. This form of group discussion attempts to discover an answer to a question or a solution to a problem. There is no known best or correct solution and the trainer uses the discussion to find an acceptable solution. The trainer's basic functions are to define the problem as he understands it, and to encourage free and full participation in a discussion whose goals include identifying the real problem, assembling and analysing data, formulating and testing hypotheses, determining and evaluating possible courses of action, arriving at conclusions and making recommendations to support these conclusions.

Some of the more important uses of discussion techniques are

- developing imaginative solutions to problems;
- stimulating interest and constructive thought, and ensuring trainee participation in situations which would otherwise permit passivity;
- emphasizing principal teaching points;
- supplementing reading, lectures, exercises, etc;
- determining the degree of trainee comprehension of concepts and principles, and trainee readiness for progression;
- preparing trainees for application of theoretical work to particular situations;
- summarizing, classifying or reviewing;
- preparing trainees for future instruction; and
- determining trainee progress and the effectiveness or otherwise of previous instruction.

The following are the major benefits of discussion methods:

- Discussions present the stimulating opportunity to both express one's own opinions and to hear those of others. In a well-planned and skilfully directed discussion, interest level is remarkably high.
- Trainees actively participate in the development of the instruction and, therefore, are more likely to accept the validity and importance of the content, and thus become more deeply committed to decisions or solutions than they would be if the subject matter was simply presented to them.
- The trainer is allowed to make use of the trainees' experience, knowledge and abilities to the benefit of everyone else in the group.
- Discussion demands a very high degree of trainee participation and should lead to better learning and retention. Learning is said to take place in direct proportion to the amount of individual participation in the learning process.

Some of the limitations of group discussion methods are:

- The trainer must remain unobtrusive and yet lead the discussion. He must be well-informed. He must be able to reduce arguments over trivia, prevent domination by a few trainees, relate comments to previous discussions, summarize, encourage full participation.
- The quality of the discussion depends directly on the thoroughness of the preparation. Little or no control can be exercised over the preparatory work of the trainees.
- Discussions demand a large time allocation and the amount of time needed is frequently the decisive factor in eliminating a discussion approach.
- Discussions are only applicable to certain forms of subject matter when the basic groundwork of the training has been completed.
- Discussions are only effective with relatively small groups, otherwise worthwhile participation is not possible.
- The group has to be carefully selected if maximum benefit is to be gained from a discussion.

Syndicate exercises

A large group can be broken into smaller subgroups for discussion or problem-solving exercises, for example, the design of a safe system of work, with the trainer available for consultation and guidance. As with various forms of group discussion, this allows for experience sharing, group decision-making and discipline in solving a particular problem. In any syndicate exercise form of training, objectives must be clearly established and the reporting back of the syndicate's findings is important.

Group dynamics (T-groups)

With this technique, situations develop, or are induced, in which trainees' behaviour is examined by other trainees. Group behaviour is also examined by the trainee. This technique is a very effective way of teaching the trainee about his behaviour and its effect on others. Knowledge of the trainee's own behaviour and that of the group as a whole increases, together with the skills to work together and communicate.

It should be appreciated that this form of training can go badly wrong if the trainer is not adequately trained in the technique. Trainees may, for instance, be badly hurt by revelations about their behaviour and may opt out causing the groups to dissolve. The trainer must be acutely aware of this problem prior to any group dynamics exercise and any losses of confidence amongst trainees must be resolved before the close of the exercise.

Programmed instruction/learning

In the glossary of training terms, programmed instruction/learning is defined as a form of instruction/teaching in which the following factors are present:

- there is a clear statement of exactly what the trainee is expected to be able to do at the end of the programme;
- the material to be learnt, which is itemized and tested, is presented serially in identifiable steps and/or frames;
- the trainee follows an actual sequence of frames which is determined for him according to his individual needs; and
- feedback of the information of the correctness or otherwise of the responses is usually given to the trainee before the next frame is presented.

Instructional programmes have many uses. Some of the most valuable are:

- the provision of remedial instruction and the retention and maintenance of levels of proficiency in infrequently practised skills;
- the filling of the gaps in instruction caused by late arrival, absence, etc.;
- the acceleration of able trainees permitting early completion of training for such trainees;
- the provision of sufficient common background among trainees to ensure a firm basis for formal classroom work;
- the consolidation of learning by review and practice of knowledge skills;
- the provision of advanced work or broader contact on the same level within a particular field of study; and
- the provision of a control in the study of learning situations.

The advantages of programmed instruction mainly stem from the meticulous preparation, testing and validation involved in the production of an instructional programme. These advantages are:

- Where such programmes can be used, the failure rate is lowered; this can be attributed to the testing and validation of programmes before use, which ensures their effectiveness, appropriate 'self-pacing' by trainees, a 'forced' response from trainees and immediate confirmation which guarantees continuous attention, indicates and rectifies incorrect responses, prevents misinterpretation and eliminates the repetition of errors.
- The self-pacing, forced attention and immediate feedback characteristics of programmes improve learning and retention, and their use usually results in improved standards of proficiency among trainees successfully completing a training course.
- The elimination of unnecessary materials made possible in the programme's development can reduce the time devoted to learning critical material. Self-pacing and forced attention can also decrease training time.
- Instruction can be almost completely standardized.
- Except for those programmes which depend upon teaching machines, the use of such programmes demands no special facilities for accommodation or for equipment.
- Although it is not recommended that programmes be used as substitutes for trainers, they are validated under conditions where they alone do the teaching and are, therefore, effective instructional vehicles, even in the absence of a trainer.
- Programmes can be designed to meet a wide range of individual needs, and also to achieve group or individual progress.
- Programmes, via their self-pacing characteristics and adaptability to any class size, can improve efficiency and economy for group or individual instruction. They also free trainers from routine and repetitive teaching tasks, permitting them to devote their time to more difficult or more demanding aspects of training.

As a form of active learning, therefore, this technique has much to offer. There are two types:

1. *basic linear*, that is the learning of small parts in a set sequence; and
2. *branching*, where the trainee is presented with a text and given a multiple choice answer, which then directs the trainee to the text, depending upon whether the original answer is correct or incorrect. If the answer is correct, he moves on to the next question. Where the answer is incorrect, he has to go back and start again, reconsidering the question asked.

The disadvantages of programmed learning stem from the same roots as its advantages.

- Programmes suitable for specific needs must generally be produced locally or by special contract. Most programmes that are available 'off the shelf' may not match particular objectives.
- Few trained programmers are available.
- Programmes are costly in terms of programmer training, writing, testing and validation or the use of outside consultants in this field.
- Programming is unsuitable for any subject matter which is unstable or subject to frequent and radical change.
- Many programmes, especially linear ones, are intrinsically boring, making heavy demands on trainers who must be able to motivate their trainees to complete programmes.
- The use of programmes poses peculiar administrative problems, due to varying completion times by trainees, such as time-tabling and the orderly assignment of trainees to jobs.

This form of indirect training is generally provided through externally purchased training packages, although some organizations actually design their own packages. Increasingly, these take the form of PC-based systems incorporating interactive and problem-solving elements. Testing can also take place at predetermined points in a programme with separate recording of the trainee's success. Failure to meet a test standard can result in the trainee having to repeat that particular part of the package.

Role playing/Simulation

These techniques increase trainee involvement in the learning process by introducing a realistic element into instruction. They all present the trainees with a situation which they have to resolve by acting out the roles of those concerned in the situation. These techniques are applicable to most instructional methods. There are a few simulation techniques which should be considered.

- **In-tray exercises**. These form part of role playing in which each trainee acts the part of, say, an executive faced with an in-tray containing a number of letters, files, memoranda, etc., usually to be actioned within a certain time limit. In-tray exercises can effectively be made part of a larger simulation, for example, by telephone calls and interruptions by visitors, and are often added or linked with other groups of individuals, each with a different in-tray and varied functions within a simulated organization. Such an exercise can be very effective, considering the broad nature of health and safety at work.
- **Case studies**. These comprise problem situations which are presented to a group of trainees, the group having to find the best solution usually in the form of a written report with an oral explanation.

- **Management games**. Management games differ from other forms of role playing exercise in that they are played to rules laid down before the game starts. Trainees are given management titles and make management decisions, using a model of the real situation. These games are competitive and usually enjoyable. However, their competitive nature and the implied criticism of the 'losers' tend to introduce unnatural reactions in the participants!

Simulation techniques vary but they are generally most often applied to give trainees practice in the application of job skills and knowledge in safe controlled conditions. They isolate part of a task and give practical training in that part. Furthermore, they introduce interest and active participation in what otherwise might be considered boring or routine instruction.

Their benefits vary according to the technique used, but simulation can add, through the high degree of commitment of participants, an extra interest and motivation to the basic instructional method employed. Limitations vary with the technique used. They do, however, require a great deal of preparation and time allocation if they are to be effective.

Individual coaching

This is an essentially personal on-the-job management training technique designed to inspire and develop individual skills. The one-to-one relationship between trainer and trainee can be extremely effective providing the trainer is fully aware of his objectives and adheres to the written training programme provided. Coaching imparts knowledge, develops skills and forms attitudes during informal but planned encounters between trainer and trainee. This technique increasingly offers sound learning situations within a management development programme.

Projects

The form taken by projects can vary immensely, but there must be objectives set for the trainee to complete, together with wide guidelines necessary to encourage initiative. A project stimulates creativity, interest and decision-making, and information on the trainee's knowledge and personality is fed back to the trainer through the way the trainee undertakes the project work.

Assignments

An assignment differs from a project in that the task or investigation is undertaken to close guidelines after a session of information absorption. An assignment encourages learning transfer to the job situation and is useful as a test for a trainee. Realistic assignments should be chosen to avoid frustration and loss of confidence by the trainee.

On the job training

This is a form of direct training carried out whilst the employee is at the workplace or site during working time. This provides familiar surroundings and the chance to make use of the workplace as a training area. On the job training can take place in several ways.

- **Training by supervisors**. Supervisors, in most cases, have a direct role in the practical training of employees, in monitoring performance of given tasks according to specified procedures. Much will depend upon the amount of time available for this task, but most supervisor job descriptions incorporate an element related to training.
- **Training by an instructor**. Some people may be designated as instructors for certain tasks and aspects of performance, such as in the use of word processing equipment, for manual handling operations, fire safety procedures and first aid. Instructors, in the main, should have attended an approved training course for this type of work.
- **Tool box talks**. These are informal talks and discussions held in the workplace with the trainees commonly grouped around the trainer. Tool box talks can cover a wide range of topics, from the safety aspects of specific types of machinery to the safe dispensing or mixing of hazardous substances. They tend to be scheduled to take place at, for instance, the commencement of a working period and concentrate on one particular aspect of health and safety. They may also incorporate demonstrations of, for example, the correct use of personal protective equipment, certain hand tools and a new safe system of work.
- **Training by fellow workers**. The skilled and experienced operator has much to offer newly appointed employees with respect to on-the-job training of people in the processes and work activities undertaken in the workplace. Selection of this particular group of trainers is important with respect to interpersonal skills, such as patience, friendliness, empathy for beginners and the clarity of explanations provided.

The advantages of informal on-the-job training are that individuals feel comfortable in their own particular work area, can relate to that area and the training tends to be more interactive than with formal training sessions. Measures should be taken to ensure trainers are not interrupted during sessions by, for example, telephone calls or having to deal with people outside the trainee group.

The training process: A model for health and safety

The training process with respect to health and safety can be summarized in the model shown in Figure 14.1.

Conclusions

Training is an essential element in the development of human reliability and in the improvement of performance and, to be effective, must take place in a series of stages. Training goes on throughout working life into retirement.

Organizations need to develop the skills and competencies of their people in order to survive. This is done by training and education. Moreover, people learn in a variety of ways and through many different training techniques.

The need for health and safety training of all employees is well established under health and safety legislation. It is a classic 'safe person' strategy for protecting people against injury and ill-health arising from work and no employer can afford to disregard these requirements.

Key points

- Careful identification of training needs, particularly with respect to health and safety, is an important task in any organization. This has the basic objective of
  - determining the content of the required training;
  - indicating the best method of undertaking the training;
  - highlighting the problem of motivating organizations to implement the recommended training methods or to use the training provided; and
  - revealing the ultimate problem of motivating organizations and people to apply the training once it has been given.
- In the case of accident prevention, there are certain basic themes which can be summarized as the importance of distinguishing between
  - accident and injury;
  - prevention and protection; and
  - techniques which are future, rather than past, orientated.
- The link between safety performance and organizational effectiveness, and between accident prevention management and organizational management as a whole.

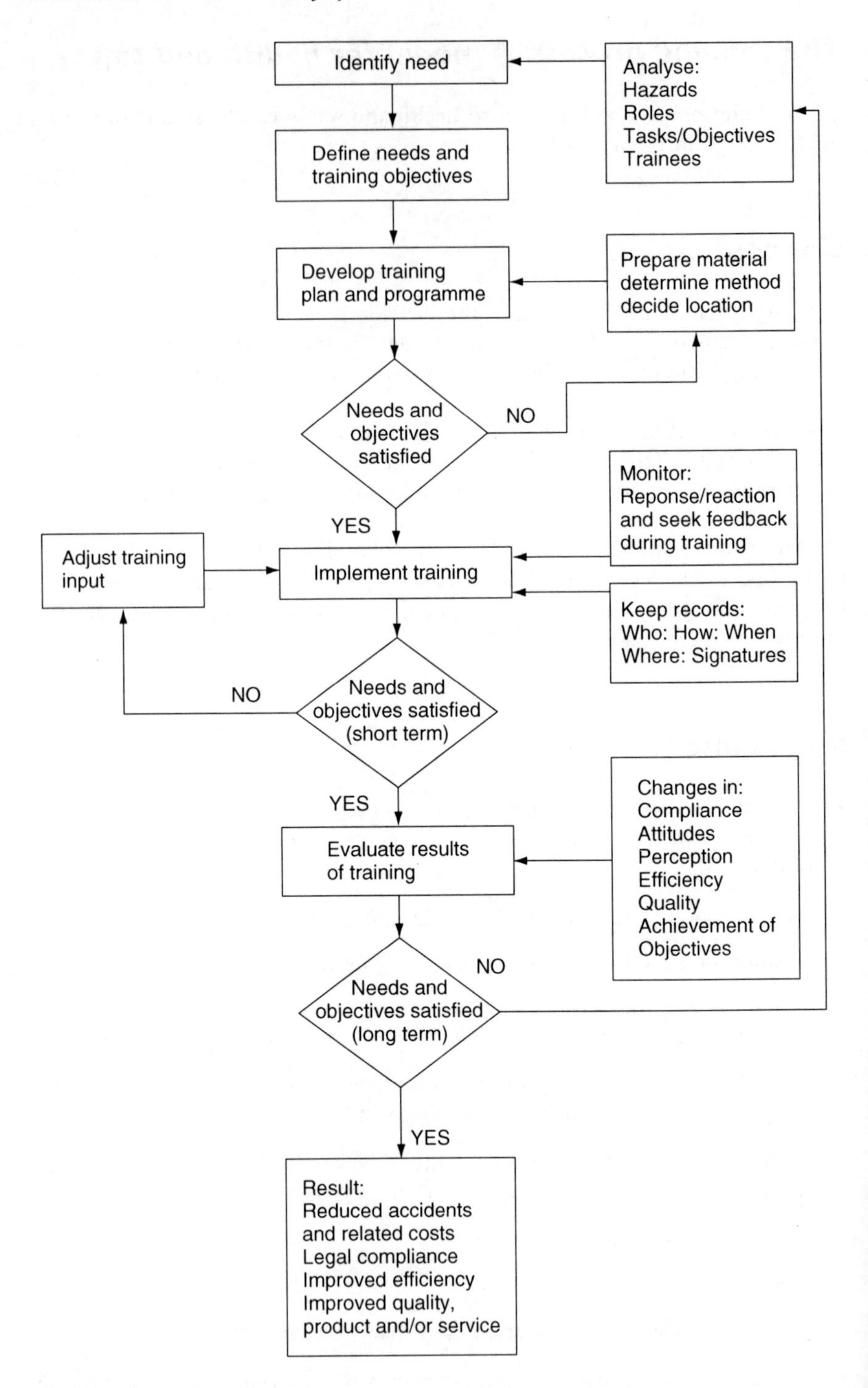

Figure 14.1 The training process: A model for health and safety (Stranks, 2005a) *The Handbook of Health and Safety Practice*: Pearson Prentice Hall

- Training must be regarded as the creation of learning situations, not simply by systematic instruction, but by better design of organizational structures and the adoption of appropriate management styles. Generally, organizations are viewed as learning systems. This applies in all situations, including health and safety.
- The concept of training must be extended to one of developing people's full potential so that organizations can genuinely satisfy human needs by effective utilization of manpower.
- Accidents 'downgrade the system'. In order to ensure effective and safe utilization of manpower, the organization's most valuable asset, there is clearly a case for effective safety training of all staff, from managing director downwards throughout the organization.

15 Presentation skills

A presentation is a fast and potentially effective method of getting things done through, or in conjunction with, other people. Presentations are used as a means for bringing people together to look at new ideas, convey information, plan future activities, monitor or assess progress in a particular project or programme.

The purpose of a presentation

As a form of communication, presentations have many purposes:

- to provide information on and explanation of, for example, a particular programme, project, new legislation and current performance;
- to obtain understanding of what the presenter is endeavouring to achieve;
- to obtain commitment and, in some cases, agreement to a particular proposition; and
- to present a particular problem and the means for overcoming that problem.

Presentation is one of the basic skills that all managers have to acquire. However, for some people, the prospect of having to make a presentation can be stressful, filling them with dread and despair. Fundamentally, presentations are about:

- personal confidence;
- knowledge of the subject to be presented;
- knowing the audience; and
- understanding of the subject to be presented.

People need to consider this aspect in a positive manner, however. For the presenter, the request to make a presentation should be seen as an opportunity for several reasons. Firstly, it puts the individual on display to those members of the organization attending the presentation. It is a chance to demonstrate skills, ability, personality and leadership qualities.

Secondly, it allows the presenter to raise issues, present problems, ask questions and initiate discussion on topics of particular significance. In particular, it enables the presenter to identify those members of the group supporting the proposal and who could provide valuable input at a later date. Finally, it enables the presenter to speak his mind and express personal opinions, whilst the group sits and listens.

Good communication

The important element of this communication exercise is not the transmission of the information forming the basis of the presentation, but the reception from the audience. On this basis, prior to planning a presentation, the identity or identities of the audience must be established. Once this has been established, the presentation should be tailored to enable or assist that audience to understand the message being put out, make decisions based on the message and to remember the message. Good communication, therefore, is concerned with holding the attention of the audience long enough to get the point or points across.

To make successful presentations, a number of factors, outlined below, must be considered.

Planning the presentation

Careful preparation is vital! It is a sad fact that many people come into a room to make a presentation to a group, and after five minutes it is patently obvious that the speaker has made no preparation whatsoever for this presentation and is, in effect, speaking 'off the top of his head'. It is not hard to understand why, within ten minutes, the majority of people attending have 'switched off', have become distracted by the weather outside or have lost interest in the topic. Preparation for a presentation entails careful attention to the following.

Objectives

What are the objectives of this exercise? These must be considered carefully and precisely. The objective should take the form of a simple, concise and straightforward statement of intent, such as the introduction of a new system following recent legislation, the people who may be affected, both directly and indirectly, an outline of cost and commitment needed from other managers. In particular, the objective should, as far as possible, link with management's objectives if it is going to receive support.

This is not just a selling job! The main objective is to focus on what needs to be done, the management involvement requested and the means for making it happen.

The audience

The nature of the audience must be identified with a view to determining how best to achieve the objective. Here it is necessary to consider the aims and objectives of those in the audience, such as engineers, line managers, directors or even contractors. It is necessary to convince these groups that such a proposal will assist them in achieving their own particular objectives.

However, this principle of matching the aims of the audience with that of the presenter, whilst being successful in some cases, may not always work on its own. The speaker must convince the audience that, not only does he understand their problems, but has the means for solving them.

A number of basic points need consideration as far as the audience is concerned:

- **Getting their attention**. It is vital that the attention of the audience is focussed on the presentation right from the beginning. It may be necessary to wait a few minutes while people settle themselves or finish conversations with other people. At this stage various techniques to establish initial rapport with the audience should be used. Getting everyone to briefly introduce themselves and tell the other people what they do is a good form of 'ice breaker'.
- **Establishing the purpose of the presentation**. The audience need to start thinking about the content of the presentation. Here it is appropriate to state the objectives and get them to consider similar experiences related to these objectives. The structure of the presentation should also be introduced at this stage. This can help to establish the theme and provide something concrete to hold their attention.
- **Establishing rapport**. The first few minutes of the presentation are crucial whereby people make a decision whether to listen. Rapport is concerned with the image the presenter wishes to present to them, as a friend, guide, expert and adviser and how he can assist them in getting things right.

Presentation structure

A good presentation structure has three main elements

1. **The introduction**. This outlines the principal objective of the presentation, the reason why the audience has been assembled and the outcome required by the presenter.
2. **The main body of the presentation**. This part covers the main areas for consideration by the audience and where a decision is required, the current situation and the proposed situation. Changes should be specified and the effect these changes will have on the organization.
3. **Recommendations**. Recommendations for future action should be summarized together with the extent of commitment, both managerial and financial.

At the end of the presentation, those present should be invited to ask questions about the proposal or recommendations being put forward.

Use of the eyes, voice and body language to communicate effectively

During the presentation, people will be watching the eyes of the presenter. The eyes say a great deal about the sincerity, honesty and confidence of the presenter, and the presenter's eye gaze should move across the audience seeking eye contact with individuals in turn for a few seconds. Eye contact should be used to maintain rapport with the audience. Much will depend upon the size of the audience in the extent of the rapport established. With large groups, regularly focussing on people at the back of the auditorium can convince them that they are the subject of the presentation. The presenter should smile at regular intervals. This assists in establishing rapport and puts people at ease.

The presenter's style and tone of language are important together with projection of the voice and variations in tone and expression. The presenter should not speak too fast and take his time in presenting the information. The voice level should be slightly louder than normal, but not shouting. After a time, the audience will adapt to a particular level of speech.

The well-established concept in public speaking of 'pitch, pace and pause' is important. The pitch of the voice can be varied in stressing important features of the presentation. The pace should be sufficient for people to take in the messages being put across and there should be regular pauses to allow the information to sink in. Repeating important messages at regular intervals is very important.

Body language, that element of presentation which is unique to the presenter, says a great deal about the sincerity and importance of the presentation. The speaker who stands at a lectern, speaks in a monotone whilst he virtually reads out his lecture notes, is not going to command attention. Hand and arm movements take the form of gestures to the audience. They are particularly important, along with variations in stance. Gestures, including movements of the head, arms and whole body are used to emphasize specific points, reinforce other points and to maintain rapport with the audience. Occasionally, the speaker should move from one side of the rostrum or presentation area to the other, seeking eye contact with those at each side of the room.

One of the problems for many people is what to do with the hands. The hands should be kept reasonably still except when using them in unison with speech and to emphasize specific points. They should not be used to fiddle with loose change in the pocket or a pencil or some other object. This creates distraction.

Effective use of visual aids

Visual aids are used to reinforce the messages being put out. A range of visual aid techniques are available, such as the use of an overhead projector, flip chart or computer lap top-driven presentation.

Whatever system is used, visual aids should attract attention and have a distinct purpose. The quantity of information on a screen shot should be limited with a view to creating maximum impact.

Flip charts can be used to stress specific points, produce diagrams and summarize points made.

Dealing with questions

The level of knowledge of people attending presentations varies considerably. On this basis, questions will be asked and the presenter must be in a position to deal with their questions. People may want a point elaborating on or seek further clarification. The way people put questions to a presenter varies considerably and the presenter must listen carefully to questions before deciding on a response.

Most presentations are short but, in some cases, a presenter may be speaking for a day or longer, running a one day seminar or a longer course. One of the problems in making presentations of this type, in particular to public audiences, is that some people attend a presentation with the main objective of getting the speaker to solve problems for them which have arisen in their own workplace. This may have absolutely nothing to do with the main theme of the presentation or its objectives. Dealing with such questions can be time consuming and can result in the speaker having to rush the rest of the presentation, in order to complete the presentation on time. The best way to deal with these questions is to indicate, in the nicest possible way, that their observations are noted and that the speaker will deal with them during a break or at the completion of the presentation.

Securing audience participation

Securing audience participation is an art in itself. This can be achieved by regularly asking questions of the audience, the use of short five minute group discussion sessions followed by a report back session, getting people to recount their own experiences of comparable situations and, in the case of certain courses, discussing questions from previous examination papers.

In some cases there will be people who wish to monopolize discussions. Whilst their contribution must be appreciated, attempts must be made to 'draw' some of the less participative members of the group into discussions. This must be done carefully without giving them the feeling of having been 'picked on'.

Some speakers tell jokes to emphasize particular points or give accounts of situations that have arisen in their own workplace relating to the points under consideration. The fact is some people are good 'story tellers' and can have the audience laughing at regular intervals, reducing tension and stimulating attention at the same time. Other people are not, for a variety of reasons, adept at this art. The rule is, 'If you can't tell stories, make jokes, etc., then don't try!' Try to obtain their participation in other ways.

Afterwards

After the presentation, the presenter should endeavour to ascertain the success or otherwise of his presentation. This may be manifested by interest shown by members of the group who approach the presenter later to discuss specific topics of the presentation, the mood of the group following the presentation and any general agreement of the matters raised or otherwise.

In the case of formal training sessions, it is common for delegates to be asked to complete an evaluation questionnaire with respect to the content of the session, its usefulness to them, etc., including the performance of the presenter, the visual aids used and the style of the presentation. Much useful feedback can be obtained from these course evaluations and presenters should study them with particular respect to presentation style, taking note of points raised for consideration in future presentations.

The presentation and the presenter

Figure 15.1 summarizes the points made.

To summarize, presentation is an art acquired over a period of time. Presentation should be geared to the needs of the audience, taking into account the type of audience, their jobs and the type of work they do.

Important points

The presenter

Appearance	Gestures and mannerisms
Body movements	Eye contact
Firmness	Empathy
Distracting habits	Pitch, pace and pause
Timing	Use of notes and visual aids
Punctuality	Conviction, confidence and sincerity

SMILE OCCASIONALLY!!

The presentation

Preparation and rehearsal	Construction – beginning, middle and end
Content	Vocabulary
Word pictures	

The environment

Room arrangements	Temperature, lighting and ventilation

Support material

Visuals	Equipment working
Handout material	

Figure 15.1 The presentation and the presenter

1. Stand up, speak up and then shut up!
2. Consider what you are going to say, say it and ensure your audience knows that you have said it.
3. Try to keep your brain ahead of your mouth.
4. Listen to what you are saying.
5. Watch for the discriminating response.
6. Be a good listener.
7. When using a word, think of how many meanings that word, as heard, can have in the minds of the audience.

Figure 15.2 The rules of verbal communication

The rules of verbal communication

In short, people making presentations should follow the rules of verbal communication. These rules are shown in Figure 15.2.

Conclusions

The presentation is a fast and effective method of getting things done through, or in conjunction with, other people. For the presenter, a presentation is concerned with personal confidence, knowledge and understanding of the subject, and knowing the audience. For presentations to be effective, they must be properly planned with defined objectives and a structure.

Presentations involve eye contact, voice control and body language. Presentation is very much an art, particularly in the case of dealing with questions, securing audience participation and obtaining general consensus with respect to the material being presented. All managers should develop their presentation skills and would do well to follow the rules of verbal communication outlined in this chapter.

Key points

- Presentations are used as a means of bringing people together to look at new ideas, to receive information and plan future activities.
- The most important aspect of a presentation is the reception from the audience and a presentation should be planned with the audience in mind.
- A number of basic points need consideration as far as audiences are concerned, namely getting their attention, establishing the purpose of the presentation and establishing rapport.

- Presentations should have a definite structure incorporating an introduction, the main body of the presentation, followed by a conclusion incorporating a summary of the recommendations made.
- Presenters need to pay attention to the nonverbal aspects of a presentation, in addition to what is actually put across verbally.
- Visual aids should be clear, simple, attract the attention of the audience and get the messages over in the shortest time possible.
- Particular attention should be paid to the environment of the presentation, in terms of room layout and factors such as temperature, lighting and ventilation.

16 Health and safety culture

'Culture' is a term that is bandied around a great deal these days. People talk about 'the right management culture', 'getting the quality culture right' and 'procuring cultural changes in organizations'. So what do we mean by 'culture'?

Culture is variously defined as

a state of manners, taste and intellectual development at a time or place;

refinement or improvement of mind, tastes, etc. by education and training.

The Confederation of British Industry (CBI) defines 'culture' quite simply as 'the way we do things around here'.

All organizations incorporate a set of cultures which have developed over a period of time. They are associated with the accepted standards of behaviour within that organization and the development of a specific culture with regard to, for instance, quality, customer service and written communication, is a continuing quest for many organizations.

Cultural indicators

Cultural indicators are used in many aspects of life and can take many forms, for example, political indicators, educational indicators and economic indicators. They are used to measure specific features of a particular culture, be it a particular group of people, such as a race, tribe or an organization, the way people spend their money or the degree of violence on television.

Cultural indicators can be linked with:

- *Key Performance Indicators* (*KPI*) within a government or organization, such as running costs, completion of a project on time or the reliability of a particular system; and
- *Success Criteria* (*Acceptance Criteria*), namely a definition in measurable terms of what must be done for work to be completed satisfactorily, products manufactured to specification or for a programme of work completed.

Within an organization, cultural indicators include

- the degree of loyalty and commitment displayed by all levels of the workforce;
- the presence of shared goals with specific indicators of achievement of these goals;
- policies and procedures to support the achievement of goals through techniques such as team-building and quality management;
- evidence of fair and effective management and well-established management systems;
- investment in people through the frequent provision of information, instruction and training at all levels;
- investment in technology to suit the needs of the organization;
- compliance with legal requirements with respect to, for example, human resources, industrial relations procedures, the health and safety of employees and others, environmental issues and consultation with the workforce;
- the personal integrity of those who direct, manage and supervise others;
- the system for communication both within and outside the organization;
- a well-written mission statement which states what the organization intends to achieve and the methods by which this achievement will take place; and
- a publicized reward structure that rewards high levels of performance.

Many more cultural indicators could be added to the list above. Much will depend upon the nature of the organization.

Safety culture

Developing strong health and safety cultures has the single greatest impact on the reduction of accidents and ill-health and should be a top priority for managers at all levels.

One definition of 'safety culture' suggested by the HSC is:

> The safety culture of an organisation is the product of the individual and group values, attitudes, competencies and patterns of behaviour that determine the commitment to, and the style and proficiency of, an organisation's health and safety programme.
>
> Organisations with a positive safety culture are characterised by communications founded on mutual trust, by shared perceptions of the importance of safety, and by confidence in the efficacy of preventative measures.

Health and safety cultures consist of shared beliefs, practices and attitudes that exist within an organization. 'Culture' is the atmosphere created by those beliefs and attitudes and which shapes individual behaviour.

An organization's safety culture is the outcome of a number of factors:

- the norms, assumptions and beliefs of directors, managers and employees;
- the attitudes to the subject held at all levels of the organization;
- polices and procedures;
- supervisor priorities, responsibilities and accountabilities;
- action, or lack of action, by management to correct unsafe behaviour;
- production and bottom line pressures versus quality issues;
- the training and motivation of employees; and
- the extent of employee involvement, consultation and actual 'buy in' of the philosophy.

Cultural factors

An organization's health and safety culture is associated with many factors:

- **Knowledge and competence of personnel**. The knowledge and competence of all personnel must be appropriate in terms of the hazards and risks arising from work. This should entail training and instruction of personnel and a certain amount of self-education.
- **Management commitment**. The management must be required to demonstrate commitment to the high priority approach and adopt the common goal of safety.
- **Compliance with legal requirements**. There must be adequate knowledge of current health and safety legislation and personnel must be motivated to comply with it. This may entail the establishment of objectives together with a system of rewards and sanctions.
- **Awareness of personnel**. Individuals must be aware of the importance of safety through the provision of information, instruction and training supported by regular supervision.
- **Supervision**. Management must supervise the work of individuals, including the regular review of work practices.
- **Responsibility**. Individual responsibilities for health and safety must be clearly stated and understood.

A strong safety culture

Where a strong safety culture exists in an organization, all employees freely accept responsibility for safety and pursue it consistently. They may go beyond what they consider their actual responsibilities to identify unsafe acts, conditions and behaviours, intervening where necessary to correct them. For example, in a strong safety culture any employee would be happy to remind a visiting senior manager of the need to wear head protection or eye protection when touring the plant. This type of behaviour would not be perceived as over-zealous, but would be supported and valued by the organization.

Organizations with strong safety cultures experience few examples of risky behaviour. As a result, they achieve low accident rates, low labour turnover, low absenteeism and high productivity. Generally, they are companies that perform well in the market place.

One of the basic requirements for the establishment of a strong safety culture is the involvement of senior members of the organization, such as directors, partners and senior managers, in the process. This has the effect of spurring line managers on to better levels of performance than might have been expected without this support.

Establishing a safety culture

With the greater emphasis on health and safety management implied in recent legislation, such as the Management of Health and Safety at Work Regulations, attention should be paid by organizations to the establishment and development of the correct safety culture within the organization. Both the HSE and the CBI have provided guidance on this issue.

The main principles, which involve the establishment of a safety culture, accepted and observed generally, are

- the acceptance of responsibility at and from the top, exercised through a clear chain of command, seen to be actual and felt through the organization;
- a conviction that high standards are achievable through proper management;
- setting and monitoring of relevant objectives/targets, based upon satisfactory internal information systems;
- systematic identification and assessment of hazards and the devising and exercise of preventive systems which are subject to audit and review; in such approaches, particular attention is given to the investigation of error;
- immediate rectification of deficiencies; and
- promotion and reward of enthusiasm and good results (Rimington, 1989).

Building and developing a safety culture

Any process that brings together all levels within an organization with a view to working together to achieve a common goal that everyone holds in high value will strengthen the organizational culture. Health and safety at work is a unique area of management activity that can achieve this end. It is one of the few initiatives that can offer significant benefits for employees compared with, for example, quality management or the means of increasing profitability. As a result, there is an employee 'buy-in' which enables the organization to bring about change without controversy and resistance from the workforce. Moreover, when the needed process improvements are put in place, safety, quality and profitability will also improve and a culture is developed that supports continuous improvement in all areas.

The process of building a safety culture incorporates many elements. Some of the more important elements are outlined below.

- **Obtaining senior management commitment or 'buy-in'**. This is the fundamental first step in the exercise and may involve a presentation at a board meeting to obtain their approval and commitment. They must be advised of the need for change and their support for such change. It may be appropriate to give an indication of current direct and indirect costs to the organization associated with, for instance, increased insurance premiums, time lost and damage to property and plant associated with accidents, emphasizing the fact that the reduction in these costs should more than pay for the changes required. This coming together with senior management to renew their commitment should be continued on a frequent basis.
- **Building trust**. To accept change, people need to have trust. Trust will occur and increase as different levels within the organization work together and begin to see success.
- **Self-assessments or benchmarking techniques**. The person running the scheme, such as a health and safety practitioner, will need to keep track, through self-assessment and techniques like benchmarking, to ensure he is being effective in stimulating progress by managers.
- **Management training**. All levels of management, employee representatives, health and safety committee members and employees will need some level of training, not only in areas such as hazard recognition, legal requirements and safety procedures, but also in communication and team building.
- **Steering committee**. A Health and Safety Steering Committee, initially chaired by a director or senior manager, and comprising management, employee representatives and specialists, such as a chief engineer, should be established. The senior health and safety specialist should act as secretary and organize the committee. The purpose of the committee should be clearly specified in a written constitution form with the principal objective of facilitating, supporting and directing the change processes being undertaken. This committee will provide guidance and direction and avoid duplication

of effort. Fundamentally, the committee should have specific authority to get things done.

- **A shared vision**. This is one of the most important features of a safety culture where everyone in the organization shares the same ambitions and feelings about the need to improve safety performance by following the policies, procedures and systems being promoted.
- **Role definition**. The role and function of everyone from the top of the organization downwards should be defined and specified. The interface between production processes and health and safety should be clarified.
- **Accountability**. A system identifying individual accountability for health and safety should be introduced. This may incorporate job safety specifications for different groups of workers. Specific groups may need training to meet the requirements of job safety specifications.
- **Feedback**. As with any system designed to bring about change, there must be feedback which gives a clear indication as to how change is proceeding. Feedback should not necessarily take the form of reduced accident and sickness rates as these are not a true measure of performance and are open to manipulation. Typical examples of feedback information are the number of workplace inspections carried out, the number of recommendations arising from inspections implemented, improved quality of risk assessment procedures, the progress of planned preventive maintenance systems, the number of people who have received training in a particular aspect, such as the safe use of hazardous substances, and the number of hazards reported by employees together with the measures taken to eliminate or control these hazards.
- **Policies for recognition**. Recognition of success by departments, sections and by individuals should feature strongly in the process. Success in achieving health- and safety-related objectives should receive publicity within the organization and recognized by the awarding of trophies at ceremonies laid on for this purpose. The public recognition of high standards of health and safety through the establishment of awards adds credence to the whole process. Typical awards could include:
  - safety officer of the year;
  - best individual initiative for improving safety;
  - best safety committee initiative;
  - best local safety initiative; and
  - most improved performance by a location, department or sector of the organization.

One of the functions of the steering committee is the determination of those who will receive awards. Rewards for high performance should take the form of local parties or other events directed at recognizing achievement. The criteria for measurement and evaluation of performance in an award scheme should be clearly defined and published. Many schemes operate on the basis of a safety sampling exercise which incorporates specific performance indicators and the maximum points awardable for each performance indicator (see Measurement techniques below).

- **Awareness training and commencement**. Everyone should be trained in the purpose of the programme, health and safety awareness and the means for measuring performance. The commencement of the scheme should receive high levels of publicity seeking the commitment of everyone to the improvements.
- **Process changes** Recommended changes arising from the various activities involved should be implemented promptly. Failure to do so results in loss of credibility of the scheme.
- **Performance measurement**. There should be continual measurement of performance and reporting back to the steering committee.
- **Communicating the results**. Results should be communicated through posters, notice boards and newsletters: progress reports should be discussed at departmental meetings.
- **Reinforcement and reassessment**. As with any scheme, there is a need for regular reinforcement, feedback, corrections to the system and reassessment of specific features.

Factors for consideration

Any process, such as that above, inevitably implies change. Change should take place in a series of steps, operating by means of a clearly defined change process or regime rather than focussing on individual tasks, e.g. the training of employees in safe handling of hazardous substances. The following aspects might be considered ripe for improvement:

- the definition of individual safety responsibilities at all levels, recognizing that safety is a line management function;
- the definition of the role and function of the health and safety specialist, with particular reference to that person's role as an adviser;
- development of positive indicators for measuring performance, e.g. the number of employees who have attended the safety awareness training, the number of hazard reports received and action taken;
- bringing senior and line management together through the establishment of a shared vision of health and safety goals and objectives as they relate to the production process;
- implementation of a process that holds managers and supervisors accountable for being visibly involved, setting the proper example and leading a positive change for health and safety;
- evaluation and reinstatement of any incentives and sanctions for health and safety;

- evaluation and reinvigoration of the joint consultation process, ensuring the safety committee is serving a useful purpose and functioning appropriately in terms of representation, role, function, the system for implementing recommendations and the conduct of meetings;
- the provision of a range of pathways for employees to make suggestions, report hazards and discuss problems with line management ensuring, that in each case, such matters are handled satisfactorily;
- the development of a system that tracks hazard correction, perhaps through a work order system;
- ensuring the reporting of injuries, first aid treatments and near misses by employees, emphasizing the significance of reporting minor incidents;
- evaluation of the accident investigation procedure to ensure it is effective, timely and complete in terms of identifying root causes of accidents and not merely that of allocating blame;
- ensuring the analysis of time lost through accidents and occupational ill-health, the provision of medical and health surveillance by occupational health practitioners in certain cases and acting on the recommendations of such practitioners.

Developing a safety culture

Several features can be identified from the CBI study (1991) which are essential for developing a sound safety culture. According to the study, a company wishing to improve its performance will need to judge its existing practices against them.

- Leadership and commitment from the top which is genuine and visible. This is the most important feature.
- Acceptance that it is a long-term strategy which requires sustained effort and interest.
- A policy statement of high expectations and conveying a sense of optimism about what is possible supported by adequate codes of practice and safety standards.
- Health and safety should be treated as other corporate aims and properly resourced.
- It must be a line management responsibility.
- 'Ownership' of health and safety must permeate at all levels of the workforce. This requires employee involvement, training and communication.
- Realistic and achievable targets should be set and performance measured against them.
- Incidents should be thoroughly investigated.
- Consistency of behaviour against agreed standards should be achieved by auditing and good safety behaviour should be a condition of employment.

- Deficiencies revealed by an investigation or audit should be remedied promptly.
- Management must receive adequate and up-to-date information to be able to assess performance.

A safety culture model

The model (Figure 16.1) shows the factors of co-operation, communication and competence within an overall framework of control.

The process commences with a system for communication of current legal requirements, standards and technical developments in the field of occupational health and safety within the organization. This communication system is an essential feature of the control framework.

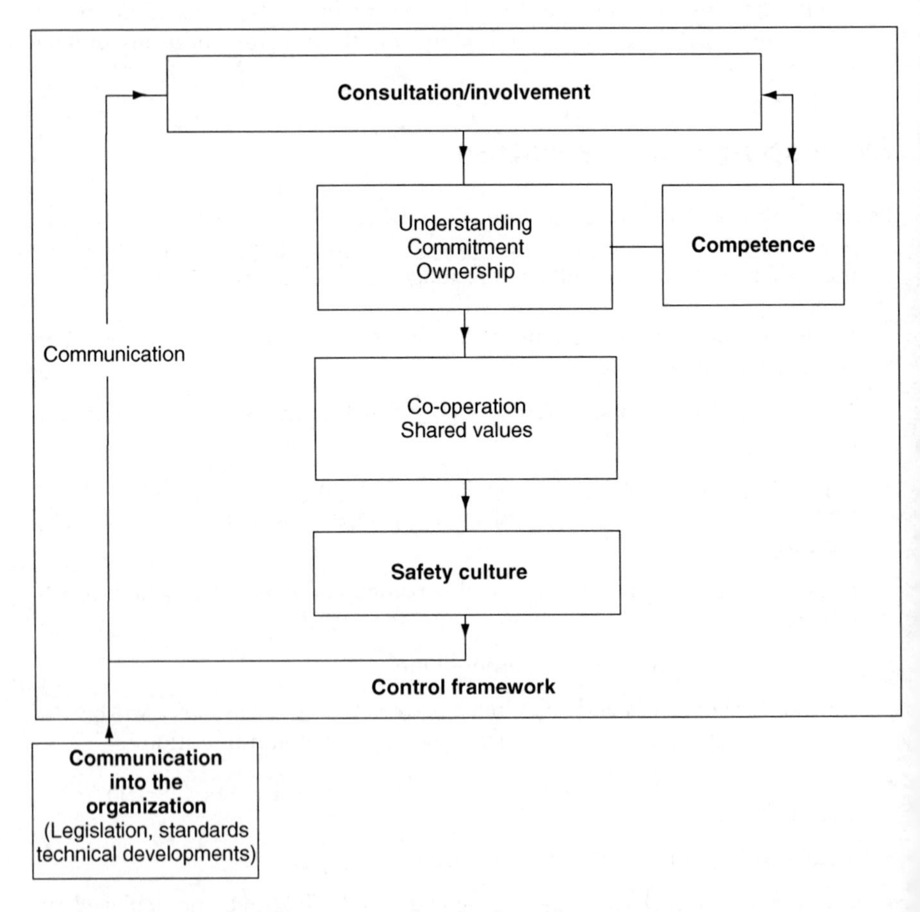

Figure 16.1 Model of a safety culture

The control framework is based on a process of consultation with, and involvement of, staff, contractors and others having a direct or indirect influence on health and safety procedures and systems. At this stage, the question of individual competence of people like health and safety practitioners, trade union safety representatives, managers and others will arise and may indicate a need for training of such people. The principal objective is to bring about a state of understanding, commitment and ownership of health and safety at all levels.

This state of ownership leads to co-operation on health and safety issues and shared values, which is the basis for a health and safety culture within the organization.

The role of senior management

As the legal requirements relating to health and safety at work move away from the concept of prescriptive standards to a more management- and human factors-orientated approach, the role of senior management in developing and sustaining an appropriate safety culture becomes increasingly significant. What must managers do, then, to encourage a positive safety culture?

Firstly, the board must clearly state their intentions, expectations and beliefs in relation to health and safety at work. In other words, they must state where they want the organization to be in terms of health and safety, and formulate action plans for achieving these objectives. Whilst these objectives may be stated in a Statement of Health and Safety Policy under the HSWA, increasingly organizations are publishing mission statements detailing these intentions, expectations and beliefs.

Adequate resources, in terms of financial resources, time and effort must be made available in order to translate these plans and objectives into effective action. In particular, managers at all levels must be made accountable and responsible for their performance, as with other areas of performance, as part of this process. This should take place through routine performance monitoring and review, such performance being related to the reward structure of the organization. On the job performance monitoring should take into account the human decision-making components of a job, in particular the potential for human error.

Above all, senior managers and directors must be seen by all concerned to be taking an active and continuing interest in the development and implementation of health and safety improvements. On this basis, they should reward positive achievement in order to reinforce their message to subordinates that health and safety is of prime importance in the activities of the organization.

In the same way, the various lower levels of management must be actively involved. They must accept their responsibilities for maintaining health and safety standards as line managers and ensure that health and safety keeps a high profile within their area of responsibility. This will entail vigilance on their part to ensure, for instance, that safe systems of work are being followed, that people under their control are wearing the appropriate personal protective equipment and that unsafe practices by workers are being adopted. They must show that deviations from recognized health and safety standards will not be tolerated but, in doing so, it is important for line managers to recognize that they will receive backing from senior managers where such deviations actually occur. On this basis, it is vital that senior management demonstrate their commitment.

Correlation between health and safety culture and health and safety performance

A well-developed safety culture in an organization can contribute significantly to improved safety performance. All those points mentioned above with respect to the development and promotion of the right safety culture – responsibility, conviction, recognition of improvement, performance indicators, the use of suggestion schemes, leadership and commitment – need to be taken on board by all levels of management if there is to be marked improvements.

Above all, health and safety needs to be driven from the top with clearly established lines of responsibility and an admission of corporate liability in the event of failure. There must be a sense of conviction and ownership, recognition that high standards of performance are achievable, that health and safety is one of the main corporate aims and that successful health and safety performance by managers is incorporated in the reward structure of the organization.

Using a range of measurement techniques, it should be possible to draw a correlation between an organization's health and safety culture and the level of performance.

Measurement techniques

A number of techniques are available for measuring the safety culture within an organization.

Attitude surveys endeavour to establish currently held attitudes to health and safety by employees (see Chapter 1, Human behaviour and safety). However, the outcome of attitude surveys may not necessarily give totally accurate results in that those participating in the survey may feel they should give responses to the survey that the researchers are seeking, sooner than what they actually feel.

In any system for assessing and measuring the safety culture within an organization, a number of general factors need consideration:

- the level of senior and line management commitment;
- the existence of documented risk assessments, safe systems of work, safety rules and procedures;

- evidence of how employees perceive risk and of personal risk taken by employees which may have been noted;
- the level of competence of individual employees in their tasks; and
- work stressors and pressures on individual employees which may vary considerably from one employee to another.

Techniques for assessing safety culture include:

- examination of a number of methods of assessment including:
  - interviews with specific individuals;
  - workshops directed at evaluating delegates' perceptions of safety and the measures being taken by the organization to improve safety performance; and
  - questionnaire surveys;
- examination and evaluation of the actual methods employed with respect to cost, timescale and effectiveness.

Safety culture elements

It is essential that the elements of safety culture are specified. These can include:

- the proposition that safety is perceived as equally important or more important than production;
- open and effective communication;
- the commitment of senior management which assists in the promotion of shared values, norms, standards and beliefs;
- provision of adequate resources by the organization to bring about necessary improvement;
- communication and frankness with respect to errors, failures and problems that have arisen; and
- clear-cut evidence that unsafe acts and unsafe behaviour are rare.

Questionnaire techniques

These take the form of questionnaire surveys or safety climate surveys directed at a representative sample of the workforce or the whole workforce. As such, they should follow a step-by-step approach.

One barrier to this technique is that certain individuals or groups may view this approach with suspicion. As with any questionnaire aimed at obtaining genuine responses from those completing it, it should be produced in a user-friendly format, with a clear explanation as to the purpose of the survey and seeking the support of those involved.

Senior management must give their commitment to the survey before commencement and representative groups should be invited to review the questionnaire and make amendments they consider appropriate.

On completion of the questionnaire, results should be assembled and analysed. The results of the questionnaire, including any significant feedback from it should also be published and made available to employees.

The next stage entails development of a plan of action based on feedback. If the action plan is to be successfully implemented, there must be consultation with employee representatives and any changes in work processes and procedures agreed.

The final stage involves implementing the recommendations to bring about change.

Interviews and workshops

One-to-one interviews are a useful method of imparting information on health and safety procedures and systems to employees and assessing their reactions. It enables individual commitment to be gained, together with confidence in working arrangements. Whilst interviews can be time consuming, much useful information can be gained.

An interview schedule should be set up and interviews should cover the basic elements of safety culture that the organization wishes to establish and develop. Interviews should be carried out in private, preferably in a location where there are no distractions and the confidential nature of the interview should be stressed. Each interviewee should be advised of the nature of the interview and, at the end of the exercise, data arising from the interviews should be analysed. In many cases, interviews will identify deficiencies in the approach to health and safety and future action should deal with these deficiencies. It is important that interviewees receive feedback on the overall outcome of the exercise and the actions intended to remedy deficiencies arising from interviews.

Workshops, on the other hand, comprising around six delegates at a time, operate on a group discussion basis. The objectives of the workshop should be explained at the commencement of the session and delegates provided with a number of key topics for discussion together with a questionnaire. Delegates should be encouraged to contribute to the discussion and to participate in the interpretation of the results of topics discussed and questionnaires completed.

Discussion can be centred on the outcome of questionnaires with each delegate discussing the elements measured by the questionnaire. Following this exercise, it should be possible to identify the positive aspects of the current safety culture, areas for improvement and actions that would genuinely make a difference.

Factors that promote a negative health and safety culture

Negative safety cultures are commonly encountered in organizations. In these organizations, health and safety is not perceived as being important. The approach is purely reactive, as opposed to proactive, and the subject is only raised following an accident or adverse event that could result in loss.

A negative safety culture is associated with a number of factors:

- an emphasis on blame wherever accidents occur, as opposed to identifying direct and indirect causes;
- a philosophy that says, 'comply with the law, but no more!';
- unclear distinction as to individual responsibilities for health and safety;
- little or no information, instruction and training for staff, with the firmly held view that 'health and safety's all a matter of common sense!';
- inadequate communication and consultation, where people who raise health and safety issues with management are perceived as 'troublemakers';
- no clear direction or guidance on what is expected of managers and employees with respect to health and safety procedures;
- payment of 'lip service' to requirements such as risk assessment, safety monitoring, accident investigation and the operation of safe systems of work;
- the absence of formal procedures for ensuring safe working, such as permit-to-work systems, and the training and appointment of competent persons, and where the manager who holds the health and safety advisory function is untrained, unclear as to his role and function as the 'health and safety person' and who merely reacts to situations sooner than taking measures to prevent these situations arising; and
- a stressful working environment where employees are expected to work in adverse conditions, such as those associated with excessive temperatures, noise, poorly maintained equipment and badly arranged work layouts;

Conclusions

Following the inquiry into the Piper Alpha incident, the last decade has seen considerable attention being paid to the concept of promoting and developing the right safety culture within organizations. In order to establish such a culture it is necessary for a range of cultural indicators, such as key performance indicators or success criteria, to be established. This enables objectives to be set which are both measurable and achievable by those areas of the organization concerned.

No one is going to bring about this transformation overnight. It needs commitment, shared vision, accountability and, most importantly, policies for recognizing success. Above all, senior management, with the support of their health and safety specialist, have to drive the process forward. This implies demonstration of support, commitment, leadership, recognition of good performance and a Statement of Health and Safety Policy that identifies the role, function and accountability of everyone from the chief executive downwards.

Key points

- 'Culture' is a term that is bandied around a great deal these days. The CBI define the term quite simply as, 'the way we do things round here'.
- An organization's safety culture is the outcome of a number of factors, such as the norms, assumptions and beliefs of everyone, the attitudes held and the need to develop and promote the right policies and procedures.
- Culture is reflected by management commitment, the awareness of people, the quality of supervision, clear-cut levels of responsibility and genuine attempts to comply with legal requirements.
- Obtaining senior management commitment or 'buy-in' is the most important starting point for promoting and developing the right safety culture.
- A health and safety steering committee, chaired by a director or senior manager, can have a significant effect on raising the profile of health and safety across the organization.
- One model of safety culture stresses the factors of co-operation, communication and competence within an overall framework of control.
- A direct correlation can be drawn between a developing health and safety culture and improved health and safety performance.
- Employers need to be aware of the factors that promote a negative health and safety culture, such as an emphasis on blame, unclear distinction of responsibilities together with inadequate information, communication and consultation.

17 Change and change management

Change is a fact of everyday life. In many cases, organizations need to change in order to survive. The need for change, and the uncertainty and insecurity that it causes some people, is one of the greatest causes of stress at work. Organizational change, if it is going to work successfully, has to be managed.

What is change?

'Change' is variously defined as:

the process of altering or making different;

putting something in the place of.

Organizations are continually changing. In many cases, they need to change in order to survive. The need for change may be associated with, for example, setting up a manufacturing process for new products, restructuring of the management system, takeover situations, downsizing, installation of new machinery and plant, the installation of new technology, customer demand, relocation of the workplace or the need to comply with new legal requirements.

People, generally, do not like change. In many cases, it raises a number of fears of, for instance, redundancy, the need to change working practices, having to train for a new job, loss of income or having to relocate to a new workplace. For most people, change can be stressful. If change is to be successful, therefore, it must be carefully managed throughout the various stages of the process.

Change also takes place on a personal level. Most people are resistant to change unless they can see both direct and indirect benefits arising from the change being introduced. It may entail the adoption of new working practices, the implementation of a particular safe system of work, revised working arrangements and the need to acquire new skills. Moreover, change may entail the need to alter behaviour with respect to attitudes held, individual motivating factors and in the relationships between people.

Change agents

Getting people to see beyond their own fears, real or perceived, is a challenge of particular relevance to senior managers. It requires a unique kind of leader, sometimes referred to as a 'change agent' or 'change manager'. Such appointments are not uncommon when introducing, for example, quality management, health and safety management and planned preventive maintenance systems in large multisite organizations.

Multipurpose change agents

A change agent is a person who has the clout, the conviction and the charisma to make things happen and to keep people engaged. He must have a strong sense of mission, good communication skills and a firm belief that change is for the good of the organization. Change agents must have a wide range of skills and should be perceived as catalysts in the management system. They must have a good understanding of an organization's culture and internal politics and be capable of analysing situations where there is a need for change. In particular, they must be capable of promoting and defending their analyses to the organization who may not want to hear what they have to say. Change agents communicate with many parts of the organization – production, engineering, distribution, finance and human resources. As such they must be able to predict typical responses to their recommendations for change and understand the financial implications for the organization. In essence, they must bring order out of chaos.

People with advisory roles, such as health and safety specialists, have been described as multipurpose change agents. Such specialists are judged on the credibility and practicability of their advice. In many cases, their advice may be based on the need to comply with new legal requirements and the best way to bring about compliance. Alternatively, health and safety specialists may need to drive forward new procedures for dealing with, for example, accident investigations by line managers, the safe use of hazardous substances or regulating the activities of contractors.

A well-trained health and safety specialist, however, should be able to see beyond the minimum standards laid down by legal requirements. As a change agent, his advice should bring about improved safety performance, a reduction in losses associated with workplace accidents and ill-health, a greater commitment on the part of senior and line management and the development and promotion of the right safety culture within the organization.

Change as the frequent outcome of safety monitoring

One of the functions of the health and safety practitioner is to bring about change and, in certain cases, to manage that change. This can bring him into conflict with managers and specialists, such as engineers, who may not be so responsive to the need for change.

Proactive monitoring, such as the undertaking of safety audits, inspections and surveys, along with the risk assessment process, together with reactive monitoring associated with the investigation of accidents and loss-producing incidents, frequently indicates the need for change. As a result of these activities, the health and safety practitioner may recommend that change take place immediately or over a period of time in terms of the elimination of control of hazards and their associated risks.

Whatever action is required, it is the health and safety practitioner's task to ensure his recommendations are implemented satisfactorily within the recommended timescale, reporting back to senior management on the relative success of the changes or, where action has not been taken, making further recommendations to bring about the necessary changes. It may be necessary to continually monitor this change management process.

Change management

Change management is defined as 'the process of guiding organizations through the process of transition'. Whilst it is frequently associated with the 'stewardship of resources', namely the prudent management of money, very frequently little attention is given to people, the human assets of the organization, without whom the business simply would not operate successfully or, in some cases, at all.

Guiding, nurturing and shepherding human capital are the skills most needed to ensure that organizational change is received and implemented enthusiastically, rather than with distrust and fear. The degree to which managers are able to manage change, develop consensus and sustain commitment will determine the success or failure of any management initiative or reform effort.

Planning change

There are two types of change that impact on organizations:

Internal change

Internal change may take the form of, for example, structured shifts or programmes that are an ongoing phenomenon within an organization. These changes may be undertaken to avoid deterioration of current performance or to improve future performance of a process or system.

They are generally managed and controlled from within the organization in an orderly, planned and systematic way, e.g. the introduction of annual performance appraisals (job and career reviews) for various levels of management, changing working patterns or the introduction of quality management systems.

External change

These changes of an environmental nature come from outside the organization. In many cases, the organization can exercise little or no control over them. External change may be associated with changing economic cycles, new competitors, advances in technology or the introduction of new legislation.

Organizational challenges

How do organizations tackle the challenges that confront them? Inevitably, this will entail some form of change from the top downwards in the organization. Change management entails aligning the organizational culture with new approaches to business, systems, processes, technology and legal requirements. Organizational culture is comprised of the current human and political dynamics, as well as the organization's history. It is dynamic in that it is continually subject to change.

Principles of change management

Change management is not a precise science due to the many variables in organizations which affect change and the organization's response, at all levels, to change. However, a number of general principles to bring about effective change management should be considered.

The human factors issues

The human factors issues should be addressed systematically. Any significant change creates people issues. New leaders will be asked or invited to come forward, jobs will be changed, new skills and capabilities must be developed, and employees will be uncertain and resistant.

A formal approach for managing change should be developed early and adapted as often as change moves through the organization. The change management approach should be fully integrated into programme design and decision-making, both informing and enabling strategic direction. It should be based on a realistic assessment of the organization's history, readiness and capacity to change.

Starting at the top

Because change is inherently stressful and unsettling for people at all levels of an organization, when it is on the horizon, all eyes will turn to the CEO and the leadership team for strength, support and direction. The leadership must embrace the new approaches first, both to challenge and to motivate the rest of the organization.

Involving every layer of the organization

As transformation programmes progress from defining the strategy and setting targets to design and implementation, they affect different levels of an organization. Change efforts must involve plans for identifying leaders throughout the organization and pushing responsibility for design and implementation down, so that changes 'cascades' through the organization. At each layer of the organization, the leaders who are identified and trained must be aligned to the organization's vision, equipped to execute the organization's vision and motivated to make change happen.

Making the formal case for change

Individuals are inherently rational and will question to what extent change is needed, whether the organization is headed in the right direction and whether they want to commit personally to making change happen. They will look to the leadership for answers. The articulation of a formal case for change and the creation of a written vision statement are invaluable opportunities to create or compel leadership team alignment.

Creating ownership

Leaders of programmes involving extensive change must overperform during the transformation and be the zealots who create a critical mass among the workforce in favour of change. Ownership is often best created by involving people in identifying problems and creating solutions. It is reinforced by incentives and rewards.

Communicating the message

The best change programmes reinforce core messages through regular timely advice that is both inspirational and practicable. Communications flow in from the bottom and out through the top and are targeted to provide employees with the right information at the right time and to solicit their input and feedback.

Assessing the cultural landscape

Successful change management programmes pick up speed and intensity as they cascade down, making it critically important that leaders understand and account for culture and behaviours at each level of the organization.

Addressing the culture explicitly

Once the culture is understood, it should be addressed as thoroughly as any other area in the change management programme. Leaders should be explicit about the culture and underlying behaviours that will best support the new way of doing things, and find opportunities to model and reward those behaviours. This entails developing a baseline, defining an explicit end state or desired culture and devising detailed plans to make the transition.

Preparing for the unexpected

No change management programme goes completely according to plan. People react in unexpected ways, areas of anticipated resistance fall away and the external environment shifts. Effectively managing change requires continual assessment of its impact and of the organization's willingness and ability to adopt the next wave of transformation.

Speaking to the individual

Individuals or groups of individuals need to know how their work will change, what is expected of them during and after the change programme, how they will be measured and what success or failure will mean to them and those around them. Team leaders should be as honest and explicit as possible. Highly visible rewards, such as promotion, recognition and bonuses, should be provided as dramatic reinforcement for embracing change.

Leadership

Leadership must come from the top of the organization – the chief executive officer (CEO) or managing director (MD). Without this leadership, the project is doomed to failure. Showing organizations how to see themselves as collaborative systems requires leadership, that is, conscious authoritative oversight that can create a sense of organizational cohesiveness and shared vision. It is the leader's job to provide managers with vision on what the organization can be and what it can achieve.

One of the most significant factors that impedes success in organizations is this lack of leadership. Leadership must be sustained, focussed, visible and demonstrable if reforms are to be successfully implemented. Evidence indicates that when innovatory projects are not mandated by individuals at executive level, the effectiveness of such projects is always compromised.

A step-by-step approach

Generally, people can only take so much change at a time. If change is to be successful, it must take a step-by-step approach. Above all, there must be a high standard of communication throughout the organization aimed at promoting the organization's values and beliefs and, in particular, the explanation of why change is necessary.

Leaders who seek to facilitate organizational change use a variety of means to communicate their organization's values and beliefs. They may often discuss these values and beliefs in meetings, team briefing exercises, internal publications and on a one-to-one basis with managers. Training is also an effective means by which organizational beliefs can be absorbed and assimilated.

The following are step-by-step ways to communicate values and beliefs and the need for change:

- the open display of top management support for changes necessary;
- preparation and publication of a mission or vision statement that captures the essence of an organization's character, purpose and aspirations and the use of illustrative slogans, stories and comparisons to convey that essence, including the need for change in order to survive;
- rewarding employees who live up to the organizational mission with incentives, awards, promotions, etc;
- the use of a management style that is compatible with these core beliefs and the need for change;
- establishment of systems, procedures and processes that are compatible with the beliefs and values, including those involving change;
- replacement or alteration of the responsibilities of employees who do not support the desired values and beliefs who, in turn, may be resistant to change; and
- assignment of a manager or group whose primary responsibility is to bring about change and/or perpetuate the culture.

Implementing change

Actually making things happen in line with recommendations made is, perhaps, the hardest part of the change management process. Before this process commences, however, there must be full support for change from senior and line management. Change can be implemented through both direct and indirect action.

Direct action

A number of tools are available for bringing about direct action to implement change. A Statement of Health and Safety Policy is one of the most important tools in this respect. A well-written statement will incorporate, in the section dealing with 'Organization and Arrangements for Health and Safety', a hierarchy of command from the most senior person in the organization downwards. This should incorporate a description of the individual role, function and accountabilities for health and safety of all levels of the hierarchy. In most cases, people will need training to ensure they fully understand their responsibilities in this respect.

Once people are clear where their responsibilities lie and are aware of the fact that they can personally be taken to task for failing to meet them, interest in the subject increases and there is a greater willingness on their part to take direct action to bring about improvements in health and safety.

Safety monitoring activities, such as safety audits and inspections, inevitably identify the need for change and specify a timescale for the improvements necessary to implement change. Similarly, the outcome of accident investigation will, in many cases, identify the need for change in a system of work or the introduction of new methods of safe working, together with breaches in legal requirements. Management may need to implement change quickly, particularly if subject to enforcement action where, for example, an improvement notice or prohibition notice under the Health and Safety at Work Act has been served on the organization.

Finally, for any direct action to be taken and sustained, the support, demonstration and commitment of senior management for bringing about change must be clearly visible. Where this is not present, the change process can falter.

Indirect action

Change can take place indirectly in many ways. In many cases, improvement in health and safety performance may be associated with the workplace culture of an organization, the quality of training and supervision and the system for measuring performance.

Cultural dividends from risk assessment

Culture is about 'the way we do things round here'. The risk assessment process provides an ideal opportunity for bringing about cultural change and recommendations from risk assessments should address this aspect. Risk assessments should examine work activities with a view to identifying the significant hazards, and specifying the precautions necessary, together with the need for information, instruction and training where appropriate. They should also involve employees in discussion on risks arising from current working practices and seek feedback from them on the way to make things safer. This level of involvement has a direct effect on the safety culture in a workplace and is a powerful tool in bringing about change.

Whilst the process of changing the safety culture is a slow one, over a period of time these cultural dividends will be noticed in terms of increased attention to the detail of safety procedures, the 'safe way' being perceived as the 'correct way' and a reduction in injuries and lost time arising from minor accidents in particular.

Training and supervision

Training and instruction should draw the attention of employees to the significant hazards arising from work activities with a view to breaking old unsafe habits and introducing employees to safer working practices. Supervision should ensure the newly introduced working practices are maintained and praise employees where they are followed.

Performance measurement

There are a number of ways of measuring performance, including the use of safety sampling exercises and check list systems. Key performance indicators should be established and individual workplaces assessed against these performance indicators.

The value and use of feedback

Feedback is created in many ways, from safety monitoring, risk assessments, training exercises, the investigation of accidents and loss-producing incidents, safety suggestion schemes (see below) and discussions with employees or their representatives on changes that have been, or need to be, introduced. It may be appropriate, in some cases, to ask employees to complete a purpose-designed health and safety questionnaire with a view to getting feedback as to what they feel needs attention. The health and safety committee is also an important tool for organizing feedback from different parts of the organization and recommendations from these meetings should be acted upon promptly.

It is important that feedback issues are properly assessed in terms of their significance with respect to changes in working practices. Recommendations from, for example, safety suggestion schemes should be acknowledged and action taken within a specific timescale.

It should be stressed that the organization values feedback from employees. However, the enthusiasm of employees for participating in this type of scheme can rapidly wane if it becomes apparent that no action has been taken, or is intended to be taken, when feedback recommendations are made.

Problems and pitfalls: Barriers to change

Generally, people are averse to change. They have their own ways of doing things and, in many cases, resent 'the new broom' approach. Prior to introducing change, therefore, it is wise to consider potential problems and pitfalls that can be linked to this in-built barrier to change.

Cultural resistance

This is one of the principal barriers to change. People are generally resistant to new approaches, they may be parochial in their approach to work and getting people to see past their own fears, whether real or perceived, is a challenge of particular relevance to all levels of management. Many processes and practices have a long, entrenched and, sometimes bureaucratic, history that has developed piecemeal over a period of many years in order to accommodate the needs of different parts of the organization and special interests. The more deeply rooted these practices, the more difficult it will be to bring about comprehensive change.

Unclear goals and performance standards

Many managers lack clear hierarchically linked indicators that offer simple illustrations of how their efforts contribute to the organization achieving its strategic goals. This situation can be complicated by poorly integrated information, reporting and setting out of expected performance standards.

Lack of incentive for change

For many organizations, management performance is measured by the extent of a department's budget, the number of people employed or amount of money spent in completion of tasks. Increased attention needs to be given to rewarding behaviour that meets strategic results-based objectives.

Bringing down the barriers to change

A number of strategies are available to enhance the process of change. These strategies should be driven by senior management.

- **Goals and performance measures**. A central principle of performance-focussed management is a clear understanding of what is to be accomplished and how progress will be measured. This means recognizing the importance of using results-orientated goals and quantifiable measures to address programme performance. Generally, people are less resistant to change if they understand the organization's goals, the changes necessary, the measures to achieve the changes and what is expected of them.
- **Building and developing the organization's human capital**. Organizational success, and the change necessary to achieve that success, is possible only with the right employees who have adequate training, the right equipment, a good workplace, incentives and accountability to work effectively. Provision of the right information, instruction, training, supervision and support is essential in order to bring about change.

- **Programmes and processes**. Successful organizations have a clear understanding of their mission and are able to articulate how day-to-day operations contribute to mission-related results. Results-orientated performance should also draw upon commercial best practice.
- **Sound data**. Many management decisions involving change have subsequently been found to be based on unsound data. This has resulted in distrust among the workforce and aggravated the process of change. Decision-making processes must be based on sound, reliable and timely data.
- **An integrated strategy**. Changes in management processes – financial, production, engineering, etc. – should not be addressed in isolation or in a piecemeal fashion. When moving through a period of change, organizations need to take an integrated approach. Failure to do so can result in uncertainty amongst those not involved.
- **Active leadership**. Strong, sustained leadership is essential to changing deeply rooted corporate cultures and successfully implementing change.
- **Clear lines of responsibility and accountability**. Successful implementation is dependent upon clear lines of decision-making authority and resource control.
- **Incentives and consequences**. Incentives should be offered that motivate decision-makers to initiate and implement efforts that are consistent with better programme outcomes.
- **Enterprise-wide management structure**. A clearly-defined, enterprise-wide management structure is essential for an organization to effectively manage any large complex modernization effort. This enables people to see where they fit into the structure and the measures they need to take personally to manage the changes to be implemented in their working lives.
- **Monitoring and oversight**. There must be periodic reports to senior management of progress in bringing about change within an organization.

Standard observation and feedback process to change unsafe behaviour

Unsafe behaviour on the part of employees and other people in the workplace, such as the employees of contractors, contributes substantially to the causes of accidents. People commonly behave unsafely because they have never been hurt in the past while doing their work in an unsafe manner. What is important is that the consequences of behaving unsafely will nearly always determine future unsafe behaviour, and people continue unsafe working practices based on their past experience. When challenged about their unsafe behaviour, such as working on a flat roof without edge protection, they will state that they have always done the job that way and that they know what they are doing.

Eliminating all hazards by means of physical and other controls may be the ideal solution, but this may not always be possible or feasible. The essence of health and safety improvement initiatives must be that of getting groups to identify the norms for safe working, helping groups reinforce these health and safety-related norms and ensuring this positive reinforcement is brought about through peer pressure. Peer pressure is an important element of changing unsafe behaviour. In other words, people who have adjusted their working practices to a safer approach will put pressure on the others who are still adopting the old unsafe working practices.

Changing unsafe behaviour commences with frequent observation of aspects of unsafe behaviour, getting people to recognize particular aspects of the identified unsafe behaviour, training and advising them in the safe way of undertaking that particular task and monitoring their subsequent performance. Feedback from the various stages of this process should be provided in terms of the way the task was originally carried out, the changes that were needed and the improvements achieved.

It is important that people receive regular praise for changing old and unsafe work practices. The level of praise should be related to the extent of change achieved. This is an important feature in the development of a proactive safety culture.

Eliminating unsafe behaviour

The observation of unsafe working practices, feedback from accident investigation and from fellow employees are typical ways of identifying unsafe behaviour by people. Any strategy directed at eliminating unsafe behaviour must be planned and take place in a series of stages, thus,

- planning the change strategy;
- identifying areas of unsafe behaviour;
- measuring the extent of the unsafe behaviour and the risks arising from it;
- putting the change strategy into operation;
- evaluating the effects of the changes made to working practices; and
- correcting any variations from the changes installed.

The timescale for this exercise should be specified and there should be regular monitoring at the various stages to ensure satisfactory completion of each stage. In some cases, where there is evidence of resistance to change by certain employees towards safer working practices, disciplinary action may be necessary. However, this should be seen as the last resort. The principal objectives are that of changing attitudes, establishing group norms for safe working and reinforcing these safe methods through praise and recognition of their implementation.

Suggestion schemes

Suggestion schemes are an important factor in bringing about change. They involve everyone irrespective of status or position. The operation of a suggestion scheme is a common feature of many organizations. When working well, they can be extremely cost-effective, resulting in massive savings in many cases. However, as with many schemes of this type, people can lose interest, particularly when their suggestions are not accepted, and the scheme can flag. They, therefore, need regular stimulation and promotion if the continuing support of employees is to be maintained.

Several questions need to be asked.

Do senior management take notice of the suggestions and ideas from people at different levels within the organization?

Is there a mechanism for processing these suggestions and acting on the recommendations made?

What are the benefits to the people whose suggestions are adopted by the organization?

Many organizations run suggestion schemes covering, for example, quality, process and product improvement with a view to achieving savings in operating costs and improvements in performance. Suggestion schemes should be perceived as the natural focal point for all good ideas lying outside immediate job responsibility. They should incorporate three basic elements:

1. a formal mechanism or procedure for reviewing suggestions put forward by employees;
2. the provision of some form of monetary reward where ideas have been found to improve performance; and
3. recognition of the achievement of the suggestion from the individual concerned.

Suggestion schemes are applicable to any size of organization and have a range of benefits. They are, fundamentally, catalysts and energy-raising schemes which encourage teamwork, involvement and reward for good ideas. To be successful, the following points need emphasizing:

- all ideas are welcome irrespective of the nature of the idea;
- everyone can put forward ideas;
- the organization will consider all ideas;
- the organization recognizes that some ideas may not be viable or will not work;
- every idea should have some form of recognition; and
- for the organization as a whole, a forum for ideas can be extremely helpful.

Successful suggestion schemes

The elements necessary for a successful suggestion scheme include:

- visible support, commitment and encouragement for the Scheme at senior management level;
- a clear indication that eligibility for the scheme includes employees at all levels;
- a simple procedure for submitting suggestions;
- a systematic review procedure which ensures all suggestions are reviewed positively;
- administrative procedures to ensure acknowledgement of suggestions, their putting forward for consideration and assessment, completion of evaluation procedures, together with monitoring and reporting on performance in respect of the flow and turnround of suggestions;
- an approved budget to support the administrative arrangements, promotional activity and the cost of rewards;
- marketing and promotional plans which treat the scheme positively, including the establishment of a brand name for the scheme, advertising, publicity and special activities to promote the scheme;
- the payment of rewards that are fair and consistent within a framework which is thoroughly understood by all those involved;
- teamwork and partnership to promote the scheme and evaluate suggestions; and
- the creation of a network of assessors who have a particular enthusiasm for the scheme.

Safety suggestion schemes

All the above elements apply in the case of a safety suggestion scheme. However, the response from employees may not be positive initially and one of the great problems is that of overcoming negative attitudes.

It is important, therefore, that the scheme has clear support and commitment from the top of the organization, that administrative procedures, marketing and promotional plans are put in place before commencement of the scheme, and that employees are aware of the rewards available. There should be a specific suggestion form which is processed by assessors and to which a response to the suggestion is produced reasonably quickly.

A team of assessors should include a senior manager, specialists, such as engineering and quality managers, workplace representatives and health and safety specialists. The role and function of the assessors should be formally specified in the suggestion scheme documentation and promotional material. The assessors, in conjunction with senior management, should be responsible for promotion and publicity, together with the procedures for distribution of rewards to successful people.

In large organizations, it may be appropriate to classify suggestions under specific groupings, for example,

- workplace health and safety procedures, such as work at a height, the operation of permit-to-work systems, electrical safety procedures and traffic movement systems;
- machinery, plant and equipment;
- hazardous substances;
- occupational health issues; and
- human factors issues, such as health and safety training.

Record-keeping is an essential feature of safety suggestion schemes. A register of all suggestions should be maintained, along with the action taken and rewards paid.

Conclusions

Health and safety specialists, as 'multipurpose change agents', are, in most cases, charged with the task of bringing about change in an organization. Some changes may be forced on the organization as a result of new legal requirements or enforcement action. Most changes, however, result from a series of interventions directed at improving performance, reducing accidents and ill-health.

For change to be effective, it must start at the top and involve every layer of the organization. Above all, there must be 'ownership' at all levels and messages should be communicated in a way that everyone understands the reason for change. It should follow a step-by-step approach and any barriers to change should be taken into account prior to its introduction.

Key points

- Most organizations go through change at regular intervals, in some cases, in order to survive.
- Change management is defined as the process of guiding organizations through the process of change.
- Much of the industrial unrest going back over a century has been associated with a failure by all levels of management to consider the impact of change on management and workforce alike.

- Change within organizations may be of an internal or external nature.
- For effective change management, there must be leadership from the top.
- Senior managers need to consider the principles of change management before endeavouring to introduce change.
- It is vital that barriers to change are identified before considering the introduction of change within the organization.
- Organizations facing the possibility of major change need to consider a change management programme directed at easing the process of change. Management and employees alike would benefit from the implementation of this exercise.
- As 'multipurpose change agents', health and safety specialists, in particular, strive to bring about change. Before this can be achieved, however, it may be necessary to change the hearts and minds of individuals at all levels of the organization through the provision of information, instruction and training, cost–benefit analysis of the necessary changes and well-intentioned advice.

18 Stress and stress management

People at work have worries and anxieties about all sorts of things – increasing competition for jobs, globalization, terrorism, 'rationalization' of the organization's operations, looking after ageing parents and relatives, the threat of redundancy, annual appraisals, new technology, outsourcing of jobs to India and other Third World countries, together with increased demands by employers for higher productivity. Moreover, they may be put under excessive pressure at certain times, for example, to meet sales targets, attend meetings in time, learn and follow new procedures and fit in with changes in the organization's culture. This can result in varying levels of stress. According to the Health and Safety Executive, workplace stress is now the fastest growing cause of absence from work.

When employees complain of stressful conditions at work, what is management's response? How often have people been told, 'If you can't stand the heat, get out of the kitchen!'? The days when such a response from employers was common are over. Employers now need to get to grips with a range of policies and procedures to deal with stress at work.

What is important is that the poor standards of performance by many employees due to the effects of stress at work represent a substantial financial loss to their organizations and the British economy. Moreover, recent cases in the civil courts, and the greater attention now being paid to the subject of stress at work by the enforcement agencies, means that employers need to consider stress in the workplace and the measures they must take to prevent employees suffering stress arising from their work. It is not uncommon for six-figure sums to be awarded as damages in civil claims for stress-induced injury.

What is stress?

Stress is variously defined as:

- the common response to attack (Selye, 1931);
- the reaction people have to excessive pressure or other types of demand placed upon them. It arises when people worry that they can't cope (Health and Safety Executive);
- a psychological response which follows failure to cope with problems;
- any influence which disturbs the natural equilibrium of the human body.

The CBI defines stress as 'that which arises when the pressures placed upon an individual exceed the perceived capacity of that individual to cope'.

According to the TUC, stress occurs where demands made on individuals do not match the resources available or meet the individual's needs and motivation. Stress will arise if the workload is too large for the number of workers and time available. Equally, a boring or repetitive task which does not use the potential skills and experience of some individuals will cause them stress.

The HSE (1995) defined work stress as 'pressure and extreme demands placed on a person beyond his ability to cope'. In 1999, the HSC stated that 'stress is the reaction that people have to excessive pressures or other types of demand placed upon them'.

According to Cox (1993), 'stress is now understood as a psychological state that results from people's perceptions of an imbalance between job demands and their abilities to cope with those demands'.

A further definition is 'work stress is a psychological state which can cause an individual to behave dysfunctionally at work and results from people's response to an imbalance between job demands and their abilities to cope'.

Fundamentally, workplace stress arises when people try to cope with tasks, responsibilities or other forms of pressure connected with their jobs, but encounter difficulty, strain, anxiety and worry in endeavouring to cope.

Defining stress

A consideration of the above definitions of 'stress' produces a number of features of stress and the stress response, for example, disturbance of the natural equilibrium, taxation of the body's resources, failure to cope, sustained anxiety, a non-specific response, pressure and extreme demands and imbalance between job demands and coping ability. Fundamentally, a stressor (or source of stress) produces stress which, in turn, produces a stress response on the part of the individual. No two people respond to the same stressor in the same way or to the same extent. What is important is that, if people are going to cope satisfactorily with the stress in their lives, they must recognize

- the existence of stress;
- their personal stress response, such as insomnia or digestive disorder;
- those events or circumstances which produce that stress response, such as dealing with aggressive clients, preparing to go on holiday or disciplining employees;
- their own personal coping strategy, such as relaxation therapy.

As stated above, a stressor causes stress. Stress is commonly associated with how well or badly people cope with changes in their lives – at home, within the family, at work or in social situations. The causes are diverse, but include:

- **environmental stressors**, such as those arising from extremes of temperature and humidity, inadequate lighting and ventilation, noise and vibration and the presence of airborne contaminants, such as dusts, fumes and gases;
- **occupational stressors**, associated with too much or too little work, overpromotion or underpromotion, conflicting job demands, incompetent superiors, working excessive hours and interactions between work and family commitments;
- **social stressors**, namely those stressors associated with family life, marital relationships, bereavement, that is, the everyday problems of coping with life.

The autonomic system

Stress has a direct association with the autonomic system, a body system which controls an individual's physiological and psychological responses. This is the *flight or fight* system, characterized by two sets of nerves, the sympathetic and parasympathetic, which are responsible for the automatic and unconscious regulation of body function. The *sympathetic* system is concerned with answering the body's call to fight involving increased heart rate, the supply of more blood to the organs, stimulation of sweat glands and the tiny muscles at the roots of the hairs, accompanied by the release of adrenalin and noradrenalin. The *parasympathetic* system is responsible for emotions and for the protection of the body, which have their physical expression in reflexes, such as enlargement of the pupils, sweating, quickened pulse, blushing, blanching and digestive disturbance. The main signs associated with the sympathetic and parasympathetic states are shown in Table 18.1.

The General Adaptation Syndrome

Stress is perceived as a mobilization of the body's defences, an ancient biochemical survival mechanism perfected during the evolutionary process, allowing human beings to adapt to threatening circumstances.

In 1931, Hans Selye, a Vienna-born endocrinologist at the University of Montreal, raised the concept of the 'general adaptation syndrome' in his book, *The Stress of Life*. According to Selye, this syndrome comprises three stages.

- **alarm reaction stage**. This stage is typified by the individual receiving some sort of shock at a time when the body's defences are down, followed by a countershock, at which stage the defences are raised. In physiological terms,

Table 18.1 The main signs associated with the parasympathetic and sympathetic states

Parasympathetic state	**Sympathetic state**
Eyes closed	Eyes open
Pupils small	Pupils enlarged
Nasal mucus increased	Nasal mucus decreased
Saliva produced	Dry mouth
Slow breathing	Rapid breathing
Slow heart rate	Rapid heart rate
Decreased heart output	Increased heart output
Surface blood vessels dilated	Surface blood vessels constricted
Skin hairs normal	Skin hairs erect ('goose pimples')
Dry skin	Sweating
Digestion increased	Digestion slowed
Muscles relaxed	Muscles tense
Slow metabolism	Increased metabolism

once a stressor is recognized, the brain sends out a biochemical messenger to the pituitary gland that secretes adrencortitrophic hormone (ACTH). ACTH causes the adrenal glands to secrete corticoids, such as adrenalin. The result is a 'call to arms' of the body's systems.

- **Resistance stage**. The resistance stage involves two responses. The body will either resist the stressor or, alternatively, adapt to the effects of the stressor. This is the opposite of the alarm reaction stage, the characteristic physiology of which fades and disperses as the body adapts to the derangement caused by the stressor.
- **Exhaustion stage**. If the stressor continues to act on the body, however, this acquired adaptation is eventually lost and a state of overload is reached. The symptoms of the initial alarm reaction stage return and, if the stress is unduly prolonged, the wear and tear will result in damage to a local area or the death of the organism as a whole (see Figure 18.1).

The evidence of stress

Research in the 1990s by Professor Cox of Nottingham University led to much of the HSE's current guidance on the subject. Following an independent review of the literature, Professor Cox indicated that there was a reasonable consensus from the literature on psychosocial hazards (or stressors) arising from work which may be experienced as stressful or otherwise, and that these stressors may carry the potential for harm. According to the research there are nine characteristics of jobs, work environments and organizations which were identified as being associated with the feeling of stress and which could damage or impair health.

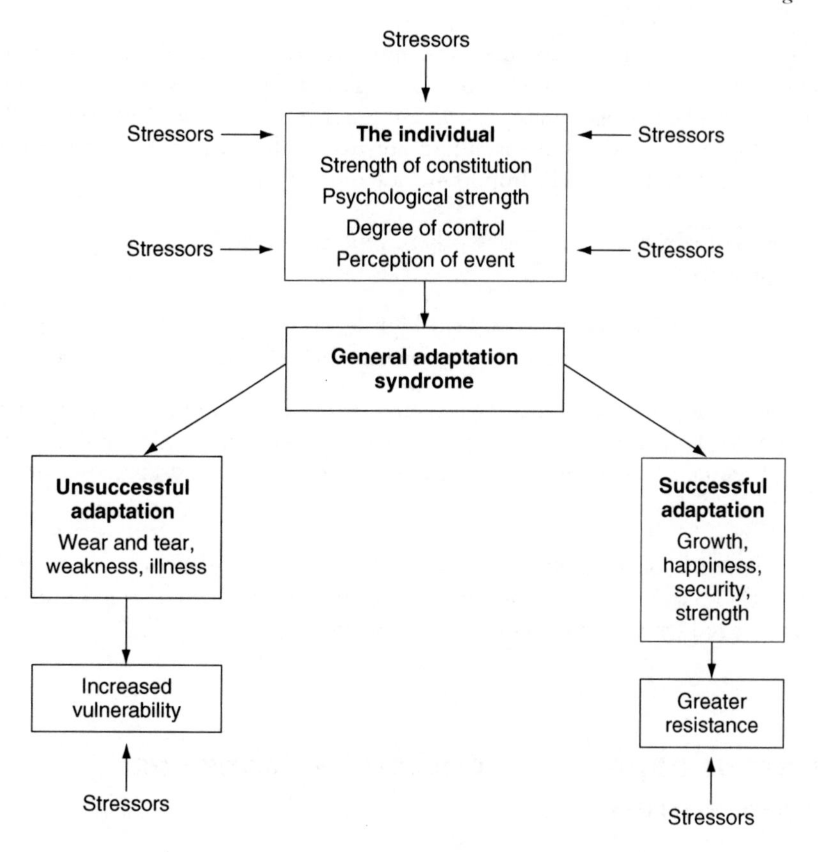

Figure 18.1 The general adaptation syndrome (Selye, 1936)

These characteristics are of two types, namely,

- those which concern the 'context or setting' in which the work takes place, namely,
  - organizational function and culture;
  - career development;
  - decision latitude/control;
  - role in organization;
  - interpersonal relationships;
  - the work/home interface; and
- the content or 'nature' of the job itself, in particular:
  - task design;
  - workload or work pace;
  - work schedule.

Further research released by the HSE gives an indication of the scale of the problem of injuries which are stress-related. In the report by Smith *et al.* (2000), *The Scale of Occupational Stress: The Bristol Stress and Health at Work Study CRR 265/2000,* it was estimated that there are five million workers suffering from high levels of stress at work. Important outcomes of this study were:

- Approximately one in five workers reported high levels of stress arising from work.
- There was an association between high levels of reported stress and specific job factors, such as excessive workloads or lack of managerial support.
- There was an association between high levels of reported stress and certain aspects of ill-health, such as poor mental health and back pain, together with certain health-related activities, such as smoking and excessive alcohol intake.

What came out of this study is that stress is now a foreseeable cause of ill-health and that employers need to take this factor into account when considering the means for reducing running costs of the undertaking.

Stress as opposed to 'pressure' – Positive and negative stress

Not all stress, however, is bad for people. Most people need a certain level of 'positive stress' or 'pressure' in order to perform well the tasks allotted to them. Some people are capable of dealing with very high levels of positive pressure. This is the classic 'fight response' or 'butterfly feeling' that people encounter before sitting an examination, running a race or attending a job interview.

Positive stress is one of the outcomes of competent management and mature leadership where everyone works together and their efforts are valued and supported. It enhances well-being and can be harnessed to improve overall performance and fuel achievement.

It is the 'negative stress', or distress, such as that arising from having to meet set deadlines or delegate responsibility, commonly leading to ill-health, that needs to be considered by employers as part of a stress management strategy. It may be the result of a bullying culture within the organization where threat, coercion and fear substitute for nonexistent management skills. With this sort of culture employees have to work twice as hard to achieve half as much to compensate for the dysfunctional and inefficient management. Negative stress diminishes quality of life and causes injury to health resulting in a range of stress-related symptoms.

Ill-health effects associated with stress

Stress affects people at work in many ways and the causes of stress are diverse. These causes can be associated with elements of the physical environment, such as open plan office layouts, the way the organization is managed, relationships within the organization and even inadequate work equipment. The causes can be classified as follows.

The physical environment

Poor working conditions associated with

- insufficient space to operate comfortably, safely and in the most efficient manner;
- lack of privacy which may be disconcerting for some people;
- open plan office layouts, resulting in distractions, noise, constant interruptions and difficulty in concentrating on the task in hand;
- inhuman workplace layouts requiring excessive bending, stretching and manual handling of materials;
- inadequate temperature and humidity control, creating excessive discomfort;
- poor levels of illumination to the extent that tasks cannot be undertaken safely;
- excessive noise levels, requiring the individual to raise his voice; and
- inadequate ventilation, resulting in discomfort, particularly in summer months, can be a frequent source of stress in the workplace.

The organization

The organization, its policies and procedures, its culture and style of operation can be a cause of stress. Culture is defined as 'a state or set of manners in a particular organization'. All organizations incorporate one or more cultures, which may be described, for example, as 'friendly', 'hostile', 'unrewarding' or 'family-style'. Stress can be associated with organizational culture and style due to, for instance,

- insufficient staff for the size of the workload, resulting in excessive overtime working;
- too many unfilled posts, with employees having to 'double up' at tasks for which they have not necessarily been trained or instructed;
- poor co-ordination between departments;
- insufficient training to do the job well, creating uncertainty and lack of confidence in undertaking tasks;
- inadequate information to the extent that people 'do not know where they stand';

- no control over the workload, the extent of which may fluctuate on a day-to-day basis;
- rigid working procedures with no flexibility in approach; and
- no time being given to adjust to change, which is one of the greatest causes of stress amongst employees.

The way the organization is managed

Management styles, philosophies, work systems, approaches and objectives can contribute to the individual stress on employees, as a result of

- inconsistency in style and approach by different managers;
- emphasis on competitiveness, often at the expense of safe and healthy working procedures;
- 'crisis management' all the time, due to management's inability, in many cases, to plan ahead and to manage sudden demands made by clients;
- information being seen as power by some people, resulting in intentional withholding of key information which is relevant to tasks, procedures and systems;
- procedures always being changed due, in many cases, to a failure by management to do the basic initial research into projects prior to commencement of them;
- overdependence on overtime working, on the presumption that employees are always amenable to the extra cash benefits to be derived from working overtime; and
- the need to operate shift work which can have a detrimental effect on the domestic lives of employees in some cases.

Role in the organization

Everyone has a role, function or purpose within the organization. Stress can be created through

- role ambiguity;
- role conflict;
- too little responsibility;
- lack of senior management support, particularly in the case of disciplinary matters dealt with by junior managers, such as supervisors; and
- responsibility for people and things which some junior managers, in particular, may not have been adequately trained to deal with.

Relations within the organization

How people relate to each other within the organizational framework and structure can be a significant cause of stress, due to, perhaps

- poor relations with the boss which may arise through lack of understanding of each other's role and responsibilities, attitudes held, and other human emotions, such as greed, envy and lack of respect;
- poor relations with colleagues and subordinates created by a wide range of human emotions;
- difficulties in delegating responsibility due, perhaps, to lack of management training, the need 'to get the job done properly', lack of confidence in subordinates and no clear dividing lines as to the individual functions of management and employees;
- personality conflicts arising from, for example, differences in language, regional accent, race, sex, temperament, level of education and knowledge; and
- no feedback from colleagues or management, creating a feeling of isolation and despair.

Career development

Stress is directly related to progression or otherwise in a career within the organization. It may be created by

- lack of job security due to continuing changes within the organization's structure;
- overpromotion due, perhaps, to incorrect selection or there being no one else available to fill the post effectively;
- underpromotion, creating a feeling of 'having been overlooked';
- thwarted ambition, where the employee's personal ambitions do not necessarily tie up with management's perception of his current and future abilities;
- the job has insufficient status; and
- not being paid as well as others who do similar jobs.

Personal and social relationships

The relationships which exist between people on a personal and social basis are frequently a cause of stress through, for instance,

- insufficient opportunities for social contact while at work due to the unremitting nature of tasks;
- sexism and sexual harassment;

- racism and racial harassment;
- conflicts with family demands; and
- divided loyalties between one's own needs and organizational demands.

Equipment

Inadequate, out-of-date, unreliable work equipment is frequently associated with stressful conditions amongst workers. Such equipment may be

- not suitable for the job or environment;
- old and/or in poor condition;
- unreliable or not properly maintained on a regular basis, resulting in constant breakdowns and down time;
- badly sited resulting, in excessive manual handling of components or the need to walk excessive distances between different parts of a processing operation;
- of such a design and sited in such a way that it requires the individual to adopt fixed and uncomfortable posture when operating it (see 'Cranfield Man' in Chapter 8, Ergonomic principles); and
- adds to noise and heat levels, increasing discomfort and reducing effective verbal communication between employees.

Individual concerns

All people are different in terms of attitudes, personality, motivation and in their ability to cope with stressors. People may experience a stress response due to

- difficulty in coping with change;
- lack of confidence in dealing with interpersonal problems, such as those arising from aggression, bullying and harassment at work;
- not being assertive enough, allowing other people to dominate in terms of deciding how to do the work;
- not being good at managing time, frequently resulting in pressure from supervisors and other employees to ensure the task is completed satisfactorily and on time; and
- lack of knowledge about managing stress.

Some of the more common occupational stressors are shown in Table 18.2.

Classification of stressors

A stressor produces stress. One of the ways of reducing stress is through changing or eliminating the cause of stress. Stressors can be classified according to type as

Table 18.2 The more common occupational stressors

New work patterns	Increased competition
New technology	Longer hours
Promotion	Redundancy
Relocation	Early retirement
Deregulation	Acquisition
Downsizing	Merger
Job design	Manning levels
Boredom	Insecurity
Noise	Lighting
Temperature	Atmosphere/ventilation

Table 18.3 Classification of stressors

Type 1	Stressors that can be changed or eliminated with minimum effort, such as hunger, thirst, inadequate lighting or ventilation, excessive noise, members of a work group and badly-fitting personal protective equipment
Type 2	Stressors that are difficult to change or eliminate, such as poor working relationships, financial problems, certain illnesses and conditions, inconsiderate managers and clients, technical difficulties with machinery and equipment and difficulties in separating work from home activities
Type 3	Stressors that are impossible to change, such as incurable illness, physical disabilities, death.

shown in Table 18.3. The category or type of stress will determine the range and scale of support provided by the organization.

Main sources of work stress

Another way of categorizing stressors is on the basis of their source. Certain stressors impact on people through their senses, such as extremes of temperature, odours, noise, light and ventilation. Other stressors cause changes in thoughts and feelings, such as fear, excitement, arousal, ambiguity, threat and worry. A third group is associated with changes in body state, such as those created by illness, inputs of drugs, chemicals and alcohol.

Irrespective of the magnitude of each of these stressors, they create some form of impact and have a cumulative effect bringing the individual closer to his tolerance level for peak performance. Excessive input of stress takes the person beyond that peak tolerance level leading to some form of stress response.

The sources of stress vary considerably from person to person. However, a number of the more common sources of stress can be considered. These are:

- **Task-related factors**. Work beyond the individual's mental capacity, information overload, boredom
- **Interpersonal factors**. Day-to-day interaction with people, abuse and harassment
- **Role ambiguity**. The individual has no clear idea of what is expected of him
- **Role conflict**. Opposing demands made on an individual by different people
- **Little or no recognition for a job well done**
- **Personal threat**. Actual threats to a person's safety, fear of redundancy or dismissal
- **Environmental factors**. Noise, excessively high or low temperatures, inadequate lighting and ventilation, dirty workplaces, inadequate work space.

Bullying and harassment at work

Many managers would perceive the problem of bullying and harassment of their employees as primarily an industrial relations issue and, as such, should be dealt with through an employer's internal grievance and disciplinary procedures long before the problem becomes a risk to the health of those employees.

Levels of bullying vary significantly. Bullying is the common denominator of harassment, discrimination, prejudice, abuse, conflict and violence. It could be said that some people are 'serial bullies', whether they be managers or employees, and simply do not recognize this fact. Employees must be in a position to report this form of behaviour confidentially to their employer with a view to seeking preventive action.

Stress caused by bullying can produce a number of symptoms in the victims, as follows.

- **Principal symptoms**. Stress, anxiety, sleeplessness, fatigue, including chronic fatigue syndrome (see below) and trauma.
- **Physical symptoms**. Aches and pains, with no clear cause, back pain, chest pains and angina, high blood pressure, headaches and migraines, sweating, palpitations, trembling and hormonal problems, etc.
- **Psychological symptoms**. Panic attacks, thoughts of suicide, stress breakdown, forgetfulness, impoverished or intermittently functioning memory, poor concentration, flashbacks and replays, excessive guilt, disbelief, confusion and bewilderment, insecurity, desperation, etc.
- **Behavioural symptoms**. Tearfulness, irritability, angry outbursts, obsessiveness (the experience takes over a person's life), hypervigilance,

hypersensitivity to any remark made, sullenness, mood swings, withdrawal, indecision, loss of humour, etc.

- **Effects on personality**. Shattered self-confidence and self-esteem, low self-image, loss of self-worth and self-love.

Other symptoms and disorders include sleep disorder, mood disorder, eating disorder, anxiety disorder, panic disorder and skin disorder.

What is important is that the traumatizing effect of bullying results in the individual being unable to state clearly what is happening to him and who is responsible. The target of bullying may be so traumatized that he is unable to articulate his experiences for a year or more after the event. This often frustrates or prevents both disciplinary action and any subsequent legal action in respect of alleged post-traumatic stress disorder (PTSD).

Bullying commonly results in feelings of fear, shame, embarrassment and guilt which are encouraged by the bully in order to prevent his target raising the issue with management. In addition, work colleagues may often withdraw their support and then join in with the bullying, which increases the stress and consequent psychiatric injury. In addition, the prospect of going to work, or the thought or sound of the bully approaching, immediately activates the stress response, but flight or fight are both inappropriate. In cases of repeated bullying, the stress response prepares the body to respond physically, but is of little avail in most cases.

Management, therefore, have a duty to do something about this problem. The starting point is a policy on bullying and harassment at work which is brought to the attention of all employees. Managers should know what motivates the bully and take disciplinary action. Those suffering bullying would benefit from assertiveness training to defend themselves against unwarranted verbal and physical harassment.

Fatigue

People subject to bullying at work commonly suffer fatigue. This is caused by the body's flight or fight mechanism being activated for long periods, for example from Sunday evening, prior to starting work on Monday morning, through to the following Saturday morning, when there is a chance to obtain some relief.

The flight or fight mechanism is, fundamentally, a brief and intermittent response. However, when activated for abnormally long periods, it can cause the body's mental, physical and emotional reserves to drain away. The body sustains damage through prolonged raised levels of glucocorticoids, which are toxic to brain cells, and excessive depletion of energy reserves resulting in loss of strength and stamina, fatigue and muscle wastage.

Chronic fatigue syndrome

People who are subject to bullying and harassment commonly suffer symptoms similar to chronic fatigue syndrome. (This is sometimes referred to as myalgic encephalomyelitis (ME), chronic fatigue immune deficiency syndrome (CFIDS) and post-viral fatigue syndrome.) This syndrome is characterized by

- overwhelming fatigue;
- joint and muscle pains;
- spasmodic bursts of energy, followed by exhaustion and joint and muscle pain;
- a general lack of the ability to concentrate for long periods;
- poor recall with respect to words and sentence construction, for example;
- mood swings, including anger and depression;
- difficulty in taking on new information;
- imbalances in the senses of smell, appetite and taste;
- a dislike of bright lights and excessive noise;
- an inability to control body temperature;
- sleep disturbances, manifested by spending the night awake and then sleeping during the day;
- disturbances in the sense of balance;
- clumsiness, e.g. inability to grasp small objects.

Psychiatric injury

The symptoms described above eventually lead to psychiatric injury, but not mental illness. In fact, it may be appropriate at this stage to distinguish between psychiatric injury and mental illness. In spite of superficial similarity, there are distinct differences between psychiatric injury and mental illness. These include

- mental illness is assumed to be inherent (internal), whereas psychiatric injury is caused by external factors, such as bullying and harassment;
- injuries tend to heal or get better;
- a person who is suffering a mental illness can exhibit a range of symptoms which are commonly associated with mental illness, such as schizophrenia, delusions and paranoia, but not those associated with psychiatric injury;
- a person suffering psychiatric injury, on the other hand, typically exhibits a range of symptoms, including obsessiveness, hypervigilance, irritability, fatigue, hypersensitivity and insomnia, commonly associated with psychiatric injury, but not mental illness.

Suicide

People who suffer bullying have many common characteristics. For instance, they are unwilling to resort to violence, or even to resort to private legal action, to resolve conflict. They have a tendency to internalise anger rather than express it outwardly which, in many cases, leads to depression. Where the bullying continues over a long period of time, say several years, the internalised anger accumulates to the point where the victim may

- start to demonstrate all the symptoms of stress as the internal pressure causes the body to go out of stasis (this happens in every case); or
- focusses the anger on to himself and harms himself, through the use of drugs and/or alcohol, or attempts to, or actually commits, suicide; or
- in rare cases, actually 'flip' and start to copy the behaviour patterns of the bully.

The lead up to suicide takes place in a series of steps. Firstly, bullying causes prolonged negative stress (psychiatric injury) which includes reactive depression. This results in a fluctuating baseline of the person's objectivity, that is, the balance of the mind is disturbed. At this stage, the victim starts to contemplate suicide culminating with the later stage where the victim actually attempts suicide. This is, of course, a cry for help and if this situation is not recognized and dealt with, actual suicide may follow.

It is conceivable that many suicides are caused by bullying. There is evidence that at least 16 children in the UK commit suicide as a result of bullying at school. In many cases, a victim of bullying is not initially aware of what is actually happening. He may be unwilling to confide with a colleague or manager what is happening to him, he may well become traumatized and unable to articulate. In many cases, the organization will deny the existence of bullying by individuals, or a culture of bullying from the top downwards, resulting in the real causes not being identified.

A recent case

In a recent case, the Law Lords established that employers have a duty to take action against bullying of employees in the workplace, whether it be by managers or other employees. This landmark judgement came after they heard the case of a former policewoman, Eileen Waters, who claimed she was raped by a colleague and then suffered four years of victimisation after she reported him to her superior officer.

Miss Waters joined the Metropolitan Police Force in 1987. She alleged that in February 1988 she was raped by a fellow officer in a police hostel in central London while they were both on duty. Even though she complained to her reporting sergeant and to other officers about what had happened, no charges were ever brought.

Following the alleged harassment, she took sick leave in July 1992 and never returned to work as a police officer, retiring from the force in May 1998 on medical grounds.

Here the Law Lords gave her permission to sue the Metropolitan Police Commissioner for compensation after being told she was subjected to poison pen letters, offensive cartoons and frequent transfers between stations. A blood-stained truncheon was said to have been placed in her locker and, on one occasion, colleagues failed to back her up during a street confrontation.

Lord Slynn of Hadley said:

> If an employer knows that acts being done by employees during their employment may cause physical or mental harm to a particular fellow employee and he does nothing to supervise or prevent such acts when it is in his power to do so, it is clearly arguable that he may be in breach of his duty to that employee.

For the claimant, the ruling was an important milestone in a 9-year-long fight for justice, having had her claims against the Metropolitan Police Force thrown out by an industrial tribunal, the High Court and the Court of Appeal.

Models of stress at work

A number of models show the relationship of stress at work to individual and organizational performance.

The organizational boundary

Studies by Cooper and Marshall (1978) into stress within the organization identified an *organizational boundary*, with the individual manager straddling that boundary and, in effect, endeavouring to cope with conflicting stressors created by external demands (the family) and internal demands (the organization) (see Figure 18.2).

Here, there is a situation where a manager's response may be affected by individual personality traits, tolerance for ambiguity, the ability to cope with change, specific motivational factors and well-established behavioural patterns. Within the organization, a number of stressors can be present. These are associated with

- *the job*, such as work overload or underload, poor physical working conditions, time pressures and the extent and difficulty of the decision-making process;
- *role in the organization*, for example, role conflict, role overload, role ambiguity, specific responsibility for people, lack of participation in the organization's decision-making process;

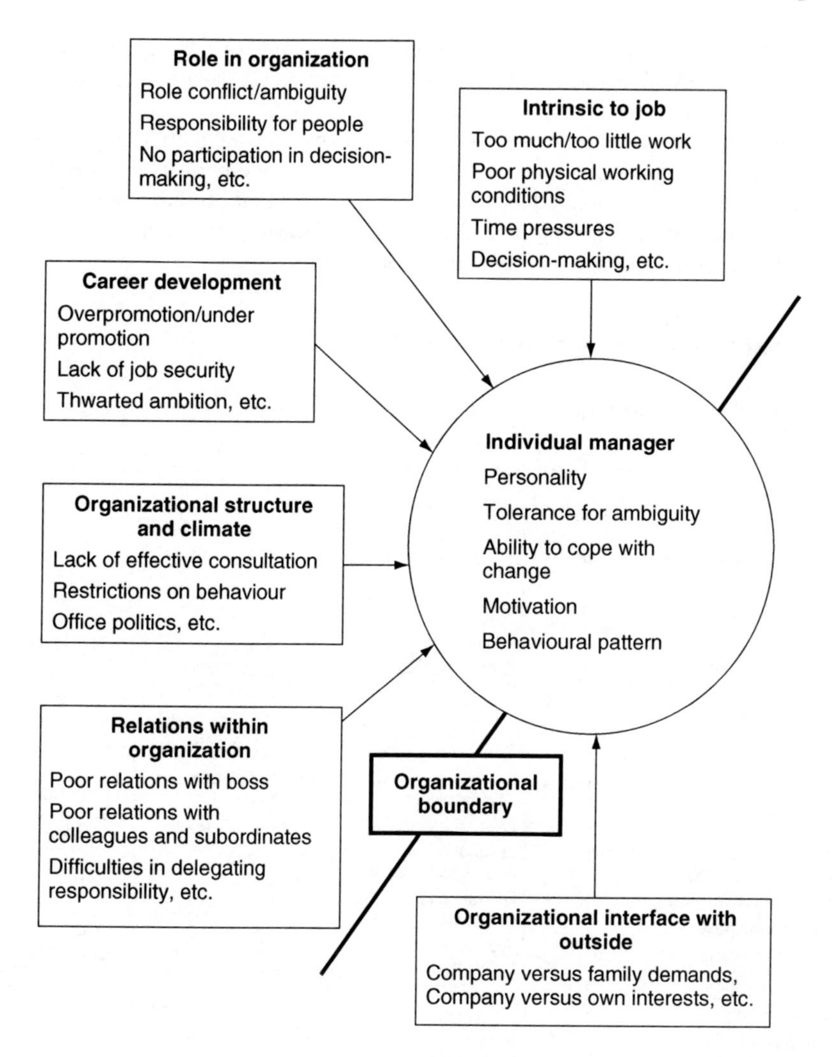

Figure 18.2 Sources of managerial stress. *Source*: Cooper and Marshall, 1978

- *career development*, for example, the stress arising from overpromotion or underpromotion, lack of job security, thwarted ambitions;
- *organizational structure and climate*, such as lack of effective consultation, restrictions on behaviour, office politics;
- *relationships within the organization*, for example, poor relationships with senior management, colleagues and subordinates, difficulties in delegating responsibility.

On the other side of the organizational boundary is the organization's interface with the outside world. Here conflict can be created where there may be competition for a manager's time between the demands of the organization and that of the family or between the organization and particular outside interests.

The conclusion drawn from these studies is that organizations should pay attention to the potentially stressful effects of their decisions, management style, consultative arrangements, environmental factors and other matters that can have repercussions for their employees and their home lives. The resulting stress arising from this conflict situation can have adverse effects on performance.

Transactional model of stress

The model shown in Figure 18.3 depicts

- the various sources of stress at work in terms of
    - those factors intrinsic to the job;
    - an individual's role in the organization;

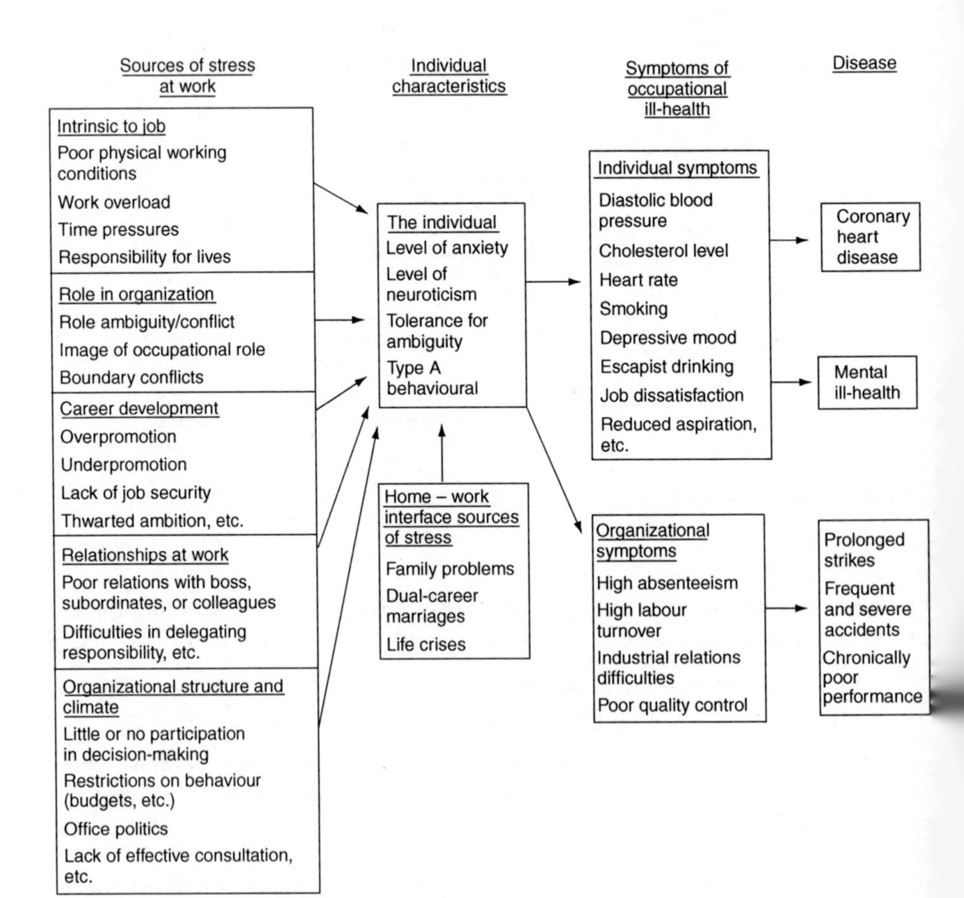

Figure 18.3 Transactional model of stress

    - career development;
    - relationships at work; and
    - the organizational structure and climate;
- individual characteristics with respect to:
    - the individual; and
    - the home–work interface;
- symptoms of occupational ill-health affecting:
    - the individual; and
    - the organization;
- the effects of stress in terms of disease:
    - coronary heart disease;
    - mental ill-health;
    - the effects on the organization.

Personality and stress

Personality is defined as 'the dynamic organization within the individual of the psychophysical systems that determine his characteristic behaviour and thought' (Allport, 1961).

Personality traits and their relationship to stress

Various types and traits of personality have been established by psychological researchers over the last 30 years. These can be classified as listed in Table 18.4. Research indicates that most people combine traits of more than one of these types and so the definitions above can only be used as a guide. The type most at risk to stress is type A (Figure 18.4).

Type A behaviour

The characteristics of type A personality can be summarized as follows:

- excessive competitiveness and striving towards advancement and achievement;
- accentuating various key words in ordinary speech without real need and tending to utter the last few words of a sentence more rapidly than the opening part;
- continual drive towards ill-defined goals;
- preoccupation with deadlines for all sorts of task;

Table 18.4 Classification of personality types

Type A, Ambitious	Active and energetic; impatient if he has to wait in a queue; conscientious; maintains high standards; time is a problem – there is never enough; often intolerant of others who may be slower.
Type B, Placid	Quiet; very little worries them; often uncompetitive; put their worries into things they can alter and leave others to worry about the rest.
Type C, Worrying	Nervous; highly strung; not very confident of self-ability; anxious about the future and of being able to cope.
Type D, Carefree	Love variety; often athletic and daring; very little worries them; not concerned about the future.
Type E, Suspicious	Dedicated and serious; very concerned with others' opinions of them; do not take criticisms kindly and remember such criticisms for a long time; distrust most people.
Type F, Dependent	Bored with their own company; sensitive to surroundings; rely on others a great deal; people who interest them most are oddly unreliable; they find the people that they really need to be boring; do not respond easily to change.
Type G, Fussy	Punctilious, conscientious and like a set routine; do not like change; any new problems throws them because there are no rules to follow; conventional and predictable; collect stamps and coins and keep them in beautifully ordered state; great believers in authority.

The potential for stress-related ill-health, such as heart disease, is common among Type A people.

Complete this questionnaire. How many YES answers did you get?

Do you walk fast?
Are you competitive?
Are you impatient?
Do you go all out?
Do you have no, or few, interests outside work?
Do you hide your feelings?
Do you talk fast?
Do you try to do more than one thing at a time?
Are you hard driving?
Do you anticipate others, e.g. interrupt, finish their sentences?
Are you emphatic in speech?
Do you want recognition from others?
Do you eat quickly?
Are you ambitious?
Do you look ahead to the next task?

Figure 18.4 Type A questionnaire

- intolerance of delays and postponements in arrangements;
- a level of mental alertness which can easily progress to aggressive behaviour;
- permanent impatience with people and situations; and
- feelings of guilt when having a rest or relaxing.

Women at work – Stress factors

Research has shown that women can be subject to many stressors at work which are not suffered by their male counterparts (Figure 18.5). Whilst sexual harassment at work is a common cause of stress amongst women, other causes of stress include

		Yes/No
1.	Do you feel undervalued at work?	
2.	Do you feel your employers don't take your concerns seriously?	
3.	Do you feel you have been overlooked with regard to promotion?	
4.	Do you suffer sexual harassment from colleagues and clients?	
5.	Do you feel male colleagues are treated more favourably by your boss?	
6.	Are there certain male colleagues you intentionally avoid at work?	
7.	If you are not in a relationship, married or otherwise, do colleagues, both male and female, look down on you?	
8.	Are you under pressure to perform to a particular standard?	
9.	Do you feel you are poorly paid compared with male colleagues doing the same type of work?	
10.	Do you have children?	
11.	Do you have dependent relatives at home?	
12.	Does your family make excessive demands on you?	
13.	Do you suffer conflict from your desire to start a family but need to work?	
14.	Do you have problems with child care, particularly during school holidays?	
15.	Do you get any help at home with housework?	
16.	Do you need 'a stiff drink' when you get home from work?	
17.	Do you suffer from insomnia?	
18.	Do you find it difficult to separate your work activities from home life?	
19.	Does you manager regularly contact you at home over work issues?	
20.	Do you feel permanently tired?	

Question: How many 'Yes' answers did you get?

Over 15: You are seriously stressed at work and need to consider alternatives, such as finding a less demanding job.

10–14: You need to discuss your work activities with your manager with a view to resolving some of your stress-related difficulties.

5–9: There are a number of aspects of your working life which you need to take in hand personally.

1–4: You have got a reasonably happy and stress-free working life. Try to do something about the aspects of work which cause you stress. Talk to your manager about it.

Figure 18.5 Stress questionnaire for women at work

- performance-related pressures;
- lower rates of pay;
- the problem of maintaining dependants at home;
- lack of encouragement from superiors, including not being taken seriously;
- discrimination in terms of advancement;
- sex discrimination and prejudice;
- pressure from dependants at home;
- career-related dilemmas, including whether to start a family or whether to marry/live with someone;
- lack of social support from colleagues;
- lack of same-sex role models;
- evidence of male colleagues being treated more favourably by management;
- being single and labelled as an oddity; and
- lack of domestic support at home.

Management should be aware of the various forms of stress to which women are exposed at work and take measures, including disciplinary measures, where evidence of, particularly, sexual harassment exists.

Responses to stress

Stress can have both short- and long-term responses. Individuals and, indeed, their employers should take the short-term symptoms seriously, bearing in mind the long-term implications, not only for individuals but for the organization as a whole. These symptoms are summarized in Table 18.5.

Responses to prolonged stress

As with any form of stressor, not only must the nature of the stressor be considered, but also the duration of exposure to it. Physical responses to prolonged stress can include a number of minor disorders which create discomfort, but which may lead to serious ill-health including headaches, migraine, allergies, skin disorders and arthritis. The following diseases have been linked to stress, but there is no clear-cut medical evidence to this effect.

- **Coronary heart disease**. Studies have identified a positive link with competitive and aggressive behaviour and coronary heart disease.
- **Cancer**. Some people, who are prone to symptoms of stress, such as anger, fear and feelings of hopelessness, may be more susceptible to cancer.

Table 18.5 Short- and long-term responses to stress

A. Short-term responses	
Physiological	**Mental, emotional, behavioural and social**
Headaches, migraine	Job dissatisfaction
Backaches	Anxiety
Eye and vision problems	Depression
Allergic skin responses	Irritability
Disturbed sleep patterns	Frustration
Digestive disorders	Breakdown of relationships at home and work
Raised heart rate	Alcohol and drug misuse
Raised blood cholesterol	Tobacco smoking
Raised adrenalin/noradrenalin levels	Inability to unwind/relax
B. Long-term responses	
Individual health	
Gastric/peptic ulcers	
Diabetes	
Arthritis	
Stroke	
High blood pressure	
Coronary heart disease	
Mental ill-health	

- **Digestive disorders**. Persistent indigestion or stomach discomfort is a classic manifestation of stress for some people. Approximately one person in ten get a stomach ulcer at some time in their life.
- **Diabetes**. Diabetes commonly follows some form of emotional or physical upset to the system.

The lessons to be learnt are that people need to be aware, firstly, of their own personal stress response(s) and, secondly, to take positive action, rather than ignoring the evidence and pushing themselves even further.

The stages of the stress response

The response to stress commonly takes place in a number of stages in which a number of symptoms may be present. These are summarized in Table 18.6. Managers should be trained to recognize some of the more obvious symptoms as part of any stress reduction programme within an organization.

Table 18.6 Stages of the stress response

Stage 1

- Speeding up
- Talking quickly
- Walking fast (head leading)
- Eating and drinking faster
- Working at high speed and for long periods of time without tiring (at the time)

Stage 2

- Irritability
- Dyspepsia and gastric symptoms
- Tension headache
- Migraine
- Insomnia, loss of energy
- Comfort seeking – alcohol, smoking
- Increased intake of food and drink

Stage 3

- Cotton wool head
- Gastric ulceration
- Palpitations, chest pain, cardiac incident
- Depression and anxiety
- Tiredness, lack of energy
- Physical and mental breakdown

Stress indicators

People

Stress indicators are the danger signals that alert people to the fact that they are subjecting themselves to stress and pressure. These indicators can be physical, emotional or mental and are characterized by exaggerated or abnormal reactions to situations compared to their normal reactions.

Physical examples include tightness in the chest, muscle tightening, headaches, skin disorders, increased physical irritations, such as asthma, fidgeting, increased alcohol intake, loss or gain of appetite and disturbed sleep patterns. It is important to recognize that no two people display the same stress indicators. Emotional examples include increased irritability, anger, anxiety, frustration, touchiness and guilt.

Mental examples include forgetfulness, trouble with thinking clearly, difficulty in forming sentences verbally and in writing, obsessiveness with petty detail and an overactive brain.

The organization

Stress indicators can also be recognized in an organization! For example, in some cases, senior management commonly fail to recognize the problem of stress amongst managers and employees. In other cases, they may take the view that stress is a manifestation of weakness on the part of an individual manager. Statements such as, 'If you can't stand the heat, get out of the kitchen' are not uncommon amongst senior managers when presented with manifestations of stress on the part of subordinates. However, stress within the organization cannot be disregarded. Typical indications of an organization manifesting high stress levels amongst employees at all levels are

- absenteeism;
- poor timekeeping;
- high labour turnover;
- high sickness absence rates;
- low productivity;
- industrial unrest;
- resistance to change in working procedures.

Failure to recognize and act on these indicators leads to diminished levels of performance, diminished profitability for the organization and, for many people, long-term ill-health.

The effects of stress on job performance

- **Absenteeism**. Absenteeism, especially on Monday mornings, or in the taking of early/extended meal breaks is a typical manifestation of stress.
- **Accidents**. People suffering stress at work can rapidly become problem drinkers. Such people have three times the average number of accidents; many accidents incorporate stress-related indirect causes.
- **Erratic job performance**. Alternating between low and high productivity due, in some cases, to changes outside the control of the individual, is a common symptom of stress within an organization.
- **Loss of concentration**. Stressful events in people's lives commonly result in a lack of the ability to concentrate, whereby a person is easily distracted, or an inability to complete one task at a time.
- **Loss of short-term memory**. This leads to arguments about who said, did or decided what?
- **Mistakes**. Stress is a classic cause of errors of judgement, which can result in accidents, wastage and rejects. Such mistakes are frequently blamed on others.

- **Personal appearance**. Becoming abnormally untidy, perhaps smelling of alcohol, is a common manifestation of a stressful state.
- **Poor staff relations**. People going through a period of stress frequently become irritable and sensitive to criticism. This may be accompanied by 'Jekyll and Hyde' mood changes, all of which have a direct effect on staff relationships and their home life.

Anxiety and depression

Anxiety and depression are the classic manifestations of stress.

Anxiety and panic disorders

'Anxiety' is defined as 'a state of tension coupled with apprehension, worry, guilt, insecurity and a constant need for reassurance'. It is accompanied by a number of psychosomatic symptoms, such as profuse perspiration, difficulty in breathing, gastric disturbances, rapid heartbeat, frequent urination, muscle tension or high blood pressure. Insomnia is a reliable indicator of a state of anxiety.

Anxiety and panic disorders arise in many different forms, varying in intensity from person to person. They may be associated with certain disruptive life events, individual genetic features and changes in neurological chemistry. Emotional and physical stress can also lead to anxiety-related disorders.

Generalized anxiety disorder

This disorder is defined as 'a period of uncontrolled worry, nervousness and anxiety for a period of six months or more'. It may initially arise as a result of a particular worry concerning, for example, relationships with people, finance or career prospects, or may present as a vague anxiety about virtually everything. It is commonly accompanied by irritability, with a number of physical symptoms, including muscle pain, gastrointestinal problems, trembling and insomnia.

Panic disorder

Some people suffer panic attacks on a regular basis which is diagnosed as panic disorder. When experiencing a panic attack, the individual suffers intense anxiety, fear or panic. Physical symptoms include an increased heart rate, perspiration, chest pains, tingling sensations and the feeling of disaster, loss of control or even imminent death. Some people who suffer panic attacks develop obsessive compulsive disorder or agoraphobia in an attempt to control the attacks.

There are three different forms of panic attack: (1) *An unexpected panic attack* can strike at any time and without warning. (2) *A situational panic attack* always occurs in a specific location or situation, such as driving at night or when entering a particular building. The particular location or circumstance always provides the cue for a panic attack. (3) *A situationally predisposed panic attack* also occurs in response to a particular situation or location, but does not occur every time the cue is present. The reaction may be delayed for some time.

Phobias

Phobias are, fundamentally, a fear response on the part of the individual to certain situations and things. People can have phobias to virtually anything or any situation, such as fear of spiders, fear of flying, fear of heights and fear of enclosed spaces. The fear reaction generated by the phobia is out of proportion to the risks actually presented by the situation, and the person is aware of this fact intellectually. However, this awareness does not prevent phobias from occurring and many people go to great lengths to avoid encountering the subject of their particular phobia.

Post-traumatic stress disorder

This is one of the few anxiety disorders where a specific trigger or cause for the disorder can be identified. It can be caused by highly stressful situations, such as those arising from war and disasters, physical attack, sight of death and other death situations, such as the death of a loved one.

People suffering PTSD often relive the event that triggered the disorder initially, either through 'flashbacks' or nightmares. Some people make considerable efforts to avoid situations, places or circumstances that remind them of the traumatic event. Many people suffering PTSD develop an emotional insensitivity or 'numbness' as the mind attempts to protect itself from this disorder.

Obsessive compulsive disorder

This disorder affects a small number of people. People experiencing the disorder suffer persistent recurring thoughts caused by fears or anxiety. These repeating thoughts can, in some cases, lead to ritualistic behaviour designed to fend off the anxiety. This includes excessive hand washing to maintain personal hygiene and prevent illness, checking and rechecking the security of the house and only using certain brands of goods.

Separation anxiety

This is a normal aspect of the development of children. Between the ages of approximately 8 and 14 months, children become aware of their need for comfortable and safe surroundings, in particular, the closeness of their mother and father. Children suffering separation anxiety cry when their parents are about to leave them in the presence of, for example, a babysitter. They cling on to their parents displaying various manifestations of anxiety. This anxiety, in some cases, persists to the age of 4 years or even later in childhood. It may re-emerge later in childhood under stressful situations, such as marital breakdown or when being left for the first time at boarding school.

When this form of behaviour becomes persistent and excessive, a child or young person may be diagnosed as suffering from separation anxiety disorder. They worry about being separated from their parents or from being taken from their homes or getting lost. They have trouble falling asleep, even in their own beds, and have a worry that their parents may die or abandon them.

Social anxiety disorder

This is a social phobia in which sufferers become irrationally concerned about being judged or ridiculed in social circumstances and situations. They feel extreme embarrassment and anxiety when exposed to the outside world, and may suffer heart palpitations, sweating and blushing.

Depression

Depression, on the other hand, is much more a mood, characterized by feelings of dejection and gloom, and other permutations, such as feelings of hopelessness, futility and guilt. The well-known American psychiatrist, David Viscott, described depression as 'a sadness which has lost its relationship to the logical progression of events'. It may be mild or severe. Its milder form may be a direct result of a crisis in work relationships. Severe forms may exhibit biochemical disturbances, and the extreme form may lead to suicide.

It is not a single disorder, however. There are different types of depression each manifesting a wide range of symptoms, each with varying degrees of severity. Symptoms of depression include sadness, anger, feelings of 'emptiness', pessimism about the future, low energy levels and sex drive and various forms of mental impairment, such as memory loss and difficulty in concentrating.

Depression may develop when one or more of the following factors are present:

- a family history of depression;
- a personal history of depression;
- chronic stress;
- death of a loved one;
- chronic pain or illness; or
- drug or alcohol misuse.

Major depression

Where several of the above symptoms are present for a period of, for example, 2 weeks or more, an individual may be diagnosed by a psychiatrist or psychologist as suffering from major depression. This form of depression may occur perhaps once in a life time or, in some cases, may occur throughout life. Various combinations of therapy and medication are used to treat this condition.

Dysthymia

Symptoms are similar to those of a major depression, but tend to be less severe. The condition generally lasts longer than major depression, diagnosis of such being based on the presence of symptoms of depression in excess of 2 years. As with a major depression, dysthymia is treated with medication, therapy or a combination of both. Dysthymia is more resistant to treatment than major depression and can last longer if left untreated.

Bipolar disorder

This is commonly known as manic depression. Bipolar disorder is characterized by periods of depression interspersed with periods of mania. While depressive periods have similar symptoms to other types of depression, mania symptoms are quite different and include euphoria, a false sense of well-being, poor judgement, an unrealistic view of personal abilities and inappropriate social behaviour.

Post-partum depression

Approximately 80 per cent of women experience the 'baby blues' after giving birth. Hormonal changes, the physical trauma of giving birth and the emotional strain of new responsibilities often make the first 3 or 4 weeks after a birth emotionally stressful. The situation eventually resolves without any need for treatment. For a small percentage of mothers, however, the baby blues do not go away, developing into post-partum depression. If not successfully treated, this can lead to major depression or dysthymia.

Post-partum psychosis

This condition is a rare complication of post-partum depression, affecting a very small number of women after childbirth. The patient will display psychotic behaviour, often directed at the new baby. Patients may suffer hallucinations and delusions and the severity of the condition can sometimes put both mother and baby at risk.

Advice to managers

Managers frequently regard depression as being of no importance. Depression is not a sign of weakness, something which is untreatable, a personal failing or just feeling sad. It is a specific condition which needs sympathetic consideration, help and a responsible approach where employees report this condition (see figure 18.6).

Stress and atypical workers

'Atypical workers' are workers who are not in normal daytime employment, together with shift workers, part-time workers and night workers.

Statement of Policy on Stress at Work

Within this Statement of Policy, 'stress' is defined as:

> The reaction people have to excessive pressure or other types of demand placed upon them. It arises when people worry that they cannot cope. (HSE)

The principal causes of stress

According to the Health and Safety Executive, the principal causes of stress at work are associated with:

- general management and culture;
- decision-making and planning;
- the employee's role in the organization;
- relationships at work;
- job design;
- work schedules;
- work load and work pace. (Stress at work: HS(G) 116: 1995)

Statement of intent

This organization recognizes:

- its common duty of care towards employees; and
- its responsibilities under the Health and Safety at Work etc Act 1974 to ensure, so far as is reasonably practicable, the health of its employees whilst at work.

However, it is hereby stated that no occupation within this organization is regarded as intrinsically dangerous to the mental health of any employee.

The organization undertakes to take steps that are reasonable in all the circumstances to protect **employees** from the risk of reasonably foreseeable mental injury caused by stress which may arise at or during the course of their work or as a result of working conditions.

Organization and arrangements for implementing this policy

All managers are responsible for preventing employees being put into stressful situations during their work activities. They will take the potential for stress into account when undertaking an assessment of the risks to health and safety of employees in accordance with the duties of employers under the Management of Health and Safety at Work Regulations.

The organization will provide, where appropriate, a range of confidential services and treatments to employees from occupational health practitioners dedicated to eliminating or reducing stress, including

---

Figure 18.6 A model statement of policy on stress at work

- counselling on stress-related issues;
- hypnotherapy; and
- any other treatment which may be advised by an occupational health practitioner.

All employees will be provided with information, instruction and training directed at enabling them to identify any causes of stress at work and the measures they must take in dealing with these causes of stress.

Responsibilities of managers

Managers are entitled to assume that an employee is able to withstand the normal pressures of his job unless they are advised of some particular problem or vulnerability by the employee concerned. Managers must receive reports or complaints of stress at work from employees with a sympathetic and caring attitude.

Any report or complaint of stress at work from an employee must be investigated and the employee concerned informed of the outcome of such investigations. Any action taken by management as a result of report or complaint will be recorded in the individual's personal file.

Where there is evidence of bullying, harassment, threatening behaviour, violence or other inappropriate behaviour:

- between individual managers and employees; or
- between individual employees,

disciplinary action will be taken by management against the persons concerned.

Responsibilities of employees

Employees must report immediately, and in confidence, to their line manager both the causes and personal symptoms of stress arising from work.

Figure 18.6 (Continued)

Approximately 29 per cent of employees in the UK work some form of shift pattern, and 25 per cent of employees undertake night shifts. Researchers have, over the years, studied the physical and psychological effects of atypical working on these groups of people, in particular factory workers and transport workers, and have reported a number of findings. For instance,

- 60–80 per cent of all shift workers experience long-standing sleeping problems;
- shift workers are 5 to 15 times more likely to experience mood disorders as a result of poor quality sleep;
- drug and alcohol abuse are much higher among shift workers;
- 80 per cent of all shift workers complain of chronic fatigue;
- approximately 75 per cent of shift workers feel isolated from family and friends;
- digestive disorders are four to five times more likely to occur in shift workers; and
- from a safety viewpoint, more serious errors and accidents, resulting from human error, occur during shift work operations.

The psychological factors which affect an individual's ability to make the adjustments required to meet varying work schedules are associated with age, individual sleep needs, sex, the type of work and the extent of desynchronization of body rhythms, this latter point being the most significant.

Reducing the stress of shiftwork

Strategies are available aimed at minimizing the desynchronization of body rhythms and other health-related effects. The principal objective is to stabilize body rhythms and to provide consistent time cues to the body.

Employers should recognize that workers must be trained to appreciate the stressful effects of shift working and that there is no perfect solution to this problem. However, they do have some control over how they adjust their lives to the working arrangements and the change in lifestyle that this implies. They need to plan their sleeping, family and social contact schedules in such a way that the stress of this adjustment is minimized. Most health effects arise as result of changing daily schedules at a rate quicker than that at which the body can adjust. This can result in desynchronization, with reduced efficiency generally due to sleep deprivation.

- **Sleep deprivation**. This can have long-term effects on the health of the shift worker. The actual environment in which sleep takes place is important.
- **Diet**. A sensible dietary regime, taking account of the difference between the time of eating and the timing of the digestive system, will assist the worker to minimize discomfort and digestive disorders.
- **Alcohol and drugs**. Avoidance of alcohol and drugs, e.g. caffeine and nicotine, can result in improved sleep quality. Occasional use of sleeping tablets may be beneficial, but should be used under medical supervision.
- **Family and friends**. They should appreciate the demands on the shift worker. Better planning of family and social events is necessary to reduce the isolation frequently experienced by shift workers.

However, a number of remedies are available to organizations. These include

- consultation prior to the introduction of shift work;
- recognition by management that shift work can be stressful for certain groups of workers and of the need to assist in their adjustment to this type of work;
- regular health surveillance of shift workers to identify any health deterioration or change at an early stage;
- training of shift workers to recognize the potentially stressful effects and of the changes in lifestyle that may be needed to reduce these stressful effects; and
- better communication between management and shift workers aimed at reducing the feeling of isolation frequently encountered amongst such workers.

Violence at work

Not all violence to people at work is of a physical nature. Many employees are subject to psychological violence arising from bullying by superiors, advances of a sexual nature, verbal threats and abuse. There is a need for organizations to have a clearly defined policy on violence at work.

What is violence?

Most people would associate violence with being physically attacked by another person or group of persons resulting in, in extreme cases, death and, in other cases, physical injury. Physical injury may take the form of cuts, abrasions, bruising, fractures, dislocation of joints, total or partial blindness, all of which may require either first aid treatment or treatment at the accident and emergency department of a local hospital. Physical disability may further arise as a result of violence at work.

However, violence may not necessarily result in physical injury alone. It may take the form of psychological violence arising from verbal threats, persistent verbal abuse, bullying, obstruction, mocking behaviour and an attempt by an individual, perhaps a senior manager or other person in authority, to belittle a victim in the presence of other people. Inevitably, the victim will feel at risk, distressed and vulnerable, may suffer shock and, in some cases, may require longstanding psychological treatment.

The HSE in their publication, *Violence at Work: A Guide for Employers*, defines work-related violence as 'any incident in which a person is abused, threatened or assaulted in circumstances relating to their work'. Clearly, violence may take many forms. The victims of violence, furthermore, may not necessarily be employees but can include customers, members of the public, bystanders and, in some cases, individual managers who may be at risk for a variety of reasons.

Potential victims of violence

Many people who have direct contact with the public, such as shop assistants, bar staff, waiters in restaurants, bank and building society employees, local authority officers and police officers, may face aggressive or violent behaviour. They may be verbally abused, sworn at, threatened with violence or actually assaulted. Organizations need to consider, therefore, the potential for violence towards their employees and adopt appropriate strategies to protect them whilst at work.

Physical violence

Certain potential victims of physical violence are obvious. These include, in particular, anyone who handles cash or valuable items, such as jewellery, namely employees in banks, building societies, money lending organizations and jewellers' shops. However, other premises are frequently subject to robberies, such as take-away food premises, off-licences, supermarkets and petrol filling stations. This risk has been increased dramatically in the last decade due to the increased opening hours operated by many organizations, including 24-hour opening by supermarkets and petrol filling stations.

Traditionally, the armed forces and police officers, together with groups such as security personnel, are trained to deal with physical violence. However, other groups, such as teachers, people employed in the delivery and collection of goods, employees in the care industry and, indeed, anyone providing a service, may need such training in order to ensure their safety at work. This aspect may well be one of the outcomes of a risk assessment required under the Management of Health and Safety at Work Regulations 1999.

Psychological violence

As commercial life becomes more competitive, many employees feel threatened and under pressure due to the intensive and often insensitive nature of the management culture in which they work. As organizations strive towards greater financial success, as managers are put under greater pressure to achieve financial and other performance-related objectives, the greater the potential for the threatening and bullying of employees by their superiors in order to ensure these objectives are achieved.

This ruthless culture is a common feature of some organizations, resulting in increased sickness absence amongst staff, the need for counselling and, in some cases, increased accidents.

The legal position

Civil liability of employers to protect vulnerable employees

Where there is clear-cut evidence to indicate that employees may be exposed to risks of both physical and psychological violence by virtue of the tasks they undertake, the services they provide and/or the groups of people with whom they may come into contact, employers could well be deemed to be negligent by a civil court if an employee were injured.

Clearly, the common duty of care that exists between an employer and an employee must take into account the nature of these risks and where it could be shown that an organization failed to take reasonable care of such persons to whom this duty is owed, then such an organization could be deemed to be negligent.

Criminal liability

- **Health and Safety at Work, etc. Act 1974**. The general and specific duties of employers towards their employees under criminal law is well established under section 2 of the Act. Similar provisions apply in the case of non-employees, e.g. members of the public, customers, delivery and security personnel.
- **Management of Health and Safety at Work Regulations 1999**. Where there is a risk to employees and, indeed, in some cases, people not in his employment, of physical violence in particular, as with a bank or building society offices, jeweller's shop or 24-hour service supermarket or petrol filling station, an employer must be particularly aware of his duties under the Management of Health and Safety at Work Regulations 1999.

Risk assessment

Firstly, an employer must undertake a suitable and sufficient risk assessment for the purpose of identifying the measures he needs to take to comply with the requirements and prohibitions imposed under the 'relevant statutory provisions'. These relevant statutory provisions may include requirements under regulations such as the Workplace (Health, Safety and Welfare) Regulations 1992 and the Personal Protective Equipment at Work Regulations 1992 (Regulation 3). This could include structural safety measures, such as grilles, physical barriers and secured workstations to prevent contact by employees with intruders, armed robbers and even irate customers. The risk assessment must take into account the principles of prevention to be applied as detailed in Schedule 1 to the regulations (Regulation 4).

Preventive and protective measures

Having carried out this risk assessment, an employer must then take such measures as are appropriate, having regard to the nature of his activities and the size of his undertaking, for the effective planning, organization, control, monitoring and review of the preventive and protective measures (Regulation 5). It is simply not good enough, however, for an employer to install 'preventive and protective measures', such as security alarms, panic buttons, protective screening, security guarding and closed circuit television monitoring to protect his employees from risk of violence. He must continually monitor and review these arrangements and make improvements where appropriate in order to provide better protection against violence.

Competent persons

Under the regulations, a competent person or persons must be appointed to assist the employer in undertaking the measures he needs to take to comply with these requirements and prohibitions imposed upon him by or under the relevant statutory provisions (Regulation 7). Competent persons, in this case, could include security managers and/or consultants who undertake regular security audits and inspections to ensure these protective measures are effective.

Other provisions

Reporting of Injuries, Diseases and Dangerous Occurrences Regulations 1995 (RIDDOR)

An employer must notify the enforcing authority in the event of any death, major injury to an employee or dangerous occurrence arising out of or in connection with work. This may arise from acts of physical violence.

Safety Representatives and Safety Committees Regulations 1997 and Health and Safety (Consultation with Employees) Regulations 1996

Where there is a risk of violence to employees by virtue of their work activities, this risk should be taken into account in any joint consultation process between employer and employees.

Risk assessment – Factors for consideration

Many factors need consideration as part of the risk assessment process considering the potential for violence towards employees. For instance,

- What group or groups of employees may be subject to violence?
- Are there particularly vulnerable jobs or tasks?
- Does the timing of opening or closing of premises, receipt of deliveries and despatch of goods, allow for incidents to be planned?
- Do employees need to retain large amounts of cash at their workstations, e.g. supermarket cashiers?
- How often is cash moved to a safe area and is there a well-controlled system for doing this?
- Is there a policy for dealing with complaints and difficult customers?
- Is there evidence of a culture of bullying, aggressive behaviour, victimization and harassment in the organization?
- Are some employees commonly working on their own?
- Where employees work away from base, is there a system for keeping in touch and, if necessary, providing an escort?
- Do employees meet clients in their own homes?
- Are employees provided with separate and safe parking areas?

- Is access to vulnerable areas, such as bank cashier stations, adequately controlled?
- Is there a well-established emergency call system in offices where employees may be meeting members of the public and customers?
- Is closed circuit television installed where employees may be handling cash or dealing with difficult customers?

Prevention and control strategies

1. Staff should be told what is expected of them if there is a robbery, for example,
   - how to raise the alarm;
   - where to go for safety; and
   - not to resist or follow violent robbers.
2. There should be clear visibility and adequate lighting so that staff can leave quickly or raise help.
3. The build-up of cash in tills should be prevented and suitable measures should be adopted to move cash safely.
4. Arrangements should be made for staff to have access to a secure location.
5. High risk entrances, exits and delivery points should be monitored.
6. Buildings should be brightly illuminated and any possible cover for assailants removed.
7. Screens or similar protective devices for areas where staff are most at risk should be provided.
8. Only experienced or less vulnerable staff should be used for high risk tasks.
9. High risk jobs should be rotated so that the same person is not always at risk; for particularly high risk tasks the number of staff should be doubled.
10. Additional staff should be provided for high risk mobile activities or communication links to base.
11. Personal alarms for high risk staff should be provided.
12. Signs asking those wearing crash helmets to remove them should be displayed in prominent positions.
13. All staff should receive training in recognizing and dealing with violence and the potential for violence.

Dealing with the public

People who deal directly with the public, such as security staff, police officers and public officials, commonly face aggressive and/or violent behaviour. In certain cases, this behaviour may arise through the effects of drugs and drug abuse. They may further come into contact with people who have behavioural disorders associated with their attitudes, personalities and perceptions. As a result, they may be verbally abused, threatened or physically assaulted. These factors should be considered in the risk assessment process.

Work-related violence

The HSE define this term as any incident in which a person is abused, threatened or assaulted in circumstances relating to their work. Verbal abuse and threats are the most common type of incident, whereas physical attacks are comparatively rare.

Who is at risk?

Any person whose job entails dealing with members of the public can be at risk. According to the HSE, those most at risk are people engaged in:

- giving a service, e.g. supermarket check-out operators, bar staff, petrol station staff;
- caring, e.g. nurses, midwives, care assistants;
- education, e.g. teachers;
- cash transactions, e.g. bank employees, those collecting cash from banks or, for instance, gaming machines in public houses;
- delivery/collection, e.g. postal service employees, deliverymen, bailiffs;
- controlling, e.g. car park attendants;
- representing authority, e.g. enforcement officers, police officers, security staff.

Effective management of violence

The HSE recommend employers take the four-stage management process set out below.

1. **Finding out if you have a problem**. Employers must undertake a risk assessment to identify the hazards and evaluate the risks arising from these hazards. This should be assisted by,
   - consultation with employees;
   - keeping detailed records of previous incidents;
   - classifying all incidents.

   Fundamentally, a risk assessment should endeavour to predict what might happen in certain situations and the measures necessary to protect staff.

2. **Deciding what action to take**. The risk assessment should take account of who might be harmed and how this harm could arise. In certain cases, potentially violent people may be known to the organization and measures taken immediately those people are encountered. Evaluation of the risk entails checking existing arrangements, including any precautions already being taken. There is also a need to consider

   - the level of training and information provided, e.g. spotting early signs of aggression, clients with a history of aggression;
   - the environment, e.g. video cameras and alarm systems, coded security locks on doors;
   - the design of the job, e.g. checking the credentials of clients prior to entry, regular radio communication with base, avoiding lone worker situations;
   - ensuring staff get home safely.

   Record the significant findings of the assessment, including the preventive and protective measures necessary to comply with legal requirements. Review and revise the risk assessment on a regular basis.

3. **Take action**. Procedures for dealing with violence should be produced and staff informed, instructed and trained in these procedures. Reference to these procedures should be incorporated in the organization's Statement of Health and Safety Policy.
4. **Check what you have done**. Regular checks and reviews of the procedures should be made, accompanied by consultation with staff on the success or otherwise of these procedures. Records of incidents should be maintained.

What about the victims?

Where there has been a violent incident involving employees, employers must act quickly to avoid any long-term distress. Employees must be provided with support, which may include

- debriefing – they will need to talk through their experience as soon as possible after the event;
- time off work – different people need differing amounts of time off to recover; some may need specific counselling;
- legal help – this may need to be offered;
- in the case of the other employees, they may need guidance and/or training to help them react appropriately.

In any situation where there is a potential for violence towards employees, employers should have a formal system for dealing with the aftermath of such an incident. This may include the provision of counselling and assistance for those who have been the subject of violence.

Occupational health initiatives

Employers have a duty to protect the health of their employees at work. To promote good health, a number of initiatives should be considered.

Smoking at work

The relationship of cigarette smoking in particular to various forms of cancer is well established. It has a direct effect on people in terms of reduced lung function and an increased potential for lung conditions, such as bronchitis. With the increasing attention that has been given to the risks associated with passive smoking, organizations should consider the development of a policy on smoking at work.

The problem has been identified mainly in poorly ventilated, open-plan offices and amongst employees who may suffer some form of respiratory complaint, e.g. asthma or bronchitis. Other people may complain of soreness of the eyes, headaches and stuffiness.

In 1988, the HSE's booklet, *Passive Smoking at Work* drew the attention of readers to the concept of passive smoking, where nonsmokers inhale environmental tobacco smoke from burning cigarettes, cigars and pipes exhaled by smokers. Work carried out by the Independent Scientific Committee on Smoking and Health identified a small but measurable increase in risk from lung cancer for passive smokers of between 10 and 30 per cent. The booklet also suggests that passive smoking could be the cause of one to three extra cases of lung cancer every year for each 100 000 nonsmokers who are exposed throughout life to other people's smoke.

A policy on smoking at work should state the intention of the organization to eliminate smoking in the workplace by a specific date, the legal requirements on the employer to provide a healthy working environment, and that smoking is bad for the health of smokers and nonsmokers alike. The organization and arrangements for implementing the policy should state the individual responsibilities of managers in supporting and implementing the various stages of the operation, and of the staff to comply with the policy.

Drugs in the workplace

Many people see drug taking as the panacea for stress, relying on tranquillizers to reduce anxiety and amphetamines (pep pills) to counter fatigue. This form of drug taking represents a major risk to health, particularly if the individual consumes alcohol. The possession of a range of drugs is a criminal offence. Evidence of drug taking or 'pushing' in the workplace should be reported to a senior manager. Where appropriate, the advice of the police should be obtained. A programme to assist identified drug users to break the habit may need to be considered and put into operation. A range of national agencies is available to assist in such situations.

Alcohol at work

The consumption of alcohol prior to or at work may be one of the outcomes of stress in certain cases. Alcoholism is a true addiction and the alcoholic must be encouraged to obtain medical help and advice.

The repeated consumption of strong spirits, especially on an empty stomach, can lead to chronic gastritis and possible inflammation of the intestines which interferes with the absorption of food substances, notably those in the vitamin B group. This, in turn, damages the nerve cells causing alcoholic neuritis, injury to the brain cells leading to certain forms of insanity and, in some cases, cirrhosis of the liver.

A healthy lifestyle

Regular exercise, such as swimming, walking and cycling taking sufficient rest and paying attention to the food one eats are all features of a healthy lifestyle. Many organizations provide health education on these aspects with a view to ensuring a healthy workforce, together with health surveillance where appropriate, in order to encourage employees towards a more healthier lifestyle.

Managing conflict

Conflict occurs when individuals or groups are not obtaining what they need or want and are seeking their own self-interest.

About conflict

In considering conflict, it is important to recognize that

- Conflict is inevitable.
- Conflict develops because managers are dealing with people's lives, jobs, self-concept, ego and sense of mission or purpose.
- Early indicators of conflict can be recognized.
- There are strategies for resolution that are available.
- Although inevitable, conflict can be minimized, diverted and/or resolved.

Common initial causes of conflict

Conflict can be caused in a number of ways:

- poor communication;
- people seeking power;

- dissatisfaction with management style;
- weak leadership;
- lack of openness;
- changes in leadership.

Indicators of conflict

Conflict can be indicated and caused in many ways, such as

- voice tone and body language;
- surprises sprung on people;
- disagreements, regardless of issue;
- strong public statements;
- airing disagreements through the media;
- desire for power;
- increasing lack of respect;
- open disagreement;
- lack of clear goals;
- no discussion on progress achieved from managers.

Destructive and constructive conflict

Conflict is *destructive* when it

- diverts attention from other important issues;
- undermines morale or self-respect;
- polarizes people and groups, reducing co-operation;
- increases or sharpens difference;
- leads to irresponsible and harmful behaviour, such as brawling, name-calling and destruction of property.

Conflict is *constructive* when it

- results in clarification of important problems and issues;
- results in solutions to problems;
- involves people in resolving issues important to them;
- causes effective and authentic communication;

- helps release emotion, anxiety and stress;
- builds co-operation amongst people through learning more about each other;
- helps individuals develop understanding and management skills.

Avoiding or resolving conflict

If conflict cannot be avoided, then it must be resolved. Managers need to consider the following recommendations in resolving conflict situations:

- meet conflict head on; don't hide or run away!
- set goals for resolving conflict situations;
- plan for conflict and communicate frequently;
- be honest about management's concerns;
- agree to disagree; understand that healthy disagreement would build better decisions;
- get individual ego out of management style;
- let your team create; people will support what they helped create;
- continually stress the importance of following policy;
- communicate honestly; avoid playing 'gotcha'-type games;
- provide more data and information;
- develop a sound management system;
- provide feedback to those involved in the conflict quickly and efficiently.

Principal aspects of stress management

Potentially stressful organizations are those

- which are large and bureaucratic;
- on which there are formally prescribed rules and regulations;
- where there is conflict between positions and people;
- where people are expected to work hard for long hours;
- where no praise is given;
- where the general culture is classified as 'unfriendly'; and
- where there is conflict between normal work and outside interests.

Why do anything?

Cost benefits

There are obvious cost benefits associated with reduced absenteeism and accidents, together with their related direct and indirect health and other costs. Generally, however, an organization will be more effective if there is conscious recognition of stress potential and efforts are made to eliminate or reduce it.

Morale

One of the standard criticisms from people at all levels is that the organization does not care about its people. This feeling is reflected in attitudes to management, the job and the organization as a whole. It is important, therefore, for the organization to show at all levels that it really does care. This will result in increased motivation and a genuine desire on the part of staff to perform better. There is clear-cut evidence throughout the world which shows that the most profitable companies are those which take an interest in their staff and promote a caring approach.

What can be done?

There is a need here to consider both organizational and individual strategies for managing stress in the workplace.

Organizational strategy

- **Employee health and welfare**. Various strategies are available for ensuring sound health and welfare of employees. These include various forms of health surveillance, health promotion activities, counselling on health-related issues and the provision of good quality welfare amenity provisions, i.e. sanitation, washing, showering facilities, facilities for taking meals, etc.
- **Management style**. Management style is frequently seen as uncaring, hostile, uncommunicative and secretive. A caring philosophy is essential, together with sound communication systems and openness on all issues that affect staff.
- **Change management**. Most organizations go through periods of change from time to time. Management should recognize that impending change, in any form, is one of the most significant causes of stress at work. It is commonly associated with job uncertainty, insecurity, the threat of redundancy, the need to acquire new skills and techniques, perhaps at a late stage in life, relocation and loss of promotion prospects. To eliminate the potentially stressful effects of change a high level of communication in terms of what is happening should be maintained and any such changes should be well managed on a stage-by-stage basis.

- **Specialist activity**. Specialist activities, such as those involving the selection and training of staff, should take into account the potential for stress in certain work activities. People should be trained to recognize the stressful elements in their work and the strategies available for coping with these stressors. Moreover, job design and work organization should be based on ergonomic principles.

Individual strategy

There may be a need for individuals to

- develop new skills for coping with the stress in their lives;
- receive support through counselling and other measures;
- receive social support;
- adopt a healthier lifestyle; and
- where appropriate, use support from prescribed drugs for a limited period.

The use of occupational health practitioners is recommended in these circumstances.

Stress management action plans

Any action plan to deal with stress at the *organizational level* should follow a number of clearly defined stages, as follows:

- Recognize the causes and symptoms of stress.
- Decide the organization needs to do something about it.
- Decide which are the group or groups of people in whom we can least afford stress, e.g. key operators, supervisors.
- Examine and evaluate by interview and/or questionnaire the specific causes of stress.
- Analyse the problem areas.
- Decide on suitable strategies, e.g. counselling, social support, training, such as time management, environmental improvement and control, redesign of jobs, ergonomic studies.

At the *individual level*, people should be encouraged to take the following action:

- Identify your work and life objectives. Re-evaluate on a regular basis or as necessary. Put them up where you can see them.
- Ensure a correct time balance.

- Identify your stress indicators. Plan how you can eliminate these sources of stress. See them as red STOP lights.
- Allow 30 minutes each day for refreshing and recharging.
- Identify crisis areas. Plan contingencies.
- Identify key tasks and priorities. Do the IMPORTANT, not necessarily the URGENT.
- Keep your eyes on your objectives. Above all, have fun!

Stress at work and the civil law

The question of stress at work has achieved considerable significance in the civil courts in the last decade. In 2002, after hearing four appeal cases, namely,

- Terence Sutherland (Chairman of the Governors of St Thomas Becket RC High School) v Penelope Hatton
- Somerset County Council v Leon Alan Barber
- Sandwell Metropolitan Borough Council v Olwen Jones
- Baker Refractories Ltd v Melvyn Edward Bishop

Lady Justice Hale, sitting with Lords Justice Brooke and Kay, outlined a number of practical propositions to assist courts in dealing with future claims for psychiatric injury arising from stress at work.

Practical propositions

1. There are no special control mechanisms applying to claims for psychiatric (or physical) illness or injury arising from the stress of doing the work the employee is required to do. The ordinary principles of employers' liability apply.
2. The threshold question is whether this kind of harm to an employee was reasonably foreseeable. This has two components:
   - an injury to health (as distinct from occupational stress); and
   - which is attributable to stress at work (as distinct from other factors).
3. Foreseeability depends upon what the employer knows (or ought reasonably to know) about an employee. Because of the nature of a mental disorder, it is harder to foresee than physical injury, but may be easier to foresee in a known individual than in the population at large. An employer is usually entitled to assume that the employee can withstand the normal pressures of the job unless they know of some particular problem or vulnerability.
4. The test is the same whatever the employment. There are no occupations that should be regarded as intrinsically dangerous to mental health.

5. Factors likely to be relevant in answering the threshold question include:

   - The nature and extent of the work done by the employee
     Is the workload much more than is normal for the particular job?

     Is the work particularly intellectually or emotionally demanding for this employee?

     Are demands being made of this employee unreasonable when compared with the demands made of others in the same or comparable jobs?

     Or are there signs that others doing this job are suffering harmful levels of stress?

     Is there an abnormal level of sickness or absenteeism in the same job or in the same department?

   - Signs from the employee of impending harm to health
     Have they a particular problem or vulnerability?

     Have they already suffered from illness attributable to stress at work?

     Have there recently been frequent or prolonged absences which are uncharacteristic?

     Is there reason to think that these are attributable to stress at work, for example because of complaints or warnings from them or others?

6. The employer is generally entitled to take what they are told by their employee at face value, unless they have good reason to think to the contrary. They do not generally have to make searching enquiries of the employee or seek permission to make further enquiries of their medical advisers.

7. To trigger a duty to take steps, the indications of impending harm to health arising from stress at work must be plain enough for any reasonable employer to realize that they should do something about it.

8. The employer is only in breach of duty if they have failed to take the steps which are reasonable in the circumstances, bearing in mind the magnitude of the risk of harm occurring, the gravity of the harm which may occur, the costs and practicability of preventing it and the justification for running the risk.

9. The size and scope of the employer's operation, its resources and the demands it faces are relevant in deciding what is reasonable; these include the interests of other employees and the need to treat them fairly, for example, in any redistribution of duties.

10. An employer can only reasonably be expected to take steps that are likely to do some good. The court is likely to need expert evidence on this.

11. An employer who offers a confidential counselling service, with referral to appropriate counselling or treatment services, is unlikely to be found in breach of duty.

12. If the only reasonable and effective step would have been to dismiss or demote the employee, the employer will not be in breach of its duty in allowing a willing employee to continue in the job.

13. In all cases it is necessary to identify the steps that the employer both could and should have taken before finding them in breach of their duty of care.
14. The claimant must show that the breach of duty has caused or materially contributed to the harm suffered. It is not enough to show that occupational stress has caused the harm.
15. Where the harm suffered has more than one cause, the employer should only pay for that proportion of the harm suffered which is attributable to their wrongdoing, unless the harm is truly indivisible. It is for the defendant to raise the question of apportionment.
16. The assessment of damages will take account of any pre-existing disorder or vulnerability and of the chance that the claimant would have succumbed to a stress-related disorder in any event.

Stress at work and the criminal law

According to the HSE, up to 13.4 million days a year are lost in the UK because of stress. These staggering figures have prompted the HSE to take a new approach to fighting the problem by writing a code which will introduce a legal basis against which companies can be assessed for their efforts to reduce stress to manageable levels. If fewer than 65–85 per cent of staff agree that each standard has been met, the organization will fail the assessment (see Table 18.7).

HSE code on stress

The code is based on evidence from the Whitehall II study of 10 000 civil servants who have been continually assessed since 1985. The study, by Professor

Table 18.7 HSE management standards for stress at work

The HSE have specified management standards for stress at work.

Under these standards

- 85% of employees must be able to say that they can cope with the demands of the job.
- 85% should consider that they have an adequate say over how they do their work.
- 85% should say that they get adequate support from colleagues and superiors.
- 65% should say that they are not subjected to unacceptable behaviour, such as bullying.
- 65% should say that they understand their role and responsibilities.
- 65% of employees should say that they are involved in organizational changes.

Sir Michael Marmot and colleagues of University College London, found that staff could cope with high pressure work environments without damage to their health provided they had a high degree of control over their working lives and good social support.

Organizations are being asked to implement the standards, which have equivalent legal status to the Highway Code.

If the HSE Code scheme is judged to be successful, the HSE said it will introduce work-related stress audits in its routine health and safety inspections using the new nationally agreed minimum standards.

Conclusions

Some managers are not prepared to recognize the problem of stress in the workplace. The common response to people complaining of stress is, 'If you can't stand the heat, get out of the kitchen!' However, it was when people, such as occupational health nurses, started to relate sickness absence levels to stress that managers eventually began to admit that the results of their decisions and actions, the environment they provided for operators and many other features of their organizational activities could be stressful.

Fortunately, the type of manager mentioned above is rapidly disappearing with the increased recognition of the existence of stress and the greater human factors-related approach to management.

The problem is still with us, however. Many people simply fail to recognize actual stressful situations or future stressful situations in their lives. These situations can arise from problems at home, in their relationships with people or as a result of a specific life event, such as bereavement. Generally, they 'bear up' and endeavour to cope. In most cases, they do cope but with varying effects on their health, some of which can be serious.

If people are to cope with stressful situations, they have got to go back to the basic principles, namely

- identify the sorts of events in their lives which create their particular stress response;
- measure and evaluate the significance of these events; and
- learn various forms of coping strategies to enable them to deal with these life events.

People would be a great deal happier if they would undertake this exercise.

Key points

- Stress is defined as the common response to attack.
- It is further defined by the HSE as the reaction people have to excessive pressure and other types of demand placed upon them. It arises when people worry and they cannot cope.
- Stressors may take a number of forms – environmental stressors, psychological stressors and social stressors.
- Stress has a direct association with the autonomic system – a body system which controls an individual's physiological and psychological responses.
- It is important to distinguish between positive and negative stress.
- Stress can be associated with both role ambiguity and role conflict.
- Stress can further be associated with other factors, such as out-of-date work equipment, inadequate career development and deteriorating personal relationships within an organization.
- Bullying and harassment are two of the principal causes of stress at work.
- The legal concept of psychiatric injury arising from stress has now been established in the civil courts and there is a body of case law dealing with this subject.
- There is evidence to show that various personality types and traits may be associated with stress.
- Anxiety and depression are two classic manifestations of stress.
- Organizations should recognize their intention to do something about stress at work through the establishment of a Statement of Policy on Stress at Work.

19 The behavioural safety approach

Professor Dominic Cooper (1999), the well-known exponent of behavioural safety, defines the term as

> the systematic application of psychological research on human behaviour to the problems of safety in the workplace.

It is the process of involving employees in defining the ways that they are most likely to be injured, seeking their involvement, obtaining their 'buy in', and asking them to observe and monitor co-employees with a view to reducing their unsafe behaviours. Fundamentally, it is a means of obtaining increased improvements in safety performance through the promotion of safe behaviours at all levels in the workplace and in an organization.

There are many approaches to behavioural safety, but the ultimate objective of this proactive approach to safety improvement is, principally, to predict and provide an early warning of accidents and loss-producing incidents arising from work activities. Most importantly, it is not a replacement for the more traditional proactive and reactive approaches to accident prevention based on safety monitoring procedures, risk assessment and the investigation of accidents, incidents and occupational ill-health. However, it does challenge these more traditional approaches which may not necessarily take into account the potential for human error as a contributory factor in accidents. This approach is directed at improving self-awareness, encouraging contribution from, and involvement by, people at work in recommending the improvements necessary, the sharing of feedback on safety performance and increasing people's perception of risk. It is also concerned with establishing and promoting the right safety culture within an organization, based on senior management commitment and demonstration of that commitment at all levels within an organization.

Unsafe behaviour

Unsafe behaviour takes many forms from simple failure to wear or use personal protective equipment to intentional actions, such as the removal of machinery guards or the defeating of a safety mechanism to a guarding system.

There are many reasons why people behave unsafely. It may be that they were taught a particular unsafe method of work at the start of employment and see it now as standard working practice. They may lack the skills or knowledge to undertake a task safely, as with certain manual handling operations, they may fail to make a load secure due to demands on time, or operate equipment, such as a lift truck, without authority to get the job finished.

The reasons for unsafe behaviour

Organizations need to ascertain the reason or reasons for unsafe behaviour. Here it is important to distinguish between errors as opposed to direct violations of established safety systems or procedures.

Errors are the unintentional violations caused by knowledge and skill deficiencies. *Critical errors* are associated with, for example, lack of attention to the task, incorrect visual perception, loss of balance and over-reaching, the latter resulting in falls. The indirect causes of these errors could be fatigue, stress, rushing to complete a task, complacency and overconfidence. Errors can be corrected by instruction, training and supervision.

Violations, on the other hand, are deliberate actions which may need correction through disciplinary procedures. Whilst most organizations have formal disciplinary procedures covering, for example, unsolicited absence from work or gross misconduct, very few organizations have such procedures with respect to unsafe behaviour which could put an employee and his fellow employees at risk.

Some people may hold inappropriate attitudes to safety, talking about 'the nanny state', 'wrapping people in cotton wool' or 'safety getting in the way of the job'. These dangerous attitudes can be taken up by other workers resulting in a culture of unsafe working practices being developed in a workplace. A behavioural safety programme targets these people, endeavours to correct their attitudes and raise their awareness of the risks.

Reason (1989) considered many of the bases for unsafe behaviours, such as poor workplace and work equipment design, supervisory factors, procedure failures and people taking shortcuts. Over a period of time, unsafe behaviour becomes accepted as the norm because the outcomes of failure are rare.

Setting the standards

If behavioural safety is going to work effectively, there must be formally established systems and operating procedures. The establishment of these systems and procedures is an on-going process commencing with those operations and activities involving the greatest danger to employees. Techniques such as job safety analysis and risk assessment are useful tools in identifying the significant hazards and the precautions necessary on the part of employees.

At this stage it may be necessary to acquire information from official sources, such as the Health and Safety Executive, Royal Society for the Prevention of Accidents

(RoSPA) and the Institution of Occupational Safety and Health (IOSH), for incorporation in systems and procedures.

This stage of the programme takes a number of stages:

- identifying and specifying safe procedures and systems of work;
- explaining to employees, perhaps on a one-to-one basis, the key aspects of the safe system of work and getting them to complete the task according to the established system;
- training new employees in the standard operating procedure;
- supervisors or trained observers use the standard operating procedures to identify employees who are deviating from, or not following, correct methods, drawing their attention to this and re-instructing them in the correct methods; and
- regular monitoring of procedures by supervisors, together with employee feedback sessions to discuss improvements in the safe system of work or problems arising from the newly introduced procedure.

Behavioural safety training

Before the start of a behavioural safety programme, it is essential that people at all levels receive the appropriate training. Such a programme incorporates a number of elements.

- **Raising safety awareness**. This is one of the principal objectives of the training programme. In addition to various forms of active and passive training directed at increasing people's perception of risk, together with the necessary knowledge of hazards and the precautions necessary, trainees should be involved in activities such as hazard spotting in the workplace and workplace inspections directed at identifying hazards, together with feedback sessions following these exercises. Examples of typical accident situations should be discussed and the direct and indirect causes analysed. Poster campaigns, targeting both safe working practices and the causes of accidents and ill-health, should be used to raise awareness. Posters should be changed regularly, perhaps weekly, and in line with a particular safety theme.
- **Coaching**. Individual line managers, particularly supervisors, should receive frequent coaching with a view to their meeting the defined standards laid down in, for example, documented safe systems of work and safe operating procedures.
- **Reviewing the consequences of unsafe behaviour**. Regular feedback sessions are necessary whereby accidents arising from unsafe behaviour are reviewed with reference to the direct and indirect causes of those accidents. There is no doubt that people learn from other people's mistakes.

Behavioural safety programmes

A well-structured programme incorporates a series of steps.

Safety observation programme

This first stage entails observation and assessment of workplace practices, review of these practices, the identification of unsafe workplace behaviours, feedback from accident and ill-health reports and reports of near misses. This assessment identifies the 'safety critical behaviours' and is undertaken through a range of techniques, such as brainstorming, hazard analysis, talking people through work processes and the hazards that can arise and reviewing the outcome of accident investigation. The most common input is a consideration of the lessons learned from accident and incident investigation. Because of the failure of many accident assessment and investigation techniques to adequately evaluate the principal or root causes of human error accidents, many of these causes still exist even after multiple accidents, and can be predicted as the cause of further accidents.

Generally, safety observation programmes incorporate a number of elements:

- observation of work operations to ensure compliance with the standard operating procedure;
- observation of behaviour that is not necessarily related to work operations, but which involves interactions between employees;
- observation of current work practices and the making of recommendations for bringing about improvements;
- examination of the latest accident and ill-health reports;
- examination of near miss reports;
- examination of hazard and unsafe condition reports submitted as part of the organization's hazard reporting procedure; and
- establishing the current situation with respect to unsafe workplace behaviours and the approach to behavioural safety by employees.

Enlisting and training the observers

Observers should be enlisted on the basis of their commitment to health and safety. As such, they may be put into a position where they may need to be critical about their superiors and colleagues. They can be managers, supervisors, individual employees and worker representatives, such as safety representatives. Alternatively, teams of employees can operate as observers in specific parts of the workplace.

Whoever is taken on as an observer will need training in observation techniques and how to recognize unsafe situations and unsafe behaviour. Training should involve recognition of the various forms of human error and unsafe working practices, specific legal requirements relating to their area of operations, 'safe place' and 'safe person' strategies, together with development of interpersonal skills in getting people to recognize their unsafe actions and in monitoring the measures necessary to avoid unsafe behaviour by people.

One of the principal objectives of observer training is that of ensuring consistency of approach to observation tasks. This may entail the trainer 'walking the workplace' with individual trainees with a view to their fully understanding the approach to observing human behaviour.

Devising checklists and observing behaviour

Observers should be assisted in their observations and assessments of behaviour by the provision of checklists and other aids for monitoring behaviour. Whilst they should be trained in the use of these checklists, they should also be encouraged to look for other aspects of unsafe behaviour that may not be incorporated in the checklist. On this basis, checklists should be subject to regular review and modification. Checklists of this type should be maintained in a manual together with instructions for completing checklists.

Safe and unsafe behaviours are observed, recorded and provided as feedback (reinforcement) to the employee. This should lead to an increase in safe behaviour with subsequent improvement and increased involvement by the employee. Sulzer-Azaroff (1978) and Sulzer-Azaroff and Santamaria (1980) demonstrated that when safety hazards are identified, and positive feedback is used following safety inspections, the number of hazards is reduced.

Creating a recording system

The observation process should be backed up by a system which extrapolates data from checklists. It should classify the various forms of unsafe behaviour and record observed instances of such behaviour according to the classification. In this way it is possible to identify trends in unsafe behaviour, such as people taking shortcuts across designated danger areas, the dangerous driving of vehicles in and around the workplace and unsafe working at height.

Analysing the data

Data can be analysed on the basis of the above classifications. Feedback from the analysis of data should identify trends and enable specific aspects of unsafe behaviour to be targeted through further one-to-one instruction or group training sessions. In certain cases, management deficiencies may be identified from the data. As a result, supervision may need to be intensified in certain work activities with a view to reminding employees about unsafe behaviour and correcting unsafe working practices.

Data and the actions arising from the analysis of data should be published and be readily available to all employees.

Promoting the programme

For the programme to be accepted by both managers and employees, the benefits of such a programme need to be established. It is a question of winning over the 'hearts and minds' of all concerned, so that it is not perceived as 'just another management initiative' which does not affect the majority of people.

Selling the programme involves publicity, spelling out the reasons for introduction of the programme and the benefits to be derived by everyone. To give the programme maximum credibility with both management and employees, a director or senior manager should be named as having overall responsibility for the programme. That person should regularly review the success of the programme, talking to managers and employees at regular intervals with a view to reinforcing the objectives of the programme.

In the case of managers, establishing the programme in their area of control and ensuring its success should feature as a continuing performance objective. When managers perceive that this is the case, there is a tendency to pay greater attention to making it work!

Implementing the programme

Once the programme has been established and is in operation, it is vital to maintain continuity. There is a great danger that, once the initial setting up stage has been completed, everyone reverts to their normal practices and disregards what is required of them under the programme. Similarly, in the case of observers, they may lose heart if they see no action being taken following their observations.

Observers need to have a formal system for reporting back progress to a dedicated director or senior manager. This may require the appointment of a senior observer who co-ordinates the programme, receives regular feedback from observers, is responsible for analysis of data produced by observers and reports back to the responsible director.

Consistent failure to change behaviour by all employees should be the subject of regular discussions between the senior observer and the responsible director. Ultimately, the responsible director may have to demand immediate action on the part of line management to bring about the necessary changes in behaviour. This may eventually lead to disciplinary action, including dismissal in blatant cases of misbehaviour.

Maintaining the behavioural safety process

This is the continuing reinforcement of the messages arising from the programme and needs sustained effort in preventing behaviour reverting to former levels. This reversion to former behaviour has been seen in organizations which have installed, for instance, Total Quality Management programmes where, once the initial drive and publicity lessened, both managers and employees reverted to 'normal' behaviour, the whole process eventually being perceived as yet another failed management initiative.

Employers should recognize that observing people by direct methods leads easily to blame. This, in turn, leads to tension between managers and employees and is not beneficial in terms of continuing safety improvement. It is, therefore, better to concentrate on physical conditions associated with behaviour rather than trying to coerce people to modify behaviour towards safer working. For example, targeting the behaviours necessary for handling materials and work equipment will make any relevant change highly visible. Whilst this behaviour may only be brief, it will stick with some employees. For example, placing a ladder at the correct pitch takes only a moment. The ladder itself remains visible and observable and most people remember to place it correctly the next time.

A behaviour change which is visible has two benefits. Firstly, it becomes obvious to everyone that improvements happen and, secondly, people learn to register their level of performance directly from the environment in which they work. In this way, the improvements in behaviour start acting as a positive consequence of correct behaviour.

Behavioural safety and risk assessment

The risk assessment process is essentially a prediction exercise. As such, a 'suitable and sufficient' risk assessment should take into account the potential for human error. Under the Management of Health and Safety at Work Regulations, employers must also take into account 'human capability' as regards health and safety when allocating tasks. They need to ensure that the demands of the job do not exceed an employee's physical and mental ability to carry out work without risk to himself or other people.

The assessment of physical capability to undertake a task may be relatively straightforward. There is plenty of official guidance on this matter. The question of mental capability, however, raises all sorts of problems. How does a manager decide whether a person has the appropriate mental capability to drive a fork lift truck? Factors such as aptitude for the work, experience, truck driving and manoeuvring skills, lifting and loading skills, attitude to safety, conscientiousness and alertness are all significant here. The complexity of the task is also important in this case and may be affected by environmental factors, such as low temperatures, inadequate space to do the job safely, fatigue, operational stress and the extent of training and education the individual has received.

Information, instruction, training and supervision

Workplace and work activity risk assessments need to take all the above human factors into account, together with the potential for error and the potentially serious effects of error. As with other forms of risk assessment, they may identify future information, instruction and training needs together with supervision requirements.

Health surveillance

One of the outcomes of the risk assessment process is the identification of health surveillance requirements for employees. This may be necessary as a direct result of risk assessment under the Control of Substances Hazardous to Health (COSHH) Regulations or the Control of Noise at Work Regulations. In this case, reference should be made to the specific requirements of the regulations with respect to health surveillance. However, not all work activities are necessarily covered by specific regulations and the health surveillance needs may need to be assessed by an occupational health practitioner, such as an occupational health nurse or occupational physician.

Health surveillance is an essential element of an occupational health strategy. It implies the regular monitoring of the state of health of individuals through the use of a number of techniques aimed at detecting exposure by employees to

- physical agents, such as noise;
- chemical agents, such as hazardous substances;
- biological agents, such as bacteria; and
- psychological agents, such as stress-induced injury

together with gathering evidence of the early stages of occupational ill-health.

The health surveillance element of risk assessment is an area that is frequently underplayed and, in many cases, ignored by employers. In terms of both mental and physical capability, employers may take the view that a person has either 'got what it takes' to do the job, or not. As far as stress-induced injury is concerned, the potential for employees sustaining stress-induced injury must be considered in the risk assessment process (see Chapter 18, Stress and stress management).

Successful behavioural safety programmes

Changing people's behaviour, in terms of attitude and motivation towards safe working, is no easy task and does not take place overnight. There are many factors which need careful consideration prior to their introduction.

Employee participation

This is a crucial feature of behavioural safety programmes giving them a say in terms of what they consider to be safe and unsafe behaviour with respect to the processes and activities undertaken at work. Without participation, there is no worker 'ownership', resulting in limited or no commitment to the process.

Targeting unsafe behaviours

In most cases, only a limited number of unsafe behaviours are identified. However, they may be responsible for a significant number of accidents. By targeting these unsafe behaviours, and the workplace factors that create or drive these behaviours, it is possible to install strategies to eliminate them. These unsafe behaviours must be directly observable by the majority of employees.

Data and decision-making

The old adage, 'What gets measured, gets done!' applies in this case. Trained observers monitor the behaviour of co-employees on a regular basis, producing data. In many cases, the very fact of observing a person's behaviour may bring about changes in that behaviour. Data are used in the decision-making process based on identified trends in behaviour and in the provision of feedback to those concerned.

Improvements

Analysis of data enables the establishment of a schedule of actions that combine to create an overall improvement intervention.

Feedback

Feedback is the key ingredient of any type of initiative directed at bringing about improvement. Feedback may take many forms, such as simple verbal feedback, graphical feedback, particularly that identifying trends, and by incorporating feedback elements in regular staff briefings.

Management support and commitment

There must be clear and visible demonstration and commitment to the programme by all levels of management. Without this commitment, the scheme is doomed to failure!

The benefits

The benefits from the behavioural safety approach can be summarized thus:

- reduced accidents, incidents, cases of occupational ill-health and their associated direct and indirect costs;
- increased skills in positive reinforcement;
- increased reporting of hazards, shortcomings in protection arrangements, accidents and near misses;
- improved levels of safety behaviour and employee commitment to safety behaviour;
- acceptance of the system by all concerned (based on the level of management commitment to the system);
- accelerated action in the case of employee recommendations, suggestions and the remedial action necessary; and
- improvements in safety culture.

Conclusions

Behavioural safety is a concept that many organizations have taken on board since the early 1990s. Whilst it is not a replacement for the more traditional approaches to occupational health and safety, it has embraced that area of health and safety management which many people still perceive as difficult to 'get to grips' with, namely getting people to look at their behaviour and, if necessary, modify their behaviour to achieve safer working conditions.

The implementation of behavioural safety programmes requires a high level of determination, commitment and demonstration of support if they are to succeed. Whilst there is no standard format for a behavioural safety programme, the principles outlined above should be considered.

Key points

- Behavioural safety is defined as the systematic application of psychological research on human behaviour to the problems of safety in the workplace.
- It is a means of obtaining increased improvements in safety performance through the promotion of safe behaviours at all levels in the workplace and in an organization.
- Unsafe behaviour takes many forms and there are many reasons why people behave unsafely.
- The principal reasons for unsafe behaviour are associated with errors and violations.

- Behavioural safety programmes take place in a series of stages commencing with training of all those concerned and the setting of standards.
- Behavioural safety relies on safety observation programmes and the assessment of safety critical behaviours.
- Behavioural safety has a direct connection with the risk assessment process and this process should identify the potential for unsafe behaviour.
- Successful behavioural safety programmes entail employee participation, a targeting of unsafe behaviours, the preparation and use of data in the decision-making process and the incorporation of feedback from which trends can be identified.

Glossary

Ability A general term referring to the potential for the acquisition of a skill or to an already acquired skill.

Accident An unforeseeable event often resulting in injury (Oxford Dictionary).

Any deviation from the normal, the expected or the planned, usually resulting in injury (Royal Society for the Prevention of Accidents).

An unintended, unplanned happening that may or may not result in personal injury, property damage, work process stoppage or interference, or any combination of these conditions under such circumstances that personal injury might have resulted (Frank Bird, American exponent of total loss control).

An unexpected, unplanned event in a sequence of events that occurs through a combination of causes, resulting in physical harm (injury or disease) to an individual, damage to property, business interruption or any combination of these effects (Health and Safety Unit, University of Aston in Birmingham).

Accident proneness The concept or notion that some people are more liable to have accidents than others or that some people are more susceptible to accidents than others.

Accountability The liability for an individual to give an account of his performance.

Achievement need A need to succeed and to strive for standards of excellence. It serves to motivate an individual to do well.

Acquisition The slow but gradual strengthening of a learned response.

Adaptation A change in the sensitivity of a sense organ due to stimulation or lack of stimulation.

Adjustment The relationship that exists between an individual and his environment, especially his social environment, in the satisfaction of his motives.

Agenda A list of things to be done or attended to.

Anthropometry A branch of ergonomic study concerned with the study and measurement of body dimensions, the orderly treatment of resulting data and the application of these data in the design of workplace and equipment control layouts.

Anxiety A state of tension coupled with apprehension, worry, guilt, insecurity and a constant need for reassurance. A vague or objectless fear. The normal response of the body to recognized danger.

Aptitude The ability or potential that an individual has to profit through a certain type of training. It indicates how well that person would be able to undertake a task after receiving the training.

Arousal An increase in alertness and muscular tension, an emotion that may be associated with fear, expectation, excitement, physical exercise and certain stressful events.

Attitude A predetermined set of responses. A tendency to behave in a particular way in a particular situation. A tendency to respond either positively or negatively to certain persons, objects or situations. A mental and neutral state of readiness, organized through experience, exerting a directive or dynamic influence upon an individual's response to all objects and situations with which it is related.

Attitude survey A means to measure and evaluate the attitudes of certain target groups in the workplace and amongst the public at large in order to be able to predict future responses of such groups. At national level, a government may wish to assess the future response of the public to proposed or impending legislation. In similar manner, an organization may wish to predict the responses of employees to the introduction of new working procedures aimed at increasing productivity.

Atypical workers Workers and others who are not in normal daytime employment, together with shift workers, part-time workers and night workers (Working Time Regulations 1998).

Authority The power or a right to command others or to act in certain situations.

Autonomic system The body system that controls an individual's physiological and psychological responses.

Avoidance A form of conditioned and specific emotional response whereby an individual avoids people and situations which may be stressful or contrary to his beliefs and opinions.

Behaviour Any observable action of a person or animal.

Behavioural intention A person's belief as to how he will respond to a given situation in the future, for example he will wear head protection during demolition work.

Behavioural safety	The systematic application of psychological research on human behaviour to the problems of safety in the workplace. A means of gaining further improvements in safety performance through promoting safe behaviours at all levels in the workplace.
Behavioural sciences	The behavioural sciences are those sciences most concerned with human and animal behaviour. Behavioural science has three main aims, namely to describe, explain and predict human behaviour. The principal behavioural sciences are psychology, sociology and social anthropology.
Belief	The acceptance of a statement or proposition. It does not necessarily involve an attitude, although it may.
Brainstorming	A method for developing creative solutions to problems by focussing on a particular problem and then deliberately proposing as many solutions and ideas to solve the problem with a view to eventually finding the right solution.
Change agent	A person who has the 'clout', the conviction and charisma to make things happen and keep people actively engaged in the change process.
Character	The ethical or moral traits of personality.
Classical conditioning	Learning that takes place when a conditioned stimulus is paired with an unconditioned stimulus.
Cognition	A thought or idea.
Cognitive dissonance	A motivational state produced by inconsistencies between simultaneously held cognitions or between a cognition and behaviour.
Cognitive overload	A state in which there is more information directed at a person that he can process in thought at a particular time.
Cognitive style	The preferred way an individual processes information.
Communication	The transfer of information, ideas, feelings, knowledge and emotions between one individual, or group of individuals, and another, the basic function of which is to convey meanings.
Communication structure	The pattern of closed and open channels of communication within a group of individuals.
Compensation	A defence mechanism in which an individual substitutes one activity for another in an attempt to satisfy frustrated motives. It commonly implies failure or loss of self-esteem in one activity and the compensation of this loss by efforts in some other aspect of endeavour.

Compliance Behaviour in accordance with group pressures without necessarily accepting the values and norms of the group.

Concept An internal process representing a common property of objects or events, usually represented by a word or name.

Conditioned response A response produced by a conditional stimulus after learning.

Conditioning A general term referring to the learning of some particular response.

Conflict A clash, struggle or trial of strength involving two or more persons or groups.

Conformity The tendency to be influenced by group pressure and to acquiesce to group norms.

Correlation The relationship between any two events.

Counselling The provision of advice, support and assistance to people with health, vocational and personal problems.

Crisis A situation when something happens that requires major decisions to be made quickly.

Culture The customs, habits, traditions and artefacts that characterize a people or social group.

Decision-making The cognitive process of selecting a course of action from among multiple alternatives.

Degradation The general slowing of human performance over a wide range of environmental conditions and associated with loss of motivation to perform well, fatigue arising from working for long periods, lack of stimulation and working in situations where there may be conflict threatening the body's homeostatic or coping mechanisms and resulting in stress.

Depression A mood characterized by feelings of dejection and gloom and other feelings, such as hopelessness, futility and guilt.

Design ergonomics A branch of ergonomics concerned with the design and specification of various features of the man–machine interface.

Deterministic view Describing the characteristics of elements within a system by means of functions and on the basis of a given input which leads to a prognosis of a clear starting and end function.

Drive A term implying an impetus to behaviour or active striving to that behaviour.

Ego	In psychoanalysis, a term referring to the self and to ways of behaving and thinking realistically. The ego delays the satisfaction of motives when necessary; it directs motives into socially acceptable channels.
Emotion	Affective states, often accompanied by facial and bodily expression, and having arousing and motivating properties.
Ergonomics	The scientific study of work. The scientific study of the interrelationships between people and their work. The scientific study of the relationship between man, the equipment with which he works and the physical environment in which this man–machine system works.
Esteem needs	According to Maslow, the need for prestige, success, self-respect and the respect of others, for competence, independence and self-confidence.
Fatigue	A general term referring to the effects of prolonged work or lack of sleep resulting in excessive tiredness.
Feedback	The situation in which some aspect of the output of an activity or event regulates or contributes to future inputs to the system. For example, feedback from the study of accident causes can contribute to future accident prevention strategies.
Formal group	A social group that has a relatively permanent structure of positions, jobs and roles.
Formal organization	An organization which incorporates a system of positions, roles and interrelated groups that is designed as the most efficient arrangement for effectively accomplishing the aims of the organization.
Functional/line organization	That part of an organization which is based on the type of work being done.
Gene	The essential element in the transmission of hereditary characteristics carried in chromosomes.
Genetics	The study of heredity.
Goal	The place, condition or object that satisfies a motive.
Group dynamics	The study of the development and functioning of groups, with special reference to the interactions between groups and the patterns of relationships between individuals within groups.
Group norm	A widely shared expectation or standard of behaviour amongst members of a group, class or culture.
Habit	A learned response. An action that has become an automatic response to a given stimulus.

Health surveillance The specific health examination at a predetermined frequency of those at risk of developing further ill-health or disability and those actually or potentially at risk by virtue of the type of work they undertake during their employment.

Homeostasis The tendency of the body to maintain a balance among internal physiological conditions, such as temperature, sugar, oxygen level and mineral level.

Human engineering The field of specialization concerned with the design of equipment and tasks performed in the operation of equipment.

Human error This is associated with limitations in human capacity to perceive, attend to, remember, process and act on information and is associated with lapses of attention, mistaken actions, misperceptions, mistaken priorities and, in some cases, wilfulness.

Human factors A term covering a wide range of issues, which include

- the perceptual, physical and mental capabilities of people and the interaction of individuals with their job and the work environment;
- the influence of equipment and system design on human performance; and
- the organizational characteristics which influence safety-related behaviour at work.

These issues are affected by

- the system for communication within the organization; and
- the training systems and procedures in operation.

All these issues are directed at preventing human error (Reducing Error and Influencing Behaviour [HS(G)48).

Environmental, organizational and job factors, together with human and individual characteristics, which influence behaviour at work in a way which can affect health and safety (HSE).

Human reliability assessment A technique which includes the identification of all the points in a sequence of operations at which incorrect human action, or the failure to act (sins of omission), may lead to adverse consequences for plant and/or people.

Id In psychoanalytic theory, the aspect of personality concerned with instinctive reactions for satisfying motives. The id seeks immediate gratification of motives with little regard for the consequences or for the realities of life.

Immunization In the case of attitudes, where a mild exposure to an opposing attitude can immunize a person against that attitude, as a result of which they are not prepared to accept further facts or arguments to support that attitude, no matter how strong they may be.

Induction The logical process by which principles or rules are derived from observed facts.

Informal group A social group having no formal or permanent structure and consisting of people who happen to be assembled together at a particular time.

Informal organization That part of a formal organization which incorporates the patterns of interpersonal and intergroup relationships that develop within the formal organization.

Intelligence A general term covering a person's abilities on a wide range of tasks involving vocabulary, numbers, problem solving, concepts, etc.

Interpersonal relationships Social associations, connections or affiliations between two or more people. They vary in differing levels of intimacy and sharing, implying the discovery or establishment of common ground and may be centred around something shared in common.

Job (task) analysis An analytical technique concerned with the identification and assessment of the skill and knowledge components of jobs. A technique allowing for the identification of important behavioural and performance qualities and for the matching of individuals to jobs.

Job description A statement of the significant characteristics of a job and of the operator characteristics necessary to perform the job satisfactorily.

Job enrichment A means of increasing satisfaction with, and the responsibilities of, a job by reducing the degree of supervision or by allocating each individual a unit of work for which they have freedom to select the method and sequence of operations.

Job impoverishment According to Herzberg, the removal of the individual interest, challenge and responsibility from a job.

Job safety analysis The identification of all the accident prevention measures appropriate to a particular job or area of work activity and the behavioural factors that most significantly influence whether or not these measures are taken.

Kinetics The study of mechanical, nervous and psychological factors which influence the function and structure of the human body as a means of producing higher standards of skill and reducing cumulative strain.

Knowledge function Attitudes are used to provide a system of standards that organize and stabilize a world of changing experiences. On this basis, people need to work within an acceptable framework, have a scale of values and generally know where they stand.

Lapses The action of forgetting to carry out a task or action, to lose the place in a task or forgetting what an individual intended to do.

Layout The space available for people working within a particular room or area and the situation of plant, equipment, machinery, furniture and stored goods in relation to those people and the tasks performed.

Leadership The process of successfully influencing the activities of a group towards the achievement of a goal or goals.

Learning The relatively permanent change in behaviour that is the result of past experience or practice.

Life change unit According to Holmes and Rahe, a unit of individual stress measurement in terms of the impact of stress on health.

Long-term memory That area of memory concerned with the ability to store and subsequently recall information. Long-term memory is developed from an early age through the repetition of items and codifying them to produce a meaning.

Macro-ergonomics Rules for the design of the organization and production methods and for working groups.

Management The effective use of resources in the pursuit of organizational goals. 'Effective' implies achieving a balance between the risk of being in business and the cost of eliminating or reducing such risks.

Man–machine interface The process whereby a machine passes information to the operator through various display elements, such as gauges and dials, and is controlled through the use of manual and pedal-operated control elements, such as steering wheels, foot pedals and switches.

Memory The process of retaining, recognizing and recalling experience (remembering) and is particularly associated with how people learn.

Mental matching A process which involves individuals' decision-making and information requirements, as well as their perception of risk.

Micro-ergonomics Rules for the technical design of work tools, equipment and workplaces.

Mistake A form of human error where an individual shows awareness of a problem, but forms a faulty plan for solving it. The situation where an individual does the wrong thing believing it to be correct.

Motivation A general term referring to behaviour instigated by needs and directed towards goals.

Motive A term implying a need and the direction of behaviour toward a goal.

Nature The genetic factors contributing to behaviour.

Need Any lack or deficit within the individual, either acquired or physiological.

Negative transfer The harmful effect on learning in one situation because of previous learning in another situation. It is due to incompatible responses being required in the two situations.

Nurture The learned factors contributing to behaviour; the factors that depend upon experience.

Off-line processing The process whereby people actually simulate in their own minds the outcomes of a different course of action prior to making any final decision as to which course of action to take.

On-line processing The spur of the moment decision-making that an individual has to take in order to survive.

Opinion Acceptance of a statement accompanied by an attitude of 'for' or 'against'; a statement of something which may be subject to change.

Organizational structure A chart depicting the working relationships and division of work within a particular organization.

Organization theory Assumptions about human motivation on which the working relationships within an organization are based.

Part learning Learning, usually by way of memorizing, in which a task is divided into smaller units, each unit being learnt separately.

Peer An equal in a given respect; an associate at roughly the same level.

Perception The awareness of objects, qualities or events stimulating a person's sense organs.The process of receiving information through the sensory channels – sight, hearing, touch, taste and smell.

Perceptual defence A feature of perception, this mechanism modifies distorts or eliminates those stimuli that appear stressful threatening or creating anxiety.

Perceptual distortion This state may take the form of perceptual defence and perceptual sensitization.

Perceptual sensitization The state where people can become 'sensitized' to certain stimuli if they are, in their opinion, relevant or meaningful to them.

Personal factors A commonly used term generally taken to include individual behavioural factors, such as attitude, motivation, memory, personality, perception, etc.

Personality The traits, modes of adjustment, defence mechanisms and ways of behaving that characterize the individual and his relation to others in his environment.

The total pattern of behaviour that is unique and manifest in a person's values, beliefs, interests, attitudes, expressions and actions.

The dynamic organization within the individual of those psychophysical systems that determine his characteristic behaviour and thought (Allport).

Personality trait An aspect of personality that is reasonably characteristic of a person and distinguishes that person in some way from other people.

A tendency to behave in a particular way according to one of the more stronger elements of an individual's personality.

Physical matching The design of the whole workplace and working environment to suit the physical needs of employees.

Physical stressor Those stressors associated with physical phenomena, such as heat, cold, radiation, noise and vibration.

Physiological needs According to Maslow, the needs of an individual for the basic necessities of life, i.e. food, water and pure air.

Physiology The study of the function of living organisms and their different parts.

Positive transfer (of training) More rapid learning in one situation because of previous learning in another situation. It is due to similarity of the stimuli and/or responses required in the two situations.

Power The ability to control or influence the behaviour of others.

Problem-solving The need to produce a solution to a problem which has never been previously encountered and which entails a creative process of

- recognizing the problem;
- defining the problem in the broadest terms;
- examining a range of solutions;
- analysing the facts; and
- selecting the best solution.

Product design ergonomics The study of the design of products adapted to the needs of the user in terms of comfort and ease of use.

Production ergonomics The study of the provision of working environments in manufacturing and service organizations that are adapted to human beings.

Propaganda The deliberate attempt to influence attitudes and beliefs.

Psychology The science that studies the behaviour of people and animals.

Reasoning Thinking in which a person attempts to solve a problem by combining two or more elements from past experience.

Reinforcement A stimulus or event that strengthens a response when it follows the response.

Retention The amount of information correctly remembered. Retention is measured on the basis of recognition, recall and savings.

Risk A chance of loss or injury.

The probability of harm, damage or injury.

The probability of a hazard leading to personal injury and the severity of that injury.

The likelihood of potential harm from a hazard being realized. The extent of the risk will depend upon

- the likelihood of that harm occurring;
- the potential severity of that harm, i.e. of any resultant injury or adverse health effect; and
- the population which might be affected by the hazard, i.e. the number of people who might be exposed (Approved Code of Practice to the Management of Health and Safety at Work Regulations).

Risk assessment The process of identifying the hazards present in an undertaking (whether arising from work activities or from other factors, e.g. the layout of the premises) and then evaluating the extent of the risks involved, taking into account whatever precautions are already being taken (Approved Code of Practice to the Management of Health and Safety at Work Regulations).

A 'suitable and sufficient' risk assessment should:

- identify the significant risks arising out of the work;
- enable the employer to identify and prioritize the measures that need to be taken to comply with the relevant statutory provisions; and
- be appropriate to the nature of the work and such that it remains in force for a reasonable period of time.

Risk threshold The maximum level of risk at which a person is prepared to operate.

Role A pattern of behaviour that a person in a particular status is expected to exhibit.

Role ambiguity The situation in which the role holder has insufficient information to adequately perform his role or where the information is open to more than one interpretation. Potentially ambiguous situations arise in jobs where there is a time lag between the action taken and visible results or where the role holder is unable to see the results of his actions.

Role conflict The simultaneous occurrence of two or more role expectations such that complying with one would make compliance with the other either more difficult or almost impossible. Role conflict arises where members of the organization, who exchange information with the role holder, have different expectations of the role holder. Each may exert pressure on the role holder and satisfying one expectation could make compliance with other expectations difficult. This is the classic 'servant of two masters' situation.

Role overload and underload This results from a combination of role ambiguity and role conflict. With *role overload*, the role holder works harder to clarify normal expectations or to satisfy conflicting priorities that are impossible to achieve within the time limits specified. *Role underload* can arise where people, who may have had a demanding job, are moved into another position where there is too much time available to complete an identified workload, resulting in boredom, excessive attention to minute detail as far as subordinates are concerned and a general feeling of isolation.

Safety climate A climate that promotes staff commitment to health and safety, emphasizing that deviation from corporate safety goals, at whatever level, is not acceptable.

Safety culture The promotion of a positive climate in which health and safety are seen by both management and employees as being fundamental to the organization's day-to-day operations.

The shared beliefs, practices and attitudes that exist within an organization with respect to safety.

The product of the individual and group values, attitudes, competencies and patterns of behaviour that determine the commitment to, and the style and proficiency of, an organization's health and safety programme. Organizations with a positive safety culture are characterized by communications founded on mutual trust, by shared perceptions of the importance of safety, and by confidence in the efficacy of preventative measures (HSE).

Safety incentive scheme A form of planned motivation, the main objectives being that of providing motivation to people by identifying targets which can be rewarded if achieved and making the rewards meaningful and attractive to the people involved in the scheme.

Safety needs According to Maslow, the needs of the individual for physical and psychological safety and security, for shelter and freedom from attack, both physically and mentally.

Self The individual's awareness or perception.

Self-actualization According to Maslow, the need to achieve what a person is capable of achieving, 'What a man can be, he must be' in terms of creativity, self-fulfilment and self-expression.

Self-defensive function This is concerned with the need for an individual to defend his self-image, both externally, in terms of how people react towards that individual, and internally, to deal with inner impulses and the individual's personal knowledge of what he is really like (Katz).

Self-image The personal image that a person wishes to project to the outside world which can incorporate a number of features, e.g. 'cool and calculating', hard to fool, constructive in thought, a friend to everyone, etc.

Shaping Teaching a desired response through a series of successive steps which lead the learner to the final response. Each small step leading to the final response is reinforced.

Short-term memory The ability to store information for a very limited period of time, e.g. a few seconds; the amount of information an individual can take in and retain.

Slips Failures in carrying out the actions of a task, that is, actions not as planned.

Social adjustive function In attitude study, this is concerned with how people relate, and adjust, to the influence of parents, teachers, friends and their superiors. Behaviour is based, to some extent, on a philosophy of 'maximum reward – minimum punishment' (Katz).

Social needs According to Maslow, the need of people to relate to other people, for friendship and affection and for belonging to a group, both at work and at home.

Social stressors Those stressors associated with, for example, family life, marriage, partnership and bereavement, the everyday problems of coping with life.

Sociotechnical factors In ergonomic study, the social relationships between people and how they work together as a group, group working practices, manning levels, working hours, the provision of meal and rest breaks, the formal and informal communication systems and the rewards and benefits available.

Organizational factors may also be included in this classification, including the structure of an organization and individual work groups, the identification of authority for certain actions and the interfaces between different work groups.

Staff organization That part of an organization represented by people who have advisory, service and control functions.

Statistics A collection of techniques in the quantitative analysis of data and used to facilitate evaluation of those data.

Status needs The needs of an individual to achieve status with respect to other people in a group, including the need for power, prestige and security.

Stereotype A fixed set of greatly oversimplified beliefs that are held generally by members of a group.

Stimulus Any object, energy or energy change in the physical environment that excites a sense organ.

Stress Any influence that disturbs the natural equilibrium of the living body. The common response to attack (Selye, 1936). The nonspecific response of the body to any demands made upon it.

Superego In psychoanalysis, that which restrains the activity of the ego and the id. It corresponds closely to that which is known as 'conscience'.

System inputs The human system receives information through a number of inputs to the sense organs – sight, hearing, touch, taste and smell. These organs have a very limited capacity for transmitting to, and receiving information from, the brain.

Task characteristics Those characteristics which feature in a task, such as freedom of operation of a keyboard, the repetitiveness of the task, the actual workload, the criticality of the task in the overall process, the duration of the task and its interaction with other tasks as part of a process.

Task demands Those features of a task that place a physical and/or mental load or demand on the operator, such as the need to be physically fit to undertake manual handling, or the need for attention and vigilance in inspection tasks.

Training The systematic development of attitude, knowledge and skill patterns required by the individual to perform adequately a given task or job. It is often integrated with further education (Department of Employment).

Training need A training need is said to exist when the optimum solution to an organization's problem is by means of some form of training.

Trait An aspect of personality that is reasonably characteristic of a person and distinguishes him in some way from many other people.

Understanding The faculty of the mind by which it apprehends the real state of things presented to it or the representation made to it; the act of comprehending or apprehending.

Unintentional error A situation where an individual fails to perform a task correctly, typical 'slips' or 'lapses' frequently associated with lack of attention to the task or carelessness.

Value expressive function In attitude study, a situation where individuals use their attitudes to present a picture of themselves that is pleasing and satisfying to them.

Violation A situation where a person deliberately carries out an action that is contrary to some rule which is organizationally required, such as an approved operating procedure.

Workstation An assembly comprising

- display screen equipment (whether provided with software determining the interface between the equipment and its operator or user, a keyboard or any other input device);
- any optional accessories to the display screen equipment;
- any disc drive, telephone, modem, printer, document holder, work chair, work desk, work surface or other item peripheral to the display screen equipment; and
- the immediate environment around the display screen equipment (Health and Safety (Display Screen Equipment) Regulations 1992).

Bibliography and further reading

Allport, GW (1961). *Pattern and Growth in Personality*. Holt, Rinehart & Winston, New York.

Anderson, JR (1996). *Cognitive Psychology and its Implications*: WH Freeman, New York.

Arbitration and Conciliation Advisory Service (2004). *Bullying and Harassment at Work: A Guide for Employers*. ACAS, London.

Argyris, C (1962). *Interpersonal Competence and Organisational Effectiveness*. Tavistock, London.

Atkinson, JW (1964). *An Introduction to Motivation*. Van Nostrand Reinhold, New York.

Atkinson, JW and Feather, NT (1966). *A Theory of Achievement Motivation*. John Wiley, New York.

Bass, BM (1965). *Organisational Psychology*. Allyn and Bacon, Boston.

Bell, CR (1974). *Men at Work*. George Allen & Unwin, London.

Bilsom International (1992). *In Defence of Hearing*. Bilsom International, Henley-on-Thames.

Binet, A and Simon, T (1905). New methods for the diagnosis of the intellectual levels of sub-normals. *L'Année Psychologique* 12, 191–244.

Bird, FE (1974). *Management Guide to Loss Control*. Institute Press, Atlanta.

Bird, FE and Loftus, RG (1984). *Loss Control Management*. Royal Society for the Prevention of Accidents, Birmingham.

Bird, FE and Germain, L (1985). *Loss Prevention Model*. International Loss Control Institute, University of Atlanta.

Bloom, P *et al.* (1956). *Taxonomy of Educational Objectives: The Classification of Educational Goals: Handbook: Cognitive Domain*. Longmans, New York.

British Standards Institution (1980). *BS 5378 Safety signs and colours*. BSI, London.

British Standards Institution (1994). *BS EN ISO 9000 Quality Systems: Specification for the Design/Development, Production, Installation and Servicing*. BSI, London.

British Standards Institution (1996). *BS 8800 Occupational Health and Safety Management Systems*. BSI, London.

British Standards Institution (2004). *Guide to Occupational Health and Safety Management Systems*. BSI, London.

Brown, JAC (1972). *The Social Psychology of Industry*. Pelican Books, Louisiana.

Cattell, RB (1946). *Personality*. McGraw-Hill, New York.

Cattell, RB (1965). *The Scientific Analysis of Personality*. Penguin Books, London.

Central Computer and Telecommunications Agency and the Council of Civil Service Unions (1988). *Ergonomic Factors Associated with the use of Visual Display Units*. CCTA, London.

Chartered Institute of Personnel and Development (2001). *Stress and Employer Liability*. CIPD, London.
Chartered Institute of Personnel and Development (2002). *A Corporate Strategy for Dealing with Stress at Work*. CIPD, London.
Confederation of British Industry (1991). *Developing a Safety Culture*. CBI, London.
Cooper, CL (1998). *Theories of Organisational Stress*. Oxford University Press, Oxford.
Cooper, CL and Marshall J (1978). Occupational Sources of Stress: A review of the literature relating to coronary heart disease and mental ill-health: *Journal of Occupational Psychology* 49, 11–28.
Cooper, CL and Palmer, S (2000). *Conquer your Stress*. Chartered Institute of Personnel and Development, London.
Cooper, CL, Cooper, RD and Eaker, IH (1988). *Living with Stress*. Penguin, London.
Cooper, DC (1999). *Behavioural Safety: A Proven Weapon in the War on Workplace Accidents*. Sheila Pantry Associates Ltd, OSH World, Sheffield.
Cox, T (1978). *Stress*. Macmillan Press, London.
Cox, T (1993). *Stress Research and Stress Management: Putting Theory to Work CRR 61/1993*. HMSO, London.
Cox, T, Griffiths, A and Barlow, C (2000). *Organisational Interventions for Work Stress*. HSE Books, Sudbury.

Dempsey, PJR (1973). *Psychology and the Manager*. Pan Books, London.
Department of Employment (1973). *Safety Training Needs and Facilities for One Industry*. HMSO, London.
Department of Employment (1975). *Health and Safety at Work etc. Act 1974*. HMSO, London.
Department of Employment and Productivity (1978). *Glossary of Training Terms*. HMSO, London.
Department of Employment (1996). *Health and Safety (Safety Signs and Signals) Regulations 1996*. HMSO, London.
Department of Employment (1995). *Reporting of Injuries, Diseases and Dangerous Occurrences Regulations 1995*. HMSO, London.
Department of Social Security (1985). *Social Security (Industrial Injuries)(Prescribed Diseases) Regulations 1985*. HMSO, London.
Department of Work and Pensions (1995). *Disability Discrimination Act 1995*. HMSO, London.
Department of Work and Pensions (1996). *Employment Rights Act 1996*. HMSO, London.
Department of Work and Pensions (2005). *Control of Noise at Work Regulations 2005*. HMSO, London.
Department of Work and Pensions (2005). *Control of Vibration at Work Regulations 2005*. HMSO, London.
Drucker, P (1999). *Practice of Management*. Butterworth-Heinemann, London.
Drucker, P (1999). *Management: Tasks, Responsibilities, Practices*. Butterworth-Heinemann, London.

Earnshaw, J and Cooper, C (2001). *Stress and Employer Liability*. Chartered Institute of Personnel and Development, London.

Edholm, OG (1967). *The Biology of Work*. World University Library, Weidenfield and Nicolson, London.

Edwards, W (1961). Behavioural decision theory. *Annual Review of Psychology* 12, 473–498.

European Agency for Safety and Health at Work (2001). *Work-Related Stress: The European Picture*. EASHW, Brussels.

Eysenck, HJ (1965). *Fact and Fiction in Psychology*. Penguin Books, London.

Festinger, FW (1957). *A Theory of Cognitive Dissonance*. Harper & Row, New York.

Flesch, R (1946). *The Art of Plain Talk*. Harper & Row, New York.

Freud, S (1949). *An Outline of Psychoanalysis*. Norton, New York.

Gilbreth, FB (1970). *Science in Management for the One Best Way to do Work: Classics in Management*. American Management Association, New York.

Glendon, AI, Clarke, GC and McKenna, EF (2006). *Human Safety and Risk Management*. Taylor & Francis, London.

Gowers, Sir E (1969). *The Complete Plain Words*. Pelican Books, Louisiana.

Grandjean, E (1980). *Fitting the Task to the Man: An Ergonomic Approach*. Taylor & Francis, London.

Hale, AR and Hale, M (1970). Accidents in perspective. *Occupational Psychology* 44, 115–122.

Hale, AR and Hale, M (1972). *A Review of Industrial Accident Research Literature*. Committee on Safety and Health at Work Research Paper: HMSO, London.

Hale, AR and Glendon, AI (1987). *Individual Behaviour in the Control of Danger*. Elsevier, Amsterdam.

Hatvany, I (1996). *Putting Pressure to Work*. Pitman Publishing, Totowa.

Health and Safety Commission (1978). *Safety Representatives and Safety Committees Regulations 1977: Approved Code of Practice and Guidance*. HMSO, London.

Health and Safety Commission (1987). *Writing a Safety Policy Statement: Advice to Employers*. HSE Books, London.

Health Safety Commission (1991). *Study Group on Human Factors: Human Reliability Assessment – A Critical Overview*. Advisory Committee on the Safety of Nuclear Installations: HMSO, London.

Health and Safety Commission (1992). *Workplace Health, Safety and Welfare: Workplace (Health, Safety and Welfare) Regulations 1992: Approved Code of Practice*. HMSO, London.

Health and Safety Commission (2000). *Management of Health and Safety at Work: Management of Health and Safety at Work Regulations 1999: Approved Code of Practice and Guidance*. HSE Books, London.

Health and Safety Executive, International Stress Management Association, ACAS (2005). *Working Together to Reduce Stress at Work: A Guide for Employees*. HSE Books, London.

Health and Safety Executive, Health and Safety Laboratory (2003). *Manual Handling Assessment Charts*. HSE Books, London.
Health and Safety Executive (1982). *Guidelines for Occupational Health Services*. HMSO, London.
Health and Safety Executive (1987a). *Lighting at Work*. HSE Books, London.
Health and Safety Executive (1987b). *Computer Control: A Question of Safety*. HSE Books, London.
Health and Safety Executive (1988a). *The Tolerability of Risk from Nuclear Power Stations*. HSE Books, London.
Health and Safety Executive (1988b). *Passive Smoking at Work*. HSE Books, London.
Health and Safety Executive (1990a). *If the Task Fits: Ergonomics at Work*. HSE Books, London.
Health and Safety Executive (1990b). *Work-related Upper Limb Disorders: A Guide to Prevention*: HSE Books, London.
Health and Safety Executive (1992a). *Lighten the Load: Guidance for Employers on Musculoskeletal Disorders*. HSE Books, London.
Health and Safety Executive (1992b). *Manual Handling: Guidance on Regulations: Manual Handling Operations Regulations 1992*. HMSO, London.
Health and Safety Executive (1992c). *Display Screen Equipment at Work: Guidance on Regulations: Health and Safety (Display Screen Equipment) Regulations 1992*. HMSO, London.
Health and Safety Executive (1992d). *Listen Up!* HSE Books, London.
Health and Safety Executive (1992e). *Successful Health and Safety Management*. HMSO, London.
Health and Safety Executive (1993). *Working with VDUs*. HSE Books, London.
Health and Safety Executive (1994a). *If the Task Fits: Ergonomics at Work*. HSE Books, London.
Health and Safety Executive (1994b). *Upper Limb Disorders: Assessing the Risks*. HSE Books, London.
Health and Safety Executive (1994c). *Officewise*. HSE Books, London.
Health and Safety Executive (1995). *Stress at Work*. HSE Books, London.
Health and Safety Executive (1996a). *Protecting your Health at Work*. HSE Books, London.
Health and Safety Executive (1996b). *Signpost to the Health and Safety (Safety Signs and Signals) Regulations 1996*. HSE Books, London.
Health and Safety Executive (1996c). *Consulting Employees on Health and Safety: A Guide to the Law*. HSE Books, London.
Health and Safety Executive (1997). *Health and Safety Climate Survey Tool*. HSE Books, London.
Health and Safety Executive (1998a). *Help on Work-related Stress: A Short Guide*. HSE Books, London.
Health and Safety Executive (1998b). *Seating at Work*. HSE Books, London.
Health and Safety Executive (1999a). *Checkouts and Musculoskeletal Disorders*. HSE Books, London.
Health and Safety Executive (1999b). *Health and Safety Benchmarking: Improving Together*. HSE Books, London.
Health and Safety Executive (2000a). *Getting to Grips with Manual Handling: A Short Guide for Employers*. HSE Books, London.

Health and Safety Executive (2000b). *Stating your Business: Guidance on Preparing a Health and Safety Policy Document for Small Firms*. HSE Books, London.

Health and Safety Executive (2000c). *Violence at Work: A Guide for Employers*. HSE Books, London.

Health and Safety Executive (2000d). *Back in Work: Managing Back Pain in the Workplace*. HSE Books, London.

Health and Safety Executive (2000e). *A Guide to Risk Assessment Requirements*. HSE Books, London.

Health and Safety Executive, Department of Health, Health Education Authority (2001). *Don't Mix It! A Guide for Employers on Alcohol at Work*. HSE Books, London.

Health and Safety Executive (2001a). *Work-related Stress: A Short Guide*. HSE Books, London.

Health and Safety Executive (2001b). *Drug Misuse at Work: A Guide for Employers*. HSE Books, London.

Health and Safety Executive (2001c). *Keep the Noise Down: Advice for Purchasers of Workplace Machinery*. HSE Books, London.

Health and Safety Executive (2001d). *An Assessment of Employee Assistance in Workplace Counselling Programmes in British Organisations: HSE Research Report*. HMSO, London.

Health and Safety Executive (2001e). *Tackling Work-Related Stress: A Manager's Guide to Improving and Maintaining Employee Health and Well-being*. HMSO, London.

Health and Safety Executive (2001f). *Reducing error and influencing behaviour*. HSE Books, London.

Health and Safety Executive (2002a). *Effective Teamwork in Reducing the Psychosocial Risks: Case Studies in Practitioner Format CRR 393*. HMSO, London.

Health and Safety Executive (2002b). *The Scale of Occupational Stress: A Further Analysis of the Impact of Demographic Factors and the Type of Job CRR 311*. HMSO, London.

Health and Safety Executive (2002c). *The Scale of Occupational Stress: Bristol Stress and Health at Work Study CRR 265*. HMSO, London.

Health and Safety Executive (2002d). *Work Environment: Alcohol Consumption and Ill-Health: The Whitehall Study CRR 422*. HMSO, London.

Health and Safety Executive (2002e). *Passport Schemes for Health, Safety and the Environment: A Good Practice Guide (INDG381)*: HSE Books

Health and Safety Executive (2002f). *Violence at Work: New Findings from the 2000 British Crime Survey*. HSE Books, London.

Health and Safety Executive (2003a). *Understanding Ergonomics at Work*. HSE Books, London.

Health and Safety Executive (2003b). *Beacons of Excellence in Stress Prevention RR133*. HSE Books, London.

Health and Safety Executive (2003c). *Best Practice in Rehabilitating Employees Following Absence due to Work-Related Stress RR138*. HSE Books, London.

Health and Safety Executive (2003d). *Home Working Guidance for Employers and Employees*. HSE Books, London.

Health and Safety Executive (2004). *Corporate Health and Safety Performance Index*. HSE Books, London.

Health and Safety Executive (2005). *HSE Management Standards for Stress at Work.* HSE Books, London.

Health and Safety Executive (2006). *Five Steps to Risk Assessment.* HSE Books, London.

Health and Safety Laboratory (2005). *Review of the Public Perception of Risk, and Stakeholder Engagement – HSL/2005/16.* Health and Safety Laboratory, Buxton.

Heinrich, HW (1931). *Unsafe Acts and Conditions.* McGraw-Hill, New York.

Heinrich, HW (1959). *Industrial Accident Prevention: A Scientific Approach.* McGraw-Hill, New York.

Herzberg, F (1957). *The Two Factor Theory.* McGraw Hill, New York.

Herzberg, F, Mansner, G and Snyderman, BB (1959). *The Motivation to Work.* John Wiley, New York.

Holmes, D and Rahe, J (1967). Scaling of life change: The social readjustment rating scale. *Journal of Psychosomatic Research* 11, 213–218.

Institution of Electrical Engineers (2003). *Behavioural Safety*: Health and Safety Briefing No. 14. IEE, London.

Institution of Occupational Safety and Health (1997). *Behavioural Safety: Kicking Bad Habits.* Technical Infosheet. IOSH, Leicester.

Jung, CG (1967). *The Development of Personality.* Routledge, London.

Katz, D and Brady, K (1933). Racial stereotypes of one hundred college students. *Journal of Abnormal and Social Psychology* 28, 280–290.

Kolb, DA (1984). *Experiential Learning.* Englewood Cliffs, NJ: Prentice-Hall.

Likert, R (1960). *New Patterns of Management.* McGraw-Hill, New York.

Lippitt, R and White, RK (1958). An experimental study of leadership and group life. In: EE Macoby, TM Newcomb and EL Hartley (eds). *Readings in Social Psychology.* Holt, New York.

Lukzak, H, Volpert, W, Racthel, A and Schwier, W (1987). *Arbeitwissienschaftliche Kern definition: Gegenstandskatalog, Forschungsbiete.* RKW, Edingen-Neckarsulm.

Mackay, C and Cox, T (1976). *A Transactional Approach to Occupational Stress.* University of Nottingham, Nottingham.

Maritain, J (1952). *The Range of Reason.* Charles Scribners' Sons, New York.

Maslow, AH (1954). *Motivation and Personality.* Harper & Row, New York.

Mayo, E (1952). *The Social Problems of an Industrial Civilisation.* Routledge & Kegan Paul, London.

McClelland, DC, Atkinson, JW, Clark, RA and Lowell, EL (1953). *The Achievement Motive.* Appleton Century Crofts, New York.

McGregor, D (1960). *The Human Side of Enterprise.* McGraw Hill, New York.

Morgan, G (1989). *Creative Organisation Theory.* Sage Publications, London.

Osborne, DJ (1982). *Ergonomics at Work.* Wiley, New York.

Pidgeon, N, Kasperson, RE *et al.* (1992). *The Social Amplification of Risk.* Cambridge University Press, Cambridge.

Pirani, M and Reynolds, J (1976). Gearing up for safety. *Personnel Management* 8, 25–29.

Pheasant, P and Stubbs, D (1992). *Lifting and Handling: An Ergonomic Approach.* National Back Pain Association, Teddington.

Powell, PI, Hale, M, Martin, P and Simon, M (1971). *2000 Accidents.* National Institute of Industrial Psychology, London.

Rasmussen, J (1983). Skills, rules and knowledge: Signals, signs and symbols and other distinctions in human performance models: *IEEE Transactions on Systems, Man and Cybernetics* SMC 13(3).

Rasmussen, J, Duncan, K and Leplat, J (1987). *New Technology and Human Error.* Wiley, Chichester.

Reason, J (1989). *Human Error.* Cambridge University Press, Cambridge.

Reason, J (1997). *Managing the Risks of Organisational Accidents.* Ashgate Publishing, London.

Ridley, J and Channing, J (1998). *Risk Management.* Butterworth-Heinemann, London.

Rimington, JR (1989). *The Onshore Safety Regime: The HSE Director General's Submission to the Piper Alpha Enquiry, December 1989.* HMSO, London.

Rodger, A (1952). *The Seven Point Plan.* NIIP, London.

Rokeach, M (1968). *The Nature of Human Values: A Theory of Organization and Change.* Jossey-Bass, San Francisco.

Sell, RG and Shipley, P (1979). *Satisfaction in Work Design: Ergonomics and Other Approaches.* Taylor & Francis, London.

Selye, H (1931). *The Stress of Life.* McGraw-Hill, New York.

Selye, H (1936). *The Stress of Work.* McGraw-Hill, New York.

Simon, HA (1947). *Administrative Behaviour: A study of Decision-Making Processes in Administrative Organizations.* MacMillan, New York.

Smith, AP *et al.* (2000). *The Scale of Occupational Stress: The Bristol Stress and Health at Work Study CRR 265/2000.* HSE, London.

Spaltro, E (1969). *Psychology of the Job.* ETAS, Rome.

Stenier, J, Jarvis, M and Parrish, J (1970). Risk taking and arousal regulation. *British Journal of Medical Psychology*, 43, 333–348.

Stranks, J (1994). *Human Factors and Safety.* Pitman Publishing, London.

Stranks, J (1996). *The Law and Practice of Risk Assessment.* Pitman Publishing, London.

Stranks, J (2005a). *The Handbook of Health and Safety Practice, 7th edn.* Pearson Prentice Hall.

Stranks, J (2005b). *Stress at Work: Management and Prevention.* Elsevier Butterworth-Heinemann.

Stranks, J. (2006). *Health and Safety Pocket Book.* Elsevier Butterworth-Heinemann.

Stress Foundation (1987). *Understanding Stress.* HMSO, London.

Sulzer-Azeroff, B (1978). Behavioural ecology and accident prevention. *Journal of Organizational Behavior Management* 2, 11–44.

Sulzer-Azeroff, B (1987). The modification of occupational safety behaviour. *Journal of Occupational Accidents* 9, 177–197.

Sulzer-Aseroff, B and Lischeid, WE (1999). Assessing the quality of behavioural safety initiatives. *Professional Safety* 44, 31–36.

Sulzer-Azeroff, B and Santamaria, MC (1980). Industrial safety reduction through performance feedback. *Journal of Applied Behavior Analysis* 13, 287–295.

Swingle, PG (1973). *Social Psychology in Everyday Life*. Penguin, London.

Taylor, FW (1911). *Principles of Scientific Management*. Harper & Row, New York.

Taylor, DH (1964). Drivers' galvanic skin response and the risk of accidents. *Ergonomics* 7, 439–451.

The Ergonomics Society (2004). *Ergonomics*. The Ergonomics Society, Loughborough.

University of Manchester Institute of Science and Technology (1987). *Understanding Stress*. HMSO, London.

Waring, AE and Glendon, AI (1998). *Managing Risk*. International Thomson Business Press, London.

Warr, PB (1971). *Psychology at Work*. Penguin Books, London.

Weber, M (1964). *The Theory of Social and Economic Organisation*. Free Press, New York.

Welford, AT (1960). The measurement of sensory-motor performance: Survey and reappraisal of twelve years' progress. *Ergonomics* 3, 189–230.

Wilde, GJS (1994). Risk homeostasis theory and its promise for improved safety. In: RM Trimpop and GJS Wilde (eds). *Challenges to Accident Prevention. The Issue of Risk Compensation Processes*. Styx Publications, Groningen, The Netherlands.

Index

Lightning Source UK Ltd.
Milton Keynes UK
UKOW03f1903190713

214088UK00002B/26/P